MATLAB®
Programming
for Engineers

MATLAB®
Programming
for Engineers

Fourth Edition

Stephen J. Chapman
BAE SYSTEMS Australia

CENGAGE
Learning™

Australia • Brazil • Japan • Korea • Mexico • Singapore • Spain • United Kingdom • United States

CENGAGE
Learning™

MATLAB® Programming for Engineers, Fourth Edition
Stephen J. Chapman

Director, Global Engineering Program:
  Chris Carson

Senior Developmental Editor: Hilda Gowans

Marketing Services Coordinator:
  Lauren Bestos

Production Manager: Renate McCloy

Production Service: RPK Editorial Services

Copyeditor: Harlan James

Proofreader: Erin Wagner

Indexer: Shelly Gerger-Knechtl

Compositor: Integra

Senior Art Director: Michelle Kunkler

Creative Director: Angela Cluer

Internal Designer: Carmela Pereira

Cover Designer: Andrew Adams

Text Permissions Researcher:
  Kristina Bowering

Senior First Print Buyer: Doug Wilke

For product information and technology assistance, contact us at **Cengage Learning Customer & Sales Support, 1-800-354-9706**.
For permission to use material from this text or product, submit all requests online at **www.cengage.com/permissions**. Further permissions questions can be emailed to **permissionrequest@cengage.com**.

Library of Congress Control Number: 2007932905

MATLAB® and SIMULINK® are registered trademarks of The MathWorks, 3 Apple Hill Drive, Natick, MA 01760

U.S. Student Edition:
ISBN-13: 978-0-495-24449-3
ISBN-10: 0-495-24449-X

Cengage Learning
200 First Stamford Place, Suite 400
Stamford, CT 06902
USA

Cengage Learning is a leading provider of customized learning solutions with office locations around the globe, including Singapore, the United Kingdom, Australia, Mexico, Brazil, and Japan. Locate your local office at: **www.cengage.com/global**.

Cengage Learning products are represented in Canada by Nelson Education Ltd.

For your course and learning solutions, visit **www.cengage.com/engineering**.

Purchase any of our products at your local college store or at our preferred online store **www.ichapters.com**.

Printed in the United States of America
3 4 5 6 7 12 11 10

This book is dedicated with love to my daughter Rachel.

Preface

MATLAB® (short for MATrix LABoratory) is a special-purpose computer program optimized to perform engineering and scientific calculations. It started life as a program designed to perform matrix mathematics, but over the years it has grown into a flexible computing system capable of solving essentially any technical problem.

The MATLAB program implements the MATLAB language and provides a very extensive library of predefined functions to make technical programming tasks easier and more efficient. This extremely wide variety of functions makes it much easier to solve technical problems in MATLAB than in other languages such as Fortran or C. This book introduces the MATLAB language and shows how to use it to solve typical technical problems.

This book teaches MATLAB as a technical programming language, showing students how to write clean, efficient, and documented programs. It makes no pretense at being a complete description of all of MATLAB's hundreds of functions. Instead, it teaches the student how to use MATLAB as a computer language and how to locate any desired function with MATLAB's extensive on-line help facilities.

The first six chapters are designed to serve as the text for an "Introduction to Programming/Problem Solving" course for freshman engineering students. This material should fit comfortably into a 9-week, 3-hour course. The remaining chapters cover advanced topics such as I/O and graphical user interfaces. These chapters may be covered in a longer course or employed as a reference by engineering students or practicing engineers who use MATLAB as a part of their coursework or employment.

Changes in the Fourth Edition

The fourth edition of this book is specifically devoted to MATLAB 7.4. MATLAB 7 and later versions contain many language and tool changes, and this book had to be revised extensively for the new version. Some of the major changes are

- Case-sensitive function and directory names on all platforms.
- Function handles.
- Nested functions.
- Math operations with single and integer data types.
- Major revision of the GUI code, including the addition of panels, button groups, and toolbars. The code auto-generated by `guide` has been totally changed. Frames have been deprecated.
- Major revisions to programming tools, such as the addition of conditional breakpoints and the `mlint` tool to check for poor programming practices within an M-file.

In addition, this fourth edition adds coverage of solving differential equations using the `ode45` function, introduces the MATLAB compiler, and provides numerous new or modified end-of-chapter exercises.

The Advantages of MATLAB for Technical Programming

MATLAB has many advantages compared with conventional computer languages for technical problem solving. These include

1. **Ease of Use**

    MATLAB is an interpreted language, as are many versions of Basic. Like Basic, it is very easy to use. The program can be used as a scratch pad to evaluate expressions typed at the command line, or it can be used to execute large prewritten programs. Programs may be easily written and modified with the built-in integrated development environment, and debugged with the MATLAB debugger. Because the language is so easy to use, it is ideal for educational use and for the rapid prototyping of new programs.

    Many program development tools are provided to make the program easy to use. They include an integrated editor/debugger, on-line documentation and manuals, a workspace browser, and extensive demos.

2. **Platform Independence**

    MATLAB is supported on many different computer systems, providing a large measure of platform independence. At the time of this writing, the language is supported on Windows 2000/XP/Vista, Linux, Unix, and the Macintosh. Programs written on any platform will run on all of the other platforms, and data files written on any platform may be read transparently on any other platform. As a result, programs written in MATLAB can migrate to new platforms when the needs of the user change.

3. **Predefined Functions**

MATLAB comes complete with an extensive library of predefined functions that provide tested and prepackaged solutions to many basic technical tasks. For example, suppose that you are writing a program that must calculate the statistics associated with an input data set. In most languages, you would need to write your own subroutines or functions to implement calculations such as the arithmetic mean, standard deviation, median, and the like. These and hundreds of other functions are built right into the MATLAB language, making your job much easier.

In addition to the large library of functions built into the basic MATLAB language, there are many special-purpose toolboxes available to help solve complex problems in specific areas. For example, a user can buy standard toolboxes to solve problems in signal processing, control systems, communications, image processing, and neural networks, among many others.

4. **Device-Independent Plotting**

Unlike other computer languages, MATLAB has many integral plotting and imaging commands. The plots and images can be displayed on any graphical output device supported by the computer on which MATLAB is running. This capability makes MATLAB an outstanding tool for visualizing technical data.

5. **Graphical User Interface**

MATLAB includes tools that allow a program to interactively construct a graphical user interface (GUI) for his or her program. With this capability, the programmer can design sophisticated data analysis programs that can be operated by relatively inexperienced users.

6. **MATLAB Compiler**

MATLAB's flexibility and platform independence is achieved by compiling MATLAB programs into a device-independent pcode, and then interpreting the pcode instructions at run-time. This approach is similar to that used by Microsoft's Visual Basic language. Unfortunately, the resulting programs can sometimes execute slowly because the MATLAB code is interpreted rather than compiled. We will point out features that tend to slow program execution when we encounter them, and suggest ways to work around this limitation.

In addition MATLAB programs can be used only on computers that have MATLAB installed. Since the program is expensive, that can be a significant limitation.

A separate MATLAB compiler is available, which converts MATLAB programs into ones that can be run on any computer without requiring a MATLAB license. With the MATLAB compiler, a programmer can create MATLAB programs and distribute then to anyone on any computer. The MATLAB compiler is described in Chapter 11.

Features of This Book

Many features of this book are designed to emphasize the proper way to write reliable MATLAB programs. These features should serve a student well when he or she is first learning MATLAB, and they should also be useful to the practitioner on the job. They include

1. **Emphasis on Top-Down Design Methodology**
   The book introduces a top-down design methodology in Chapter 3, and uses it consistently throughout the rest of the book. This methodology encourages a student to think about the proper design of a program *before* beginning to code. It emphasizes the importance of clearly defining the problem to be solved and the required inputs and outputs before any other work is begun. Once the problem has been properly defined, it teaches the student to employ stepwise refinement to break the task down into successively smaller sub-tasks and to implement the subtasks as separate subroutines or functions. Finally, it teaches the importance of testing at all stages of the process—both unit testing of the component routines and exhaustive testing of the final product.

   The formal design process taught by the book may be summarized as follows:

   1. *Clearly state the problem that you are trying to solve.*
   2. *Define the inputs required by the program and the outputs to be produced by the program.*
   3. *Describe the algorithm that you intend to implement in the program.* This step involves top-down design and stepwise decomposition, using pseudocode or flow charts.
   4. *Turn the algorithm into MATLAB statements.*
   5. *Test the MATLAB program.* This step includes unit testing of specific functions as well as exhaustive testing of the final program with many different data sets.

2. **Emphasis on Functions**
   The book emphasizes the use of functions to logically decompose tasks into smaller sub-tasks. It teaches the advantages of functions for data hiding. It also emphasizes the importance of unit testing functions before they are combined into the final program. In addition, the book teaches about the common mistakes made with functions and how to avoid them.

3. **Emphasis on MATLAB Tools**
   The book teaches the proper use of MATLAB's built-in tools to make programming and debugging easier. The tools covered include the Editor / Debugger, the Workspace Browser, the Help Browser, and GUI design tools.

4. **Good Programming Practice Boxes**

   These boxes highlight good programming practices when they are introduced for the convenience of the student. In addition, the good programming practices introduced in a chapter are summarized at the end of the chapter. An example Good Programming Practice Box is shown here.

☀ Good Programming Practice

Always indent the body of an `if` construct by two or more spaces to improve the readability of the code.

---

5. **Programming Pitfalls Boxes**

   These boxes highlight common errors so that they can be avoided. An example Programming Pitfalls Box is shown here.

💣 Programming Pitfalls

Make sure that your variable names are unique in the first 31 characters. Otherwise, MATLAB will not be able to tell the difference between them.

---

6. **Emphasis on Data Structures**

   Chapter 7 contains a detailed discussion of MATLAB data structures, including sparse arrays, cell arrays, and structure arrays. The proper use of these data structures is illustrated in the chapters on handle graphics and graphical user interfaces.

Pedagogical Features

The first six chapters of this book are specifically designed to be used in a freshman "Introduction to Program/Problem Solving" course. It should be possible to cover this material comfortably in a 9-week, 3-hour-per-week course. If there is insufficient time to cover all of the material in a particular Engineering program, Chapter 6 may be bypassed, the remaining material will still teach the fundamentals of programming and using MATLAB to solve problems. This feature should appeal to harassed engineering educators trying to cram ever more material into a finite curriculum.

The remaining chapters cover advanced material that will be useful to the engineer and engineering students as they progress in their careers. This material includes advanced I/O and the design of graphical user interfaces for programs.

The book includes several features designed to aid student comprehension. A total of 15 quizzes appear scattered throughout the chapters, with answers to all questions included in Appendix B. These quizzes can serve as a useful self-test of comprehension. In addition, there are approximately 160 end-of-chapter exercises. Answers to all exercises are included in the Instructor's Manual. Good programming practices are highlighted in all chapters with special Good Programming Practice boxes, and common errors are highlighted in Programming Pitfalls boxes. End-of-chapter materials include Summaries of Good Programming Practice and Summaries of MATLAB Commands and Functions.

The book is accompanied by an Instructor's Manual, containing the solutions to all end-of-chapter exercises. The source code for all examples in the book is available from the book's Web site, and the source code for all solutions in the Instructor's Manual is available separately to instructors.

A Final Note to the User

No matter how hard I try to proofread a document like this book, it is inevitable that some typographical errors will slip through and appear in print. If you should spot any such errors, please drop me a note via the publisher, and I will do my best to get them eliminated from subsequent printings and editions. Thank you very much for your help in this matter.

I will maintain a complete list of errata and corrections at the book's World Wide Web site, which is www.cengage.com/engineering. Please check that site for any updates and/or corrections.

Acknowledgments

I would like to thank Chris Carson, Hilda Gowans, and the crew at Cengage Learning for the support they have given me in getting this book to market. I would also like to thank the reviewers of this text for their valuable input. Those willing to be acknowledged include:

David Dux, Purdue University
Mani Mina, Iowa State University
John E. Parsons, North Carolina State University
Leonardo Perez y Perez, Long Beach State University
Kenneth Rouse, Auburn University
Matthew Whiteacre, Texas A&M University

In addition, I would like to thank my wife Rosa, and our children Avi, David, Rachel, Aaron, Sarah, Naomi, Shira, and Devorah for their help and encouragement.

STEPHEN J. CHAPMAN
Melbourne, Australia

Contents

Chapter 2   MATLAB Basics   **21**

Chapter 3   Branching Statements and Program Design 87

Chapter 4   Loops   149

Chapter 8   Input/Output Functions   371

Chapter 11 The MATLAB Compiler 521

CHAPTER 1

Introduction to MATLAB

MATLAB (short for MATrix LABoratory) is a special-purpose computer program optimized to perform engineering and scientific calculations. It started life as a program designed to perform matrix mathematics, but over the years it has grown into a flexible computing system capable of solving essentially any technical problem.

The MATLAB program implements the MATLAB programming language and provides a very extensive library of predefined functions to make technical programming tasks easier and more efficient. This book introduces the MATLAB language as it is implemented in MATLAB Version 7.4 (Release 2007a) and shows how to use it to solve typical technical problems.

MATLAB is a huge program with an incredibly rich variety of functions. Even the basic version of MATLAB without any toolkits is much richer than other technical programming languages. There are more than 1000 functions in the basic MATLAB product alone, and the toolkits extend this capability with many more functions in various specialties. This book makes no attempt to introduce the user to all of MATLAB's functions. Instead, it teaches a user the basics of how to write, debug, and optimize good MATLAB programs and presents a subset of the most important functions. Just as importantly, it teaches the programmer how to use MATLAB's own tools to locate the right function for a specific purpose from the enormous number of choices available.

1.1 The Advantages of MATLAB

MATLAB has many advantages compared with conventional computer languages for technical problem solving. These include

1. **Ease of Use**

   MATLAB is an interpreted language, like many versions of Basic. Like Basic, it is very easy to use. The program can be used as a scratch pad to evaluate expressions typed at the command line, or it can be used to execute large prewritten programs. Programs may be easily written and modified with the built-in integrated development environment, and debugged with the MATLAB debugger. Because the language is so easy to use, it is ideal for the rapid prototyping of new programs.

   Many program development tools are provided to make the program easy to use. They include an integrated editor/debugger, on-line documentation and manuals, a workspace browser, and extensive demos.

2. **Platform Independence**

   MATLAB is supported on many different computer systems, providing a large measure of platform independence. At the time of this writing, the language is supported on Windows 2000/XP/Vista, Linux, several versions of Unix, and the Macintosh. Programs written on any platform will run on all of the other platforms, and data files written on any platform may be read transparently on any other platform. As a result, programs written in MATLAB can migrate to new platforms when the needs of the user change.

3. **Predefined Functions**

   MATLAB comes complete with an extensive library of predefined functions that provide tested and prepackaged solutions to many basic technical tasks. For example, suppose that you are writing a program that must calculate the statistics associated with an input data set. In most languages, you would need to write your own subroutines or functions to implement calculations such as the arithmetic mean, standard deviation, median, and so on. These and hundreds of other functions are built right into the MATLAB language, making your job much easier.

   In addition to the large library of functions built into the basic MATLAB language, there are many special-purpose toolboxes available to help solve complex problems in specific areas. For example, a user can buy standard toolboxes to solve problems in signal processing, control systems, communications, image processing, and neural networks, among many others. There is also an extensive collection of free user-contributed MATLAB programs that are shared through the MATLAB Web site.

4. **Device-Independent Plotting**

   Unlike most other computer languages, MATLAB has many integral plotting and imaging commands. The plots and images can be displayed on any graphical output device supported by the computer on which MATLAB is running. This capability makes MATLAB an outstanding tool for visualizing technical data.

5. **Graphical User Interface**

MATLAB includes tools that allow a programmer to interactively construct a Graphical User Interface (GUI) for his or her program. With this capability, the programmer can design sophisticated data-analysis programs that can be operated by relatively inexperienced users.

6. **MATLAB Compiler**

MATLAB's flexibility and platform independence is achieved by compiling MATLAB programs into a device-independent p-code and then interpreting the p-code instructions at runtime. This approach is similar to that used by Microsoft's Visual Basic language. Unfortunately, the resulting programs can sometimes execute slowly because the MATLAB code is interpreted rather than compiled. We will point out features that tend to slow program execution when we encounter them.

A separate MATLAB compiler is available. This compiler can compile a MATLAB program into a true executable that runs faster than the interpreted code. It is a great way to convert a prototype MATLAB program into an executable suitable for sale and distribution to users.

1.2   Disadvantages of MATLAB

MATLAB has two principal disadvantages. The first is that it is an interpreted language and therefore may execute more slowly than compiled languages. This problem can be mitigated by properly structuring the MATLAB program.

The second disadvantage is cost: a full copy of MATLAB is five to ten times more expensive than a conventional C or Fortran compiler. This relatively high cost is more than offset by the reduced time required for an engineer or scientist to create a working program, so MATLAB is cost-effective for businesses. However, it is too expensive for most individuals to consider purchasing. Fortunately, there is also an inexpensive Student Edition of MATLAB, which is a great tool for students wishing to learn the language. The Student Edition of MATLAB is essentially identical to the full edition.

1.3   The MATLAB Environment

The fundamental unit of data in any MATLAB program is the **array.** An array is a collection of data values organized into rows and columns and known by a single name. Individual data values within an array may be accessed by including the name of the array, followed by subscripts in parentheses that identify the row and column of the particular value. Even scalars are treated as arrays by MATLAB—they are simply arrays with only one row and one column. We will learn how to create and manipulate MATLAB arrays in Section 1.4.

When MATLAB executes, it can display several types of windows that accept commands or display information. The three most important types of windows are Command Windows, where commands may be entered; Figure Windows, which display plots and graphs; and Edit Windows, which permit a user to create and modify MATLAB programs. We will see examples of all three types of windows in this section.

In addition, MATLAB can display other windows that provide help and that allow the user to examine the values of variables defined in memory. We will examine some of these additional windows here, examine the others when we discuss how to debug MATLAB programs.

1.3.1 The MATLAB Desktop

When you start MATLAB Version 7.4, a special window called the MATLAB desktop appears. The desktop is a window that contains other windows showing MATLAB data, plus toolbars and a "Start" button similar to that used by Windows XP. By default, most MATLAB tools are "docked" to the desktop so that they appear inside the desktop window. However, the user can choose to "undock" any or all tools, making them appear in windows separate from the desktop.

The default configuration of the MATLAB desktop is shown in Figure 1.1. It integrates many tools for managing files, variables, and applications within the MATLAB environment.

The major tools within or accessible from the MATLAB desktop are

- The Command Window
- The Command History Window
- The Start Button
- The Documents Window, including the Editor/Debugger and the Array Editor
- Figure Windows
- Workspace Browser
- Help Browser
- Path Browser

We will discuss the functions of these tools in later sections of this chapter.

1.3.2 The Command Window

The right-hand side of the default MATLAB desktop contains the **Command Window**. A user can enter interactive commands at the command prompt (») in the Command Window, and they will be executed on the spot.

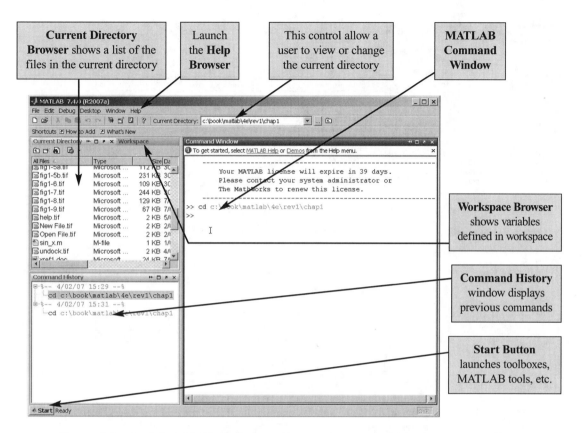

Figure 1.1   The default MATLAB desktop. The exact appearance of the desktop may differ slightly on different types of computers.

As an example of a simple interactive calculation, suppose that you want to calculate the area of a circle with a radius of 2.5 m. This can be done in the MATLAB Command Window by typing

```
» area = pi * 2.5^2
area =
   19.6350
```

MATLAB calculates the answer as soon as the Enter key is pressed and stores the answer in a variable (really a 1×1 array) called `area`. The contents of the variable are displayed in the Command Window as shown in Figure 1.2, and the variable can be used in further calculations. (Note that π is predefined in MATLAB, so we can just use `pi` without first declaring it to be 3.141592 . . .).

If a statement is too long to type on a single line, it may be continued on successive lines by typing an **ellipsis** (. . .) at the end of the first line and then continuing on the next line. For example, the following two statements are identical.

Figure 1.2 The Command Window appears on the right side of the desktop. Users enter commands and see responses here.

```
x1 = 1 + 1/2 + 1/3 + 1/4 + 1/5 + 1/6
```

and

```
x1 = 1 + 1/2 + 1/3 + 1/4 ...
     + 1/5 + 1/6
```

Instead of typing commands directly in the Command Window, a user can place a series of commands into a file, and the entire file can be executed by typing its name in the Command Window. Such files are called **script files**. Script files (and functions, which we will see later) are also known as **M-files**, because they have a file extension of ".m".

1.3.3 The Command History Window

The Command History Window displays a list of the commands that a user has entered in the Command Window. The list of previous commands can extend back to previous executions of the program. Commands remain in the list until they are deleted. To reexecute any command, simply double-click it with the left mouse

Figure 1.3   The Command History Window, showing two commands being deleted.

button. To delete one or more commands from the Command History window, select the commands and right-click them with the mouse. A popup menu will be displayed that allows the user to delete the items (see Figure 1.3).

1.3.4   The Start Button

The Start Button (see Figure 1.4) allows a user to access MATLAB tools, desktop tools, help files, and so on. It works just like the Start button on a Windows desktop. To start a particular tool, just click on the Start Button and select the tool from the appropriate submenu.

1.3.5   The Edit/Debug Window

An **Edit Window** is used to create new M-files, or to modify existing ones. An Edit Window is created automatically when you create a new M-file or open an existing one. You can create a new M-file by selecting "File/New/M-file" from the desktop menu or by clicking the ▯ toolbar icon. You can open an existing M-file file by selecting "File/Open" from the desktop menu or by clicking the 📂 toolbar icon.

Figure 1.4 The Start Button, which allows a user to select from a wide variety of MATLAB and desktop tools.

An Edit Window displaying a simple M-file called `calc_area.m` is shown in Figure 1.5. This file calculates the area of a circle, given its radius, and displays the result. By default, the Edit Window is an independent window not docked to the desktop, as shown in Figure 1.5*(a)*. The Edit Window can also be docked to the MATLAB desktop. In that case, it appears within a container called the Documents Window, as shown in Figure 1.5*(b)*. We will learn how to dock and undock a window later in this chapter.

The Edit Window is essentially a programming text editor, with the MATLAB languages features highlighted in different colors. Comments in an M-file file appear in green, variables and numbers appear in black, complete character strings appear in magenta, incomplete character strings appear in red, and language keywords appear in blue.

After an M-file is saved, it may be executed by typing its name in the Command Window. For the M-file in Figure 1.5, the results are

```
» calc_area
The area of the circle is 19.635
```

The Edit Window also doubles as a debugger, as we shall see in Chapter 2.

1.3.6 Figure Windows

A **Figure Window** is used to display MATLAB graphics. A figure can be a two- or three-dimensional plot of data, an image, or a graphical user interface (GUI).

(a)

(b)

Figure 1.5 *(a)* The MATLAB Editor, displayed as an independent window. *(b)* The MATLAB Editor, docked to the MATLAB desktop.

A simple script file that calculates and plots the function sin x is shown as follows:

```
% sin_x.m: This M-file calculates and plots the
% function sin(x) for 0 <= x <= 6.
x = 0:0.1:6
y = sin(x)
plot(x,y)
```

If this file is saved under the name `sin_x.m`, then a user can execute the file by typing "`sin_x`" in the Command Window. When this script file is executed, MATLAB opens a Figure Window and plots the function sin x in it. The resulting plot is shown in Figure 1.6.

1.3.7   Docking and Undocking Windows

MATLAB windows such as the Command Window, the Edit Window, and Figure Windows can either be *docked* to the desktop, or they can be *undocked*. When a window is docked, it appears as a pane within the MATLAB desktop. When it is undocked, it appears as an independent window on the computer screen separate from the desktop. When a window is docked to the desktop, the upper right-hand corner contains a small button with an arrow pointing up and to the right (⬈). If this button is clicked, the window will become an independent window. When

Figure 1.6   MATLAB plot of sin x versus x.

the window is an independent window, the upper right-hand corner contains a small button with an arrow pointing down and to the right (⬓). If this button is clicked, the window will be re-docked with the desktop. Figure 1.5 shows the Edit Window in both its docked and undocked state. Note the undock and dock arrows in the upper right-hand corner.

1.3.8   The MATLAB Workspace

A statement such as

```
z = 10
```

creates a variable named z, stores the value 10 in it, and saves it in a part of computer memory known as the **workspace**. A workspace is the collection of all the variables and arrays that can be used by MATLAB when a particular command, M-file, or function is executing. All commands executed in the Command Window (as well as all script files executed from the Command Window) share a common workspace, so they can all share variables. As we will see later, MATLAB functions differ from script files in that each function has its own separate workspace.

A list of the variables and arrays in the current workspace can be generated with the whos command. For example, after M-files calc_area and sin_x are executed, the whos command will display the following list of variables:

```
» whos
  Name      Size      Bytes    Class       Attributes
  area      1x1          8     double
  radius    1x1          8     double
  string    1x32        64     char
  x         1x61       488     double
  y         1x61       488     double
```

Script file calc_area created variables area, radius, and string, and script file sin_x created variables x and y. Note that all of the variables are in the same workspace, so if two script files are executed in succession, the second script file can use variables created by the first script file.

The contents of any variable or array may be determined by typing the appropriate name in the Command Window. For example, the contents of string can be found as follows:

```
» string
string =
The area of the circle is 19.635
```

A variable can be deleted from the workspace with the clear command. The clear command takes the form

```
clear var1 var2 ...
```

where `var1` and `var2` are the names of the variables to be deleted. The command `clear variables` or simply `clear` deletes all variables from the current workspace.

1.3.9 The Workspace Browser

The contents of the current workspace can also be examined with a GUI-based Workspace Browser. The Workspace Browser appears by default in the upper left-hand corner of the desktop. It provides a graphic display of the same information as the `whos` command, and it also shows the actual contents of each array if the information is short enough to fit within the display area. The Workspace Browser is dynamically updated whenever the contents of the workspace change.

A typical Workspace Browser window is shown in Figure 1.7. As you can see, it displays the same information as the `whos` command. Double-clicking on

Figure 1.7 The Workspace Browser and the Array Editor. The Array Editor is invoked by double-clicking a variable in the Workspace Browser. It allows a user to change the values contained in a variable or array.

any variable in the window will bring up the Array Editor, which allows the user to modify the information stored in the variable.

One or more variables may be deleted from the workspace by selecting them in the Workspace Browser with the mouse and pressing the delete key, or by right-clicking with the mouse and selecting the delete option.

1.3.10   Getting Help

There are three ways to get help in MATLAB. The preferred method is to use the Help Browser. The Help Browser can be started by selecting the ![?] icon from the desktop toolbar or by typing `helpdesk` or `helpwin` in the Command Window. A user can get help by browsing the MATLAB documentation, or he or she can search for the details of a particular command. The Help Browser is shown in Figure 1.8.

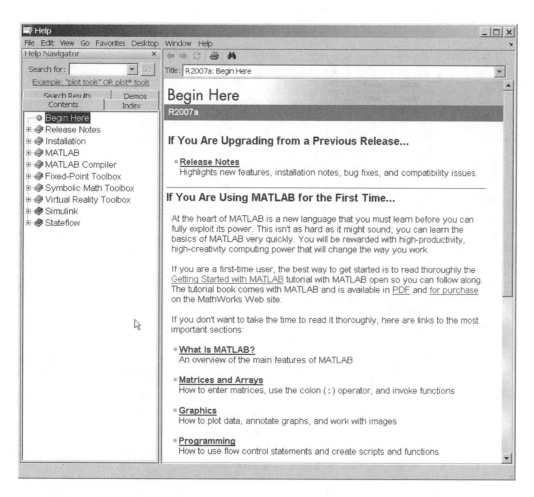

Figure 1.8   The Help Browser.

There are also two command-line–oriented ways to get help. The first way is to type help or help followed by a function name in the Command Window. If you just type help, MATLAB will display a list of possible help topics in the Command Window. If a specific function or a toolbox name is included, help will be provided for that particular function or toolbox.

The second way to get help is the lookfor command. The lookfor command differs from the help command in that the help command searches for an exact function name match, while the lookfor command searches the quick summary information in each function for a match. This makes lookfor slower than help, but it improves the chances of getting back useful information. For example, suppose that you were looking for a function to take the inverse of a matrix. Since MATLAB does not have a function named inverse, the command "help inverse" will produce nothing. On the other hand, the command "lookfor inverse" will produce the following results:

```
» lookfor inverse
INVHILB    Inverse Hilbert matrix.
ACOS       Inverse cosine.
ACOSH      Inverse hyperbolic cosine.
ACOT       Inverse cotangent.
ACOTH      Inverse hyperbolic cotangent.
ACSC       Inverse cosecant.
ACSCH      Inverse hyperbolic cosecant.
ASEC       Inverse secant.
ASECH      Inverse hyperbolic secant.
ASIN       Inverse sine.
ASINH      Inverse hyperbolic sine.
ATAN       Inverse tangent.
ATAN2      Four quadrant inverse tangent.
ATANH      Inverse hyperbolic tangent.
ERFINV     Inverse error function.
INV        Matrix inverse.
PINV       Pseudoinverse.
IFFT       Inverse discrete Fourier transform.
IFFT2      Two-dimensional inverse discrete Fourier transform.
IFFTN      N-dimensional inverse discrete Fourier transform.
IPERMUTE   Inverse permute array dimensions.
```

From this list, we can see that the function being sought is named inv.

1.3.11   A Few Important Commands

If you are new to MATLAB, a few demonstrations may help to give you a feel for its capabilities. To run MATLAB's built-in demonstrations, type demo in the Command Window, or select "demos" from the Start Button.

The contents of the Command Window can be cleared at any time using the `clc` command, and the contents of the current Figure Window can be cleared at any time using the `clf` command. The variables in the workspace can be cleared with the `clear` command. As we have seen, the contents of the workspace persist between the executions of separate commands and M-files, so it is possible for the results of one problem to have an effect on the next one that you may attempt to solve. To avoid this possibility, it is a good idea to issue the `clear` command at the start of each new independent calculation.

Another important command is the **abort** command. If an M-file appears to be running for too long, it may contain an infinite loop and it will never terminate. In this case, the user can regain control by typing control-c (abbreviated ^c) in the Command Window. This command is entered by holding down the control key while typing a "c". When MATLAB detects a ^c, it interrupts the running program and returns a command prompt.

The exclamation point (!)is another important special character. Its special purpose is to send a command to the computer's operating system. Any characters after the exclamation point will be sent to the operating system and executed as though they had been typed at the operating system's command prompt. This feature lets you embed operating system commands directly into MATLAB programs.

Finally, it is possible to keep track of everything done during a MATLAB session with the **diary** command. The form of this command is

```
diary filename
```

After this command is typed, a copy of all input and most output typed in the Command Window will be echoed in the diary file. This is a great tool for recreating events when something goes wrong during a MATLAB session. The command "`diary off`" suspends input into the diary file, and the command "`diary on`" resumes input again.

1.3.12   The MATLAB Search Path

MATLAB has a search path that it uses to find M-files. MATLAB's M-files are organized in directories on your file system. Many of these directories of M-files are provided along with MATLAB, and users may add others. If a user enters a name at the MATLAB prompt, the MATLAB interpreter attempts to find the name as follows:

1. It looks for the name as a variable. If it is a variable, MATLAB displays the current contents of the variable.
2. It checks to see if the name is an M-file in the current directory. If it is, MATLAB executes that function or command.
3. It checks to see if the name is an M-file in any directory in the search path. If it is, MATLAB executes that function or command.

Note that MATLAB checks for variable names first, so *if you define a variable with the same name as a MATLAB function or command, that function or command becomes inaccessible*. This is a common mistake made by novice users.

💣 Programming Pitfalls

Never use a variable with the same name as a MATLAB function or command. If you do so, that function or command will become inaccessible.

Also, if there is more than one function or command with the same name, the *first* one found on the search path will be executed and all of the others will be inaccessible. This is a common problem for novice users, since they sometimes create M-files with the same names as standard MATLAB functions, making them inaccessible.

💣 Programming Pitfalls

Never create an M-file with the same name as a MATLAB function or command.

MATLAB includes a special command (`which`) to help you find out just which version of a file is being executed and where it is located. This can be

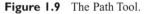

Figure 1.9   The Path Tool.

useful in finding filename conflicts. The format of this command is `which functionname`, where `functionname` is the name of the function that you are trying to locate. For example, the cross-product function `cross.m` can be located as follows:

```
» which cross
C:\Program Files\MATLAB\R2007a\toolbox\matlab\specfun\cross.m
```

The MATLAB search path can be examined and modified at any time by selecting "Desktop Tools/Path" from the Start Button, or by typing `editpath` in the Command Window. The Path Tool is shown in Figure 1.9. It allows a user to add, delete, or change the order of directories in the path.

Other path-related functions include

- `addpath`—Add directory to MATLAB search path.
- `path`—Display MATLAB search path.
- `path2rc`—Add current directory to MATLAB search path.
- `rmpath`—Remove directory from MATLAB search path.

1.4   Using MATLAB as a Scratch Pad

In its simplest form, MATLAB can be used as a scratch pad to perform mathematical calculations. The calculations to be performed are typed directly into the Command Window, using the symbols +, –, *, /, and ^ for addition, subtraction, multiplication, division, and exponentiation, respectively. After an expression is typed, the results of the expression will be automatically calculated and displayed. For example, suppose we would like to calculate the volume of a cylinder of radius r and length l. The area of the circle at the base of the cylinder is given by the equation

$$A = \pi r^2 \tag{1-1}$$

and the total volume of the cylinder will be

$$V = Al \tag{1-2}$$

If the radius of the cylinder is 0.1 m and the length is 0.5 m, the volume of the cylinder can be found using the MATLAB statements (user inputs are shown in boldface):

```
» A = pi * 0.1^2
A =
    0.0314
» V = A * 0.5
V =
    0.0157
```

Note that `pi` is predefined to be the value 3.141592 . . . . Also, note that the value stored in A was saved by MATLAB and reused when we calculated V.

Quiz 1.1

This quiz provides a quick check to see if you have understood the concepts introduced in Chapter 1. If you have trouble with the quiz, reread the sections, ask your instructor, or discuss the material with a fellow student. The answers to this quiz are found in the back of the book.

1. What is the purpose of the MATLAB Command Window? The Edit Window? The Figure Window?
2. List the different ways that you get help in MATLAB.
3. What is a workspace? How can you determine what is stored in a MATLAB workspace?
4. How can you clear the contents of a workspace?
5. The distance traveled by a ball falling in the air is given by the equation

$$x = x_0 + v_0 t + \frac{1}{2}at^2$$

   Use MATLAB to calculate the position of the ball at time $t = 5$ s if $x_0 = 10$ m, $v_0 = 15$ m/s, and $a = -9.81$ m/sec^2.
6. Suppose that $x = 3$ and $y = 4$. Use MATLAB to evaluate the following expression:

$$\frac{x^2 y^3}{(x - y)^2}$$

   The following questions are intended to help you become familiar with MATLAB tools.

7. Execute the M-files `calc_area.m` and `sin_x.m` in the Command Window (these M-files are available from the book's Web site). Then use the Workspace Browser to determine which variables are defined in the current workspace.
8. Use the Array Editor to examine and modify the contents of variable x in the workspace. Then type the command `plot(x,y)` in the Command Window. What happens to the data displayed in the Figure Window?

1.5 Summary

In this chapter, we have learned about the basic types of MATLAB windows, the workspace, and how to get on-line help. The MATLAB desktop appears when the program is started. It integrates many of the MATLAB tools in a single location. These tools include the Command Window, the Command History Window, the

Start Button, the Workspace Browser, the Array Editor, and the Current Directory Viewer. The Command Window is the most important of the windows. It is the one in which all commands are typed and results are displayed.

The Edit/Debug window is used to create or modify M-files. It displays the contents of the M-file with the contents of the file color-coded according to function: comments, keywords, strings, and so forth. This window can be docked to the desktop, but by default it is independent.

The Figure Window is used to display graphics.

A MATLAB user can get help by either using the Help Browser or the command-line help functions help and lookfor. The Help Browser allows full access to the entire MATLAB documentation set. The command-line function help displays help about a specific function in the Command Window. Unfortunately, you must know the name of the function in order to get help about it. The function lookfor searches for a given string in the first comment line of every MATLAB function and displays any matches.

When a user types a command in the Command Window, MATLAB searches for that command in the directories specified in the MATLAB path. It will execute the *first* M-file in the path that matches the command; any further M-files with the same name will never be found. The Path Tool can be used to add, delete, or modify directories in the MATLAB path.

1.5.1 MATLAB Summary

The following summary lists all of the MATLAB special symbols described in this chapter, along with a brief description of each one.

Special Symbols

+	Addition
–	Subtraction
*	Multiplication
/	Division
^	Exponentiation

1.6 Exercises

1.1 The following MATLAB statements plot the function $y(x) = 2e^{-0.2x}$ for the range $0 \le x \le 10$.

```
x = 0:0.1:10;
y = 2 * exp(-0.2 * x);
plot(x,y);
```

Use the MATLAB Edit Window to create a new empty M-file, type these statements into the file, and save the file with the name test1.m. Then, execute the program by typing the name test1 in the Command Window. What result do you get?

1.2 Get help on the MATLAB function exp using: *(a)* The "help exp" command typed in the Command Window and *(b)* the Help Browser.

1.3 Use the lookfor command to determine how to take the base-10 logarithm of a number in MATLAB.

1.4 Suppose that $u = 1$ and $v = 3$. Evaluate the following expressions using MATLAB:

(a) $\dfrac{4u}{3v}$

(b) $\dfrac{2v^{-2}}{(u + v)^2}$

(c) $\dfrac{v^3}{v^3 - u^3}$

(d) $\dfrac{4}{3}\pi v^2$

1.5 Use the MATLAB Help Browser to find the command required to show MATLAB's current directory. What is the current directory when MATLAB starts up?

1.6 Use the MATLAB Help Browser to find out how to create a new directory from within MATLAB. Then, create a new directory called mynewdir under the current directory. Add the new directory to the top of MATLAB's path.

1.7 Change the current directory to mynewdir. Then open an Edit Window and add the following lines:

```
% Create an input array from -2*pi to 2*pi
t = -2*pi:pi/10:2*pi;

% Calculate |sin(t)|
x = abs(sin(t));

% Plot result
plot(t,x);
```

Save the file with the name test2.m and execute it by typing test2 in the Command Window. What happens?

1.8 Close the Figure Window and change back to the original directory that MATLAB started up in. Next type "test2" in the Command Window. What happens, and why?

CHAPTER **2**

MATLAB Basics

In this chapter, we will introduce some basic elements of the MATLAB language. By the end of the chapter, you will be able to write simple but functional MATLAB programs.

2.1   Variables and Arrays

The fundamental unit of data in any MATLAB program is the **array**. An array is a collection of data values organized into rows and columns, and known by a single name (see Figure 2.1). Individual data values within an array are accessed by including the name of the array followed by subscripts in parentheses that identify the row and column of the particular value. Even scalars are treated as arrays by MATLAB—they are simply arrays with only one row and one column.

Arrays can be classified as either **vectors** or **matrices**. The term "vector" is usually used to describe an array with only one dimension, while the term "matrix" is usually used to describe an array with two or more dimensions. In this text, we will use the term "vector" when discussing one-dimensional arrays, and the term "matrix" when discussing arrays with two or more dimensions. If a particular discussion applies to both types of arrays, we will use the generic term "array".

The **size** of an array is specified by the number of rows and the number of columns in the array, with the number of rows mentioned first. The total number of elements in the array will be the product of the number of rows and the number of columns. For example, the sizes of the following arrays are

Array	Size
$a = \begin{bmatrix} 1 & 2 \\ 3 & 4 \\ 5 & 6 \end{bmatrix}$	This is 3 × 2 matrix, containing 6 elements.
b = [1 2 3 4]	This is a 1 × 4 array containing 4 elements, known as a **row vector**.
$c = \begin{bmatrix} 1 \\ 2 \\ 3 \end{bmatrix}$	This is a 3 × 1 array containing 3 elements, known as a **column vector**.

Individual elements in an array are addressed by the array name followed by the row and column of the particular element. If the array is a row or column vector, then only one subscript is required. For example, in the preceding arrays a(2,1) is 3 and c(2) = 2.

A MATLAB **variable** is a region of memory containing an array, which is known by a user-specified name. The contents of the array may be used or modified at any time by including its name in an appropriate MATLAB command.

MATLAB variable names must begin with a letter, followed by any combination of letters, numbers, and the underscore (_) character. Only the first

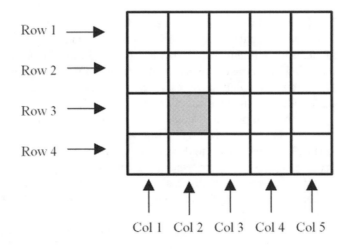

array arr

Figure 2.1 An array is a collection of data values organized into rows and columns.

63 characters are significant; if more than 63 are used, the remaining characters will be ignored. If two variables are declared with names that only differ in the 64th character, MATLAB will treat them as the same variable. MATLAB will issue a warning if it has to truncate a long variable name to 63 characters.

☄ Programming Pitfalls

Make sure that your variable names are unique in the first 63 characters. Otherwise, MATLAB will not be able to tell the difference between them.

When writing a program, it is important to pick meaningful names for the variables. Meaningful names make a program *much* easier to read and to maintain. Names such as day, month, and year are quite clear even to a person seeing a program for the first time. Since spaces cannot be used in MATLAB variable names, underscore characters can be substituted to create meaningful names. For example, *exchange rate* might become exchange_rate.

✳ Good Programming Practice

Always give your variables descriptive and easy-to-remember names. For example, a currency exchange rate could be given the name exchange_rate. This practice will make your programs clearer and easier to understand.

It is also important to include a **data dictionary** in the header of any program that you write. A data dictionary lists the definition of each variable used in a program. The definition should include both a description of the contents of the item and the units in which it is measured. A data dictionary may seem unnecessary while the program is being written, but it is invaluable when you or another person have to go back and modify the program at a later time.

✳ Good Programming Practice

Create a data dictionary for each program to make program maintenance easier.

The MATLAB language is case sensitive, which means that uppercase and lowercase letters are not the same. Thus the variables name, NAME, and Name are all different in MATLAB. You must be careful to use the same capitalization every time that variable name is used. While it is not required, it is customary to use all lowercase letters for ordinary variable names.

> ✳ **Good Programming Practice**
>
> Be sure to capitalize a variable exactly the same way each time that it is used. It is good practice to use only lowercase letters in variable names.

The most common types of MATLAB variables are `double` and `char`. Variables of type `double` consist of scalars or arrays of 64-bit double-precision floating-point numbers. They can hold real, imaginary, or complex values. The real and imaginary components of each variable can be positive or negative numbers in the range 10^{-308} to 10^{308}, with 15 to 16 significant decimal digits of accuracy. Type `double` is the principal numerical data type in MATLAB.

A variable of type `double` is automatically created whenever a numerical value is assigned to a variable name. The numerical values assigned to `double` variables can be real, imaginary, or complex. A real value is just a number. For example, the following statement assigns the real value 10.5 to the `double` variable `var`:

```
var = 10.5
```

An imaginary number is defined by appending the letter i or j to a number[1]. For example, `10i` and `-4j` are both imaginary values. The following statement assigns the imaginary value $4i$ to the `double` variable `var`:

```
var = 4i
```

A complex value has both a real and an imaginary component. It is created by adding a real and an imaginary number together. For example, the following statement assigns the complex value $10 + 10i$ to variable `var`:

```
var = 10 + 10i
```

Variables of type `char` consist of scalars or arrays of 16-bit values, each representing a single character. Arrays of this type are used to hold character strings. They are automatically created whenever a single character or a character string is assigned to a variable name. For example, the following statement creates a variable of type `char` whose name is `comment`, and stores the specified string in it. After the statement is executed, `comment` will be a 1×26 character array.

```
comment = 'This is a character string'
```

In a language such as C, the type of every variable must be explicitly declared in a program before it is used. These languages are said to be **strongly typed**. In contrast, MATLAB is a **weakly typed** language. Variables may be created at any

---

[1]An imaginary number is a number multiplied by $\sqrt{-1}$. The letter i is the symbol for $\sqrt{-1}$ used by most mathematicians and scientists. The letter j is the symbol for $\sqrt{-1}$ used by electrical engineers, because the letter i is usually reserved for currents in that discipline.

time by simply assigning values to them, and the type of data assigned to the variable determines the type of variable that is created.

2.2 Creating and Initializing Variables in MATLAB

MATLAB variables are automatically created when they are initialized. There are three common ways to initialize a variable in MATLAB:

1. Assign data to the variable in an assignment statement.
2. Input data into the variable from the keyboard.
3. Read data from a file.

The first two ways will be discussed here, and the third will be discussed in Section 2.6.

2.2.1 Initializing Variables in Assignment Statements

The simplest way to initialize a variable is to assign it one or more values in an **assignment statement**. An assignment statement has the general form

```
var = expression;
```

where `var` is the name of a variable, and *expression* is a scalar constant, an array, or a combination of constants, other variables, and mathematical operations (+, −, etc.). The value of the expression is calculated using the normal rules of mathematics, and the resulting values are stored in named variable. The semicolon at the end of the statement is optional. If the semicolon is absent, the value assigned to var will be echoed in the Command Window. If it is present, nothing will be displayed in the Command Window even though the assignment has occurred. Simple examples of initializing variables with assignment statements include

```
var = 40i;
var2 = var/5;
x = 1; y = 2;
array = [1  2  3  4];
```

The first example creates a scalar variable of type `double` and stores the imaginary number 40*i* in it. The second example creates a scalar variable and stores the result of the expression `var/5` in it. The third example shows that multiple assignment statements can be placed on a single line, provided that they are separated by semicolons or commas. Note that if any of the variables had already existed when the statements were executed, then their old contents would have been lost.

The last example shows that variables can also be initialized with arrays of data. Such arrays are constructed using brackets (`[ ]`) and semicolons. All of the elements of an array are listed in **row order**. In other words, the values in each row are listed from left to right, with the topmost row first and the bottommost row last. Individual values within a row are separated by blank spaces or commas, and the rows themselves are separated by semicolons or new lines. The following expressions are all legal arrays that can be used to initialize a variable:

`[3.4]` This expression creates a 1×1 array (a scalar) containing the value 3.4. The brackets are not required in this case.

`[1.0 2.0 3.0]` This expression creates a 1×3 array containing the row vector [1 2 3].

`[1.0; 2.0; 3.0]` This expression creates a 3×1 array containing the column vector $\begin{bmatrix} 1 \\ 2 \\ 3 \end{bmatrix}$.

`[1, 2, 3; 4, 5, 6]` This expression creates a 2×3 array containing the matrix $\begin{bmatrix} 1 & 2 & 3 \\ 4 & 5 & 6 \end{bmatrix}$.

`[1, 2, 3`
`4, 5, 6]` This expression creates a 2×3 array containing the matrix $\begin{bmatrix} 1 & 2 & 3 \\ 4 & 5 & 6 \end{bmatrix}$. The end of the first line terminates the first row.

`[]` This expression creates an **empty array**, which contains no rows and no columns. (Note that this is not the same as an array containing zeros.)

The number of elements in every row of an array must be the same, and the number of elements in every column must be the same. An expression such as

```
[1 2 3; 4 5];
```

is illegal because row 1 has three elements while row 2 has only two elements.

☄ Programming Pitfalls

The number of elements in every row of an array must be the same, and the number of elements in every column must be the same. Attempts to define an array with different numbers of elements in its rows or different numbers of elements in its columns will produce an error when the statement is executed.

The expressions used to initialize arrays can include algebraic operations and all of or portions of previously defined arrays. For example, the assignment statements

```
a = [0  1+7];
b = [a(2)  7 a];
```

will define an array a = [0 8] and an array b = [8 7 0 8].

Also, not all of the elements in an array must be defined when it is created. If a specific array element is defined and one or more of the elements before it are not, then the earlier elements will automatically be created and initialized to zero. For example, if c is not previously defined, the statement

```
c(2,3)  =  5;
```

will produce the matrix $c = \begin{bmatrix} 0 & 0 & 0 \\ 0 & 0 & 5 \end{bmatrix}$. Similarly, an array can be extended by specifying a value for an element beyond the currently defined size. For example, suppose that array d = [1 2]. Then the statement

```
d(4)  =  4;
```

will produce the array d = [1 2 0 4].

The semicolon at the end of each assignment statement shown above has a special purpose: it *suppresses the automatic echoing of values* that normally occurs whenever an expression is evaluated in an assignment statement. If an assignment statement is typed without the semicolon, the result of the statement is automatically displayed in the Command Window:

```
»  e  =  [1,  2,  3;  4,  5,  6]
e  =
    1   2   3
    4   5   6
```

If a semicolon is added at the end of the statement, the echoing disappears. Echoing is an excellent way to quickly check your work, but it seriously slows down the execution of MATLAB programs. For that reason, we normally suppress echoing at all times by ending each line with a semicolon.

However, echoing the results of calculations makes a great quick-and-dirty debugging tool. If you are not certain what the results of a specific assignment statement are, just leave off the semicolon from that statement, and the results will be displayed in the Command Window as the statement is executed.

✳ Good Programming Practice

Use a semicolon at the end of all MATLAB assignment statements to suppress echoing of assigned values in the Command Window. This greatly speeds program execution.

✳ Good Programming Practice

If you need to examine the results of a statement during program debugging, you may remove the semicolon from that statement only so that its results are echoed in the Command Window.

2.2.2   Initializing with Shortcut Expressions

It is easy to create small arrays by explicitly listing each term in the array, but what happens when the array contains hundreds or even thousands of elements? It is just not practical to write out each element in the array separately!

MATLAB provides a special shortcut notation for these circumstances using the **colon operator**. The colon operator specifies a whole series of values by specifying the first value in the series, the stepping increment, and the last value in the series. The general form of a colon operator is

```
first:incr:last
```

where `first` is the first value in the series, `incr` is the stepping increment, and `last` is the last value in the series. If the increment is one, it may be omitted. This operator will generate an array containing the values `first`, `first+incr`, `first+2*incr`, `first+3*incr`, and so forth as long as the values are less than or equal to last. The list stops when the next value in the series is greater than the value of last.

For example, the expression 1:2:10 is a shortcut for a 1×5 row vector containing the values 1, 3, 5, 7, and 9. The next value in the series would be 11, which is greater than 10, so the series terminates at 9.

```
» x = 1:2:10
x =
    1   3   5   7   9
```

With colon notation, an array can be initialized to have the hundred values $\frac{\pi}{100}$, $\frac{2\pi}{100}$, $\frac{3\pi}{100}$, . . . , π as follows:

```
angles = (0.01:0.01:1.00) * pi;
```

Shortcut expressions can be combined with the **transpose operator** (`'`) to initialize column vectors and more complex matrices. The transpose operator swaps the row and columns of any array that it is applied to. Thus the expression

```
f = [1:4]';
```

generates a 4-element row vector [1 2 3 4], and then transposes it into the 4-element column vector $f = \begin{bmatrix} 1 \\ 2 \\ 3 \\ 4 \end{bmatrix}$. Similarly, the expressions

```
g = 1:4;
h = [g' g'];
```

will produce the matrix $h = \begin{bmatrix} 1 & 1 \\ 2 & 2 \\ 3 & 3 \\ 4 & 4 \end{bmatrix}$.

2.2.3   Initializing with Built-In Functions

Arrays can also be initialized using built-in MATLAB functions. For example, the function `zeros` can be used to create an all-zero array of any desired size. There are several forms of the `zeros` function. If the function has a single scalar argument, it will produce a square array using the single argument as both the number of rows and the number of columns. If the function has two scalar arguments, the first argument will be the number of rows, and the second argument will be the number of columns. Since the `size` function returns two values containing the number of rows and columns in an array, it can be combined with the `zeros` function to generate an array of zeros that is the same size as another array. Some examples using the `zeros` function follow:

```
a = zeros(2);
b = zeros(2,3);
c = [1 2; 3 4];
d = zeros(size(c));
```

These statements generate the following arrays:

$$a = \begin{bmatrix} 0 & 0 \\ 0 & 0 \end{bmatrix} \qquad b = \begin{bmatrix} 0 & 0 & 0 \\ 0 & 0 & 0 \end{bmatrix}$$

$$c = \begin{bmatrix} 1 & 2 \\ 3 & 4 \end{bmatrix} \qquad d = \begin{bmatrix} 0 & 0 \\ 0 & 0 \end{bmatrix}$$

Similarly, the `ones` function can be used to generate arrays containing all ones, and the `eye` function can be used to generate arrays containing **identity matrices**, in which all on-diagonal elements are one, while all off-diagonal elements are zero. Table 2.1 contains list of common MATLAB functions useful for initializing variables.

2.2.4   Initializing Variables with Keyboard Input

It is also possible to prompt a user and initialize a variable with data that he or she types directly at the keyboard. This option allows a script file to prompt a user for input data values while it is executing. The `input` function displays a prompt string in the Command Window and then waits for the user to type in a response. For example, consider the following statement:

```
my_val = input('Enter an input value:');
```

Table 2.1 MATLAB Functions Useful for Initializing Variables

Function	Purpose
zeros(n)	Generates an n × n matrix of zeros.
zeros(m,n)	Generates an m × n matrix of zeros.
zeros(size(arr))	Generates a matrix of zeros of the same size as arr.
ones(n)	Generates an n × n matrix of ones.
ones(m,n)	Generates an m × n matrix of ones.
ones(size(arr))	Generates a matrix of ones of the same size as arr.
eye(n)	Generates an n × n identity matrix.
eye(m,n)	Generates an m × n identity matrix.
length(arr)	Returns the length of a vector, or the longest dimension of a 2-D array.
size(arr)	Returns two values specifying the number of rows and columns in arr.

When this statement is executed, MATLAB prints out the string `'Enter an input value: '`, and then waits for the user to respond. If the user enters a single number, it may just be typed in. If the user enters an array, it must be enclosed in brackets. In either case, whatever is typed will be stored in variable `my_val` when the return key is entered. If only the return key is entered, then an empty matrix will be created and stored in the variable.

If the `input` function includes the character `'s'` as a second argument, then the input data is returned to the user as a character string. Thus, the statement

```
» in1 = input('Enter data: ');
Enter data: 1.23
```

stores the value 1.23 into `in1`, while the statement

```
» in2 = input('Enter data: ','s');
Enter data: 1.23
```

stores the character string `'1.23'` into `in2`.

Quiz 2.1

This quiz provides a quick check to see if you have understood the concepts introduced in Sections 2.1 and 2.2. If you have trouble with the quiz, reread the sections, ask your instructor, or discuss the material with a fellow student. The answers to this quiz are found in the back of the book.

1. What is the difference between an array, a matrix, and a vector?

2. Answer the following questions for the array shown below.

$$c = \begin{bmatrix} 1.1 & -3.2 & 3.4 & 0.6 \\ 0.6 & 1.1 & -0.6 & 3.1 \\ 1.3 & 0.6 & 5.5 & 0.0 \end{bmatrix}$$

(*a*) What is the size of c?

(*b*) What is the value of c(2,3)?

(*c*) List the subscripts of all elements containing the value 0.6.

3. Determine the size of the following arrays. Check your answers by entering the arrays into MATLAB and using the whos command or the Workspace Browser. Note that the later arrays may depend on the definitions of arrays defined earlier in this exercise.

(*a*) u = [10 20*i 10+20];

(*b*) v = [-1; 20; 3];

(*c*) w = [1 0 -9; 2 -2 0; 1 2 3];

(*d*) x = [u' v];

(*e*) y(3,3) = -7;

(*f*) z = [zeros(4,1) ones(4,1) zeros(1,4)'];

(*g*) v(4) = x(2,1);

4. What is the value of w(2,1) above?

5. What is the value of x(2,1) above?

6. What is the value of y(2,1) above?

7. What is the value of v(3) after statement (*g*) is executed?

2.3   Multidimensional Arrays

As we have seen, MATLAB arrays can have one or more dimensions. One-dimensional arrays can be visualized as a series of values laid out in a row or column, with a single subscript used to select the individual array elements (Figure 2.2*a*). Such arrays are useful to describe data that is a function of one independent variable, such as a series of temperature measurements made at fixed intervals of time.

Some types of data are functions of more than one independent variable. For example, we might wish to measure the temperature at five different locations at four different times. In this case, our 20 measurements could logically be grouped into five different columns of four measurements each, with a separate column for each location (Figure 2.2*b*). In this case, we will use two subscripts to access a given element in the array—the first one to select the row and the second one to select the column. Such arrays are called **two-dimensional arrays**. The number of elements in a two-dimensional array will be the product of the number of rows and the number of columns in the array.

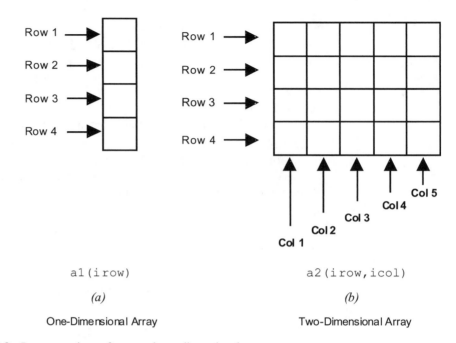

al(irow)

a2(irow,icol)

(a)

(b)

One-Dimensional Array

Two-Dimensional Array

Figure 2.2 Representations of one- and two-dimensional arrays.

MATLAB allows us to create arrays with as many dimensions as necessary for any given problem. These arrays have one subscript for each dimension, and an individual element is selected by specifying a value for each subscript. The total number of elements in the array will be the product of the maximum value of each subscript. For example, the following two statements create a $2 \times 3 \times 2$ array c:

```
» c(:,:,1)=[1 2 3; 4 5 6];
» c(:,:,2)=[7 8 9; 10 11 12];
» whos c

   Name    Size    Bytes   Class    Attributes

    c      2x3x2      96    double
```

This array contains 12 elements ($2 \times 3 \times 2$). Its contents can be displayed just like any other array.

```
» c
c(:,:,1) =
        1    2    3
        4    5    6
c(:,:,2) =
        7    8    9
       10   11   12
```

2.3.1 Storing Multidimensional Arrays in Memory

A two-dimensional array with m rows and n columns will contain m × n elements, and these elements will occupy m × n successive locations in the computer's memory. How are the elements of the array arranged in the computer's memory? MATLAB always allocates array elements in **column major order**. That is, MATLAB allocates the first column in memory, then the second, then the third, and so on., until all of the columns have been allocated. Figure 2.3 illustrates this memory allocation scheme for a 4 × 3 array a. As we can see, element a(1,2) is really the fifth element allocated in memory. The order in which elements are allocated in memory will become important when we discuss single-subscript addressing in the following section, and low-level I/O functions in Chapter 8.

This same allocation scheme applies to arrays with more than two dimensions. The first array subscript is incremented most rapidly, the second subscript is incremented less rapidly, and so on, and the last subscript in incremented most slowly. For example, in a 2 × 2 × 2 array, the elements would be allocated in the following order: (1,1,1), (2,1,1), (1,2,1), (2,2,1), (1,1,2), (2,1,2), (1,2,2), (2,2,2).

2.3.2 Accessing Multidimensional Arrays with One Dimension

One of MATLAB's peculiarities is that it will permit a user or programmer to treat a multidimensional array as though it were a one-dimensional array whose length is equal to the number of elements in the multidimensional array. If a multidimensional array is addressed with a single dimension, then the elements will be accessed in the order in which they were allocated in memory.

For example, suppose that we declare the 4 × 3 element array a as follows:

```
» a = [1 2 3; 4 5 6; 7 8 9; 10 11 12]
a =
     1     2     3
     4     5     6
     7     8     9
    10    11    12
```

Then the value of a(5) will be 2, which is the value of element a(1,2), because a(1,2) was allocated fifth in memory.

Under normal circumstances, you should never use this feature of MATLAB. Addressing multidimensional arrays with a single subscript is a recipe for confusion.

✳ Good Programming Practice

Always use the proper number of dimensions when addressing a multidimensional array.

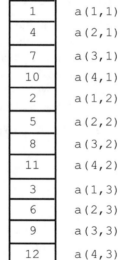

a

(a)

Arrangement in Computer Memory

(b)

Figure 2.3 *(a)* Data values for array a. *(b)* Layout of values in memory for array a.

2.4   Subarrays

It is possible to select and use subsets of MATLAB arrays as though they were separate arrays. To select a portion of an array, just include a list of all of the elements to be selected in the parentheses after the array name. For example, suppose array `arr1` is defined as follows:

```
arr1 = [1.1  -2.2  3.3  -4.4  5.5];
```

Then `arr1(3)` is just 3, `arr1([1  4])` is the array `[1.1 -4.4]`, and `arr1(1:2:5)` is the array `[1.1 3.3 5.5]`.

For a two-dimensional array, a colon can be used in a subscript to select all of the values of that subscript. For example, suppose

```
arr2 = [1  2  3;  -2  -3  -4;  3  4  5];
```

This statement would create an array `arr2` containing the values $\begin{bmatrix} 1 & 2 & 3 \\ -2 & -3 & -4 \\ 3 & 4 & 5 \end{bmatrix}$. With this definition, the subarray `arr2(1,:)` would be `[1 2 3]`, and the subarray `arr2(:,1:2:3)` would be $\begin{bmatrix} 1 & 3 \\ -2 & -4 \\ 3 & 5 \end{bmatrix}$.

2.4.1   The end Function

MATLAB includes a special function named `end` that is very useful for creating array subscripts. When used in an array subscript, `end` *returns the highest value taken on by that subscript*. For example, suppose that array `arr3` is defined as follows:

```
arr3 = [1  2  3  4  5  6  7  8];
```

Then `arr3(5:end)` would be the array `[5 6 7 8]`, and `array(end)` would be the value 8.

The value returned by `end` is always the *highest value* of a given subscript. If `end` appears in different subscripts, it can return *different* values within the same expression. For example, suppose that the 3 × 4 array `arr4` is defined as follows:

```
arr4 = [1  2  3  4;  5  6  7  8;  9  10  11  12];
```

Then the expression `arr4(2:end,2:end)` would return the array $\begin{bmatrix} 6 & 7 & 8 \\ 10 & 11 & 12 \end{bmatrix}$. Note that the first `end` returned the value 3, while the second `end` returned the value 4!

2.4.2 Using Subarrays on the Left-Hand Side of an Assignment Statement

It is also possible to use subarrays on the left-hand side of an assignment statement to update only some of the values in an array, as long as the **shape** (the number of rows and columns) of the values being assigned matches the shape of the subarray. If the shapes do not match, then an error will occur. For example, suppose that the 3×4 array `arr4` is defined as follows:

```
» arr4 = [1 2 3 4; 5 6 7 8; 9 10 11 12]
arr4 =
      1    2    3    4
      5    6    7    8
      9   10   11   12
```

Then the following assignment statement is legal, since the expressions on both sides of the equal sign have the same shape (2×2):

```
» arr4(1:2,[1 4]) = [20 21; 22 23]
arr4 =
     20    2    3   21
     22    6    7   23
      9   10   11   12
```

Note that the array elements (1,1), (1,4), (2,1), and (2,4) were updated. In contrast, the following expression is illegal because the two sides do not have the same shape.

```
» arr5(1:2,1:2) = [3 4]
??? In an assignment A(matrix,matrix) = B, the num-
ber of rows in B and the number of elements in the
A row index matrix must be the same.
```

☀ Programming Pitfalls

For assignment statements involving subarrays, the *shapes of the subarrays on either side of the equal sign must match*. MATLAB will produce an error if they do not match.

There is a major difference in MATLAB between assigning values to a subarray and assigning values to an array. If values are assigned to a subarray, *only those values are updated, while all other values in the array remain unchanged.* On the other hand, if values are assigned to an array, *the entire contents of the*

array are deleted and replaced by the new values. For example, suppose that the
3 × 4 array `arr4` is defined as follows:

```
» arr4 = [1 2 3 4; 5 6 7 8; 9 10 11 12]
arr4 =
      1    2    3    4
      5    6    7    8
      9   10   11   12
```

Then the following assignment statement replaces the *specified elements* of `arr4`:

```
» arr4(1:2,[1 4]) = [20 21; 22 23]
arr4 =
     20    2    3   21
     22    6    7   23
      9   10   11   12
```

In contrast, the following assignment statement replaces the *entire contents* of
`arr4` with a 2 × 2 array:

```
» arr4 = [20 21; 22 23]
arr4 =
     20   21
     22   23
```

Good Programming Practice

Be sure to distinguish between assigning values to a subarray and assigning values to an array. MATLAB behaves differently in these two cases.

2.4.3 Assigning a Scalar to a Subarray

A scalar value on the right-hand side of an assignment statement always matches the shape specified on the left-hand side. The scalar value is copied into every element specified on the left-hand side of the statement. For example, assume that the 3 × 4 array `arr4` is defined as follows:

```
arr4 = [1  2  3  4;  5  6  7  8;  9  10  11  12];
```

Then the expression shown below assigns the value one to four elements of the array.

```
» arr4(1:2,1:2) = 1
arr4 =
      1    1    3    4
      1    1    7    8
      9   10   11   12
```

2.5 Special Values

MATLAB includes a number of predefined special values. These predefined values may be used at any time in MATLAB without initializing them first. A list of the most common predefined values is given in Table 2.2.

These predefined values are stored in ordinary variables, so they can be overwritten or modified by a user. If a new value is assigned to one of the predefined variables, then that new value will replace the default one in all later calculations. For example, consider the following statements that calculate the circumference of a 10-cm circle:

```
circ1 = 2 * pi * 10
pi = 3;
circ2 = 2 * pi * 10
```

In the first statement, pi has its default value of 3.14159. . . , so circ1 is 62.8319, which is the correct circumference. The second statement redefines pi to be 3, so in the third statement circ2 is 60. Changing a predefined value in the program has created an incorrect answer, and also introduced a subtle and hard-to-find bug. Imagine trying to locate the source of such a hidden error in a 10,000-line program!

Table 2.2   Predefined special values

Function	Purpose
pi	Contains π to 15 significant digits
i, j	Contain the value $i(\sqrt{-1})$.
inf	This symbol represents machine infinity. It is usually generated as a result of a division by 0.
nan	This symbol stands for Not-a-Number. It is the result of an undefined mathematical operation, such as the division of zero by zero.
clock	This special variable contains the current date and time in the form of a 6-element row vector containing the year, month, day, hour, minute, and second.
date	Contains the current data in a character string format, such as 24-Nov-1998.
eps	This variable name is short for "epsilon". It is the smallest difference between two numbers that can be represented on the computer.
ans	A special variable used to store the result of an expression if that result is not explicitly assigned to some other variable.

💣 **Programming Pitfalls**

Never redefine the meaning of a predefined variable in MATLAB. It is a recipe for disaster, producing subtle and hard-to-find bugs.

Quiz 2.2

This quiz provides a quick check to see if you have understood the concepts introduced in Sections 2.3 through 2.5. If you have trouble with the quiz, reread the sections, ask your instructor, or discuss the material with a fellow student. The answers to this quiz are found in the back of the book.

1. Assume that array c is defined as shown, and determine the contents of the following subarrays:

$$
c = \begin{bmatrix} 1.1 & -3.2 & 3.4 & 0.6 \\ 0.6 & 1.1 & -0.6 & 3.1 \\ 1.3 & 0.6 & 5.5 & 0.0 \end{bmatrix}
$$

 (a) c(2,:)
 (b) c(:,end)
 (c) c(1:2,2:end)
 (d) c(6)
 (e) c(4:end)
 (f) c(1:2,2:4)
 (g) c([1 3],2)
 (h) c([2 2],[3 3])

2. Determine the contents of array a after the following statements are executed.

 (a) a = [1 2 3; 4 5 6; 7 8 9];
     a([3 1],:) = a([1 3],:);

 (b) a = [1 2 3; 4 5 6; 7 8 9];
     a([1 3],:) = a([2 2],:);

 (c) a = [1 2 3; 4 5 6; 7 8 9];
     a = a([2 2],:);

3. Determine the contents of array a after the following statements are executed.

 (a) a = eye(3,3);
     b = [1 2 3];
     a(2,:) = b;

```
(b) a = eye(3,3);
    b = [4 5 6];
    a(:,3) = b';

(c) a = eye(3,3);
    b = [7 8 9];
    a(3,:) = b([3 1 2]);
```

2.6 Displaying Output Data

There are several ways to display output data in MATLAB. This simplest way is one we have already seen—just leave the semicolon off of the end of a statement and it will be echoed to the Command Window. We will now explore a few other ways to display data.

2.6.1 Changing the Default Format

When data is echoed in the Command Window, integer values are always displayed as integers, character values are displayed as strings, and other values are printed using a **default format**. The default format for MATLAB shows four digits after the decimal point, and it may be displayed in scientific notation with an exponent if the number is too large or too small. For example, the statements

```
x = 100.11
y = 1001.1
z = 0.00010011
```

produce the following output

```
x =
    100.1100

y =
    1.0011e+003

z =
    1.0011e-004
```

This default format can be changed in one of two ways: from the main MATLAB Window menu, or using the **format** command. You can change the format by selecting the "File / Preferences" menu option. This option will pop up the preferences window, and the format can be selected from the Command Window item in the preferences list.

Alternatively, a user can use the `format` command to change the preferences. The format command changes the default format according to the values given in Table 2.3. The default format can be modified to display more significant digits of data, to force the display to be in scientific notation, to display data to two decimal

Table 2.3   Output Display Formats

Format Command	Results	Example[1]
format short	4 digits after decimal (default format)	12.3457
format long	14 digits after decimal	12.34567890123457
format short e	5 digits plus exponent	1.2346e+001
format short g	5 total digits with or without exponent	12.346
format long e	15 digits plus exponent	1.234567890123457e+001
format long g	15 total digits with or without exponent	12.3456789012346
format bank	"dollars and cents" format	12.35
format hex	hexadecimal display of bits	4028b0fcd32f707a
format rat	approximate ratio of small integers	1000/81
format compact	suppress extra line feeds	
format loose	restore extra line feeds	
format +	Only signs are printed.	+

[1]The data value used for the example is 12.345678901234567 in all cases.

digits, or to eliminate extra line feeds to get more data visible in the Command Window at a single time. Experiment with the commands in Table 2.3 for yourself.

Which of these ways to change the data format is better? If you are working directly at the computer, it is probably easier to use the menu item. On the other hand, if you are writing programs, it is probably better to use the format command, because it can be embedded directly into a program.

2.6.2   The `disp` function

Another way to display data is with the disp function. The disp function accepts an array argument and displays the value of the array in the Command Window. If the array is of type char, then the character string contained in the array is printed out.

This function is often combined with the functions num2str (convert a number to a string) and int2str (convert an integer to a string) to create messages to be displayed in the Command Window. For example, the following MATLAB statements will display "The value of pi = 3.1416" in the Command Window. The first statement creates a string array containing the message, and the second statement displays the message.

```
str = ['The value of pi = ' num2str(pi)];
disp (str);
```

2.6.3   Formatted Output with the `fprintf` Function

An even more flexible way to display data is with the fprintf function. The fprintf function displays one or more values together with related text and lets

(a)

(b)

Figure 2.4 *(a)* Selecting preferences on the MATLAB Menu. *(b)* Selecting the desired numeric format within the Command Window Preferences.

Table 2.4   Common Special Characters in `fprintf` Format Strings

Format String	Results
%d	Display value as an integer.
%e	Display value in exponential format.
%f	Display value in floating-point format.
%g	Display value in either floating-point or exponential format, whichever is shorter.
\n	Skip to a new line.

the programmer control the way that the displayed value appears. The general form of this function when it is used to print to the Command Window is

```
fprintf(format,data)
```

where `format` is a string describing the way the data is to be printed, and `data` is one or more scalars or arrays to be printed. The `format` is a character string containing text to be printed plus special characters describing the format of the data. For example, the function

```
fprintf('The value of pi is %f \n',pi)
```

will print out `'The value of pi is 3.141593'` followed by a line feed. The characters `%f` are called **conversion characters**; they indicate that a value in the data list should be printed out in floating-point format at that location in the format string. The characters `\n` are **escape characters**; they indicate that a line feed should be issued so that the following text starts on a new line. There are many types of conversion characters and escape characters that may be used in an `fprintf` function. A few of them are listed in Table 2.4, and a complete list can be found in Chapter 8.

It is also possible to specify the width of the field in which a number will be displayed and the number of decimal places to display. This is done by specifying the the width and precision after the `%` sign and before the `f`. For example, the function

```
fprintf('The value of pi is %6.2f \n',pi)
```

will print out `'The value of pi is   3.14'` followed by a line feed. The conversion characters `%6.2f` indicate that the first data item in the function should be printed out in floating-point format in a field six characters wide, including two digits after the decimal point.

The `fprintf` function has one very significant limitation: *it displays only the real portion of a complex value*. This limitation can lead to misleading results when calculations produce complex answers. In those cases, it is better to use the `disp` function to display answers.

For example, the following statements calculate a complex value x and display it using both `fprintf` and `disp`.

```
x = 2 * ( 1 - 2*i )^3;
str = ['disp: x = ' num2str(x)];
```

```
disp(str);
fprintf('fprintf: x = %8.4f\n',x);
```

The results printed out by these statements are

```
disp: x = -22+4i
fprintf: x = -22.0000
```

Note that the fprintf function ignored the imaginary part of the answer.

Programming Pitfalls

The fprintf function displays only the *real* part of a complex number, which can produce misleading answers when working with complex values.

2.7 Data Files

There are many ways to load and save data files in MATLAB, most of which will be addressed in Chapter 8. For the moment, we will consider only the **load** and **save** commands, which are the simplest ones to use.

The **save** command saves data from the current MATLAB workspace into a disk file. The most common form of this command is

```
save filename var1 var2 var3
```

where filename is the name of the file where the variables are saved, and var1, var2, etc. are the variables to be saved in the file. By default, the file name will be given the extension "mat," and such data files are called MAT-files. If no variables are specified, then the entire contents of the workspace are saved.

MATLAB saves MAT-files in a special compact format that preserves many details, including the name and type of each variable, the size of each array, and all data values. A MAT-file created on any platform (PC, Mac, Unix, or Linux) can be read on any other platform, so MAT-files are a good way to exchange data between computers if both computers run MATLAB. Unfortunately, the MAT-file is in a format that cannot be read by other programs. If data must be shared with other programs, then the -ascii option should be specified, and the data values will be written to the file as ASCII character strings separated by spaces. However, the special information such as variable names and types are lost when the data is saved in ASCII format, and the resulting data file will be much larger.

For example, suppose the array x is defined as

```
x=[1.23 3.14 6.28; -5.1 7.00 0];
```

The the command "`save -ascii x.dat x`" will produce a file named `x.dat` containing the following data:

```
1.2300000e+000   3.1400000e+000   6.2800000e+000
-5.1000000e+000   7.0000000e+000   0.0000000e+000
```

This data is in a format that can be read by spreadsheets or by programs written in other computer languages, so it makes it easy to share data between MATLAB programs and other applications.

✳ Good Programming Practice

If data must be exchanged between MATLAB and other programs, save the MATLAB data in ASCII format. If the data will be used only in MATLAB, save the data in MAT-file format.

MATLAB doesn't care what file extension is used for ASCII files. However, it is better for the user if a consistent naming convention is used, and an extension of "`dat`" is a common choice for ASCII files.

✳ Good Programming Practice

Save ASCII data files with a "`dat`" file extension to distinguish them from MAT-files, which have a "`mat`" file extension.

The **load** command is the opposite of the `save` command. It loads data from a disk file into the current MATLAB workspace. The most common form of this command is

```
load filename
```

where `filename` is the name of the file to be loaded. If the file is a MAT-file, then all of the variables in the file will be restored, with the names and types the same as before. If a list of variables is included in the command, then only those variables will be restored. If the given `filename` has no extent, or if the file extent is `.mat`, then the `load` command will treat the file as a MAT-file.

MATLAB can load data created by other programs in comma- or space-separated ASCII format. If the given `filename` has any file extension other than `.mat`, then the load command will treat the file as an ASCII file. The contents of an ASCII file will be converted into a MATLAB array having the same name as the file (without the file extension) that the data was loaded from. For example, suppose that an ASCII data file named `x.dat` contains the following data:

```
1.23    3.14    6.28
-5.1    7.00    0
```

Then the command "load x.dat" will create a 2 × 3 array named x in the current workspace, containing these data values.

The load statement can be forced to treat a file as a MAT-file by specifying the −mat option. For example, the statement

```
load −mat x.dat
```

would treat file x.dat as a MAT-file even though its file extent is not .mat. Similarly, the load statement can be forced to treat a file as an ASCII file by specifying the −ascii option. These options allow the user to load a file properly even if its file extent doesn't match the MATLAB conventions.

Quiz 2.3

This quiz provides a quick check to see if you have understood the concepts introduced in Sections 2.6 and 2.7. If you have trouble with the quiz, reread the sections, ask your instructor, or discuss the material with a fellow student. The answers to this quiz are found in the back of the book.

1. How would you tell MATLAB to display all real values in exponential format with 15 significant digits?

2. What do the following sets of statements do? What is the output from them?

   (*a*)
   ```
   radius = input('Enter circle radius:\n');
   area = pi * radius^2;
   str = ['The area is ' num2str(area)];
   disp(str);
   ```

   (*b*)
   ```
   value = int2str(pi);
   disp(['The value is ' value '!']);
   ```

3. What do the following sets of statements do? What is the output from them?
   ```
   value = 123.4567e2;
   fprintf('value = %e\n',value);
   fprintf('value = %f\n',value);
   fprintf('value = %g\n',value);
   fprintf('value = %12.4f\n',value);
   ```

2.8   Scalar and Array Operations

Calculations are specified in MATLAB with an assignment statement, whose general form is

```
variable_name = expression;
```

The assignment statement calculates the value of the expression to the right of the equal sign, and *assigns* that value to the variable named on the left of the equal sign. Note that the equal sign does not mean equality in the usual sense of the word. Instead, it means: *store the value of* expression *into location* variable_name. For this reason, the equal sign is called the **assignment operator**. A statement such as

```
ii = ii + 1;
```

is complete nonsense in ordinary algebra, but it makes perfect sense in MATLAB. It means take the current value stored in variable ii, add one to it, and store the result back into variable ii.

2.8.1   Scalar Operations

The expression to the right of the assignment operator can be any valid combination of scalars, arrays, parentheses, and arithmetic operators. The standard arithmetic operations between two scalars are given in Table 2.5.

Parentheses may be used to group terms whenever desired. When parentheses are used, the expressions inside the parentheses are evaluated before the expressions outside the parentheses. For example, the expression 2 ^ ((8+2)/5) is evaluated as follows:

```
2 ^ ((8+2)/5) = 2 ^ (10/5)
              = 2 ^ 2
              = 4
```

2.8.2   Array and Matrix Operations

MATLAB supports two types of operations between arrays, known as *array operations* and *matrix operations*. **Array operations** are operations performed between arrays on an **element-by-element basis**. That is, the operation is performed on corresponding elements in the two arrays. For example, if $a = \begin{bmatrix} 1 & 2 \\ 3 & 4 \end{bmatrix}$ and $b = \begin{bmatrix} -1 & 3 \\ -2 & 1 \end{bmatrix}$, then $a + b = \begin{bmatrix} 0 & 5 \\ 1 & 5 \end{bmatrix}$. Note that for these operations to work, *the number of rows and columns in both arrays must be the same.* If not, MATLAB will generate an error message.

Table 2.5   Arithmetic Operations Between Two Scalars

Operation	Algebraic Form	MATLAB Form
Addition	$a + b$	a + b
Subtraction	$a - b$	a - b
Multiplication	$a \times b$	a * b
Division	$\frac{a}{b}$	a / b
Exponentiation	a^b	a ^ b

Array operations may also occur between an array and a scalar. If the operation is performed between an array and a scalar, the value of the scalar is applied to every element of the array. For example, if $a = \begin{bmatrix} 1 & 2 \\ 3 & 4 \end{bmatrix}$, then $a + 4 = \begin{bmatrix} 5 & 6 \\ 7 & 8 \end{bmatrix}$.

In contrast, **matrix operations** follow the normal rules of linear algebra, such as matrix multiplication. In linear algebra, the product $c = a \times b$ is defined by the equation

$$c(i, j) = \sum_{k=1}^{n} a(i, k)b(k, j)$$

For example, if $a = \begin{bmatrix} 1 & 2 \\ 3 & 4 \end{bmatrix}$ and $b = \begin{bmatrix} -1 & 3 \\ -2 & 1 \end{bmatrix}$, then $a \times b = \begin{bmatrix} -5 & 5 \\ -11 & 13 \end{bmatrix}$. Note that for matrix multiplication to work, *the number of columns in matrix* a *must be equal to the number of rows in matrix* b.

MATLAB uses a special symbol to distinguish array operations from matrix operations. In the cases where array operations and matrix operations have a different definition, MATLAB uses a period before the symbol to indicate an array operation (for example, . *). A list of common array and matrix operations is given in Table 2.6.

New users often confuse array operations and matrix operations. In some cases, substituting one for the other will produce an illegal operation, and MATLAB will report an error. In other cases, both operations are legal, and MATLAB will perform the wrong operation and come up with a wrong answer. The most common problem happens when working with square matrices. Both array multiplication and matrix multiplication are legal for two square matrices of the same size, but the resulting answers are totally different. Be careful to specify exactly what you want!

⊛ Programming Pitfalls

Be careful to distinguish between array operations and matrix operations in your MATLAB code. Confusing array multiplication with matrix multiplication is a very common mistake.

▶

Example 2.1

Assume that a, b, c, and d are defined as follows:

$$a = \begin{bmatrix} 1 & 0 \\ 2 & 1 \end{bmatrix} \qquad\qquad b = \begin{bmatrix} -1 & 2 \\ 0 & 1 \end{bmatrix}$$

$$c = \begin{bmatrix} 3 \\ 2 \end{bmatrix} \qquad\qquad d = 5$$

What is the result of each of the following expressions?

(a) a + b   (e) a + c
(b) a .* b   (f) a + d
(c) a * b   (g) a .* d
(d) a * c   (h) a * d

SOLUTION

(a) This is array or matrix addition: a + b = $\begin{bmatrix} 0 & 2 \\ 2 & 2 \end{bmatrix}$

(b) This is element-by-element array multiplication: a .* b = $\begin{bmatrix} -1 & 0 \\ 0 & 1 \end{bmatrix}$

(c) This is matrix multiplication: a * b = $\begin{bmatrix} -1 & 2 \\ -2 & 5 \end{bmatrix}$

(d) This is matrix multiplication: a * c = $\begin{bmatrix} 3 \\ 8 \end{bmatrix}$

(e) This operation is illegal, since a and c have different numbers of columns.

(f) This is addition of an array to a scalar: a + d = $\begin{bmatrix} 6 & 5 \\ 7 & 6 \end{bmatrix}$

(g) This is array multiplication: a .* d = $\begin{bmatrix} 5 & 0 \\ 10 & 5 \end{bmatrix}$

(h) This is matrix multiplication: a * d = $\begin{bmatrix} 5 & 0 \\ 10 & 5 \end{bmatrix}$ ◄

The matrix left division operation has a special significance that we must understand. A 3 × 3 set of simultaneous linear equations takes the form

$$a_{11}x_1 + a_{12}x_2 + a_{13}x_3 = b_1$$
$$a_{21}x_1 + a_{22}x_2 + a_{23}x_3 = b_2 \qquad (2\text{-}1)$$
$$a_{31}x_1 + a_{32}x_2 + a_{33}x_3 = b_3$$

which can be expressed as

$$Ax = B \qquad (2\text{-}2)$$

where $A = \begin{bmatrix} a_{11} & a_{12} & a_{13} \\ a_{21} & a_{22} & a_{23} \\ a_{31} & a_{32} & a_{33} \end{bmatrix}$, $B = \begin{bmatrix} b_1 \\ b_2 \\ b_3 \end{bmatrix}$, and $x = \begin{bmatrix} x_1 \\ x_2 \\ x_3 \end{bmatrix}$.

Equation (2-2) can be solved for x using linear algebra. The result is

$$x = A^{-1}B \qquad (2\text{-}3)$$

Since the left division operator A \ B is defined to be inv(A) * B, the left division operator solves a system of simultaneous equations in a single statement!

※ **Good Programming Practice**

Use the left division operator to solve systems of simultaneous equations.

Table 2.6 Common Array and Matrix Operations

Operation	MATLAB Form	Comments
Array Addition	a + b	Array addition and matrix addition are identical.
Array Subtraction	a – b	Array subtraction and matrix subtraction are identical.
Array Multiplication	a .* b	Element-by-element multiplication of a and b. Both arrays must be the same shape, or one of them must be a scalar.
Matrix Multiplication	a * b	Matrix multiplication of a and b. The number of columns in a must equal the number of rows in b.
Array Right Division	a ./ b	Element-by-element division of a and b: a(i,j) / b(i,j). Both arrays must be the same shape, or one of them must be a scalar.
Array Left Division	a .\ b	Element-by-element division of a and b, but with b in the numerator: b(i,j) / a(i,j). Both arrays must be the same shape, or one of them must be a scalar.
Matrix Right Division	a / b	Matrix division defined by a * inv(b), where inv(b) is the inverse of matrix b.
Matrix Left Division	a \ b	Matrix division defined by inv(a) * b, where inv(a) is the inverse of matrix a.
Array Exponentiation	a .^ b	Element-by-element exponentiation of a and b: a(i,j) ^ b(i,j). Both arrays must be the same shape, or one of them must be a scalar.

2.9   Hierarchy of Operations

Often, many arithmetic operations are combined into a single expression. For example, consider the equation for the distance traveled by an object starting from rest and subjected to a constant acceleration:

```
distance = 0.5 * accel * time ^ 2
```

There are two multiplications and an exponentiation in this expression. In such an expression, it is important to know the order in which the operations are evaluated. If exponentiation is evaluated before multiplication, this expression is equivalent to

```
distance = 0.5 * accel * (time ^ 2)
```

But if multiplication is evaluated before exponentiation, this expression is equivalent to

```
distance = (0.5 * accel * time) ^ 2
```

These two equations have different results, and we must be able to unambiguously distinguish between them.

To make the evaluation of expressions unambiguous, MATLAB has established a series of rules governing the hierarchy or order in which operations are evaluated within an expression. The rules generally follow the normal rules of algebra. The order in which the arithmetic operations are evaluated is given in Table 2.7.

Table 2.7   Hierarchy of Arithmetic Operations

Precedence	Operation
1	The contents of all parentheses are evaluated, starting from the innermost parentheses and working outward.
2	All exponentials are evaluated, working from left to right.
3	All multiplications and divisions are evaluated, working from left to right.
4	All additions and subtractions are evaluated, working from left to right.

▶

Example 2.2

Variables a, b, c, and d have been initialized to the following values:

```
a = 3;     b = 2;     c = 5;     d = 3;
```

Evaluate the following MATLAB assignment statements:

(a) `output = a*b+c*d;`
(b) `output = a*(b+c)*d;`
(c) `output = (a*b)+(c*d);`
(d) `output = a^b^d;`
(e) `output = a^(b^d);`

SOLUTION

(*a*) Expression to evaluate: `output = a*b+c*d;`
    Fill in numbers: `output = 3*2+5*3;`
    First, evaluate multiplications
    and divisions from left to right: `output = 6 +5*3;`
                               `output = 6 + 15;`
    Now evaluate additions: `output = 21`

(*b*) Expression to evaluate: `output = a*(b+c)*d;`
    Fill in numbers: `output = 3*(2+5)*3;`
    First, evaluate parentheses: `output = 3*7*3;`
    Now, evaluate multiplications
    and divisions from left to right: `output = 21*3;`
                               `output = 63;`

(*c*) Expression to evaluate: `output = (a*b)+(c*d);`
    Fill in numbers: `output = (3*2)+(5*3);`
    First, evaluate parentheses: `output = 6 + 15;`
    Now evaluate additions: `output = 21`

(*d*) Expression to evaluate: `output = a^b^d;`
    Fill in numbers: `output = 3^2^3;`
    Evaluate exponentials
    from left to right: `output = 9^3;`
                               `output = 729;`

(*e*) Expression to evaluate: `output = a^(b^d);`
    Fill in numbers: `output = 3^(2^3);`
    First, evaluate parentheses: `output = 3^8;`
    Now, evaluate exponential: `output = 6561;`

◀

As we see in the preceding example, the order in which operations are performed has a major effect on the final result of an algebraic expression.

It is important that every expression in a program be made as clear as possible. Any program of value must not only be written but must also be maintained and modified when necessary. You should always ask yourself: "Will I easily understand this expression if I come back to it in six months? Can another programmer look at my code and easily understand what I am doing?" If there is any doubt in your mind, use extra parentheses in the expression to make it as clear as possible.

✳ Good Programming Practice

Use parentheses as necessary to make your equations clear and easy to understand.

If parentheses are used within an expression, then the parentheses must be balanced. That is, there must be an equal number of open parentheses and close parentheses within the expression. It is an error to have more of one type than the other. Errors of this sort are usually typographical, and they are caught by the MATLAB interpreter when the command is executed. For example, the expression

```
(2 + 4) / 2)
```

produces an error during when the expression is executed.

Quiz 2.4

This quiz provides a quick check to see if you have understood the concepts introduced in Sections 2.8 and 2.9. If you have trouble with the quiz, reread the sections, ask your instructor, or discuss the material with a fellow student. The answers to this quiz are found in the back of the book.

1. Assume that a, b, c, and d are defined as follows, and calculate the results of the following operations if they are legal. If an operation is, explain why it is illegal.

$$a = \begin{bmatrix} 2 & 1 \\ -1 & 2 \end{bmatrix} \qquad b = \begin{bmatrix} 0 & -1 \\ 3 & 1 \end{bmatrix}$$

$$c = \begin{bmatrix} 1 \\ 2 \end{bmatrix} \qquad d = -3$$

  (*a*) `result = a .* c;`
  (*b*) `result = a * [c c];`
  (*c*) `result = a .* [c c];`

(*d*) `result = a + b * c;`
(*e*) `result = a + b .* c;`

2. Solve for *x* in the equation $Ax = B$, where $A = \begin{bmatrix} 1 & 2 & 1 \\ 2 & 3 & 2 \\ -1 & 0 & 1 \end{bmatrix}$ and

$B = \begin{bmatrix} 1 \\ 1 \\ 0 \end{bmatrix}$.

2.10 Built-in MATLAB Functions

In mathematics, a **function** is an expression that accepts one or more input values and calculates a single result from them. Scientific and technical calculations usually require functions that are more complex than the simple addition, subtraction, multiplication, division, and exponentiation operations that we have discussed so far. Some of these functions are very common and are used in many different technical disciplines. Others are rarer and specific to a single problem or a small number of problems. Examples of very common functions are the trigonometric functions, logarithms, and square roots. Examples of rarer functions include the hyperbolic functions, Bessel functions, and so forth. One of MATLAB's greatest strengths is that it comes with an incredible variety of built-in functions ready for use.

2.10.1 Optional Results

Unlike mathematical functions, MATLAB functions can return *more than one result* to the calling program. The function `max` is an example of such a function. This function normally returns the maximum value of an input vector, but it can also return a second argument containing the location in the input vector where the maximum value was found. For example, the statement

```
maxval = max ([1  -5  6  -3])
```

returns the result `maxval = 6`. However, if two variables are provided to store results in, the function returns *both* the maximum value *and* the location of the maximum value.

```
[maxval index] = max ([1  -5  6  -3])
```

produces the results `maxval = 6` and `index = 3`.

2.10.2 Using MATLAB Functions with Array Inputs

Many MATLAB functions are defined for one or more scalar inputs and produce a scalar output. For example, the statement `y = sin(x)` calculates the sine of `x` and stores the result in `y`. If these functions receive an array of input values, then they will calculate an array of output values on an element-by-element basis. For example, if x = [0 pi/2 pi 3*pi/2 2*pi], then the statement

    y = sin(x)

will produce the result y = [0 1 0 -1 0].

2.10.3   Common MATLAB Functions

A few of the most common and useful MATLAB functions are shown in Table 2.8. These functions will be used in many examples and homework problems. If you need to locate a specific function not on this list, you can search for the function alphabetically or by subject using the MATLAB Help Browser.

Note that unlike most computer languages, many MATLAB functions work correctly for both real and complex inputs. MATLAB functions automatically calculate the correct answer, even if the result is imaginary or complex. For example, the function sqrt(-2) will produce a runtime error in languages such as C++, Java, or Fortran. In contrast, MATLAB correctly calculates the imaginary answer:

```
» sqrt(-2)
ans =
    0 + 1.4142i
```

2.11   Introduction to Plotting

MATLAB's extensive, device-independent plotting capabilities are one of its most powerful features. They make it very easy to plot any data at any time. To plot a data set, just create two vectors containing the x and y values to be plotted and use the plot function.

For example, suppose that we wish to plot the function $y = x^2 - 10x + 15$ for values of x between 0 and 10. It takes only three statements to create this plot. The first statement creates a vector of x values between 0 and 10 using the colon operator. The second statement calculates the y values from the equation (note that we are using array operators here so that this equation is applied to each x value on an element-by-element basis). Finally, the third statement creates the plot.

```
x = 0:1:10;
y = x.^2 - 10.*x + 15;
plot(x,y);
```

When the plot function is executed, MATLAB opens a Figure Window and displays the plot in that window. The plot produced by these statements is shown in Figure 2.5.

Table 2.8   Common MATLAB Functions

Function	Description
Mathematical Functions	
abs(x)	Calculates $|x|$.
acos(x)	Calculates $\cos^{-1}x$.
angle(x)	Returns the phase angle of the complex value x, in radians.
asin(x)	Calculates $\sin^{-1}x$.
atan(x)	Calculates $\tan^{-1}x$.
atan2(y,x)	Calculates $\tan^{-1}\frac{y}{x}$ over all four quadrants of the circle (results in *radians* in the range $-\pi \leq \tan^{-1}\frac{y}{x} \leq \pi$).
cos(x)	Calculates $\cos x$, with x in radians.
exp(x)	Calculate e^x.
log(x)	Calculates the natural logarithm $\log_e x$
[value,index] = max(x)	Returns the maximum value in vector x, and optionally the location of that value.
[value,index] = min(x)	Returns the minimum value in vector x, and optionally the location of that value.
mod(x,y)	Remainder or modulo function.
sin(x)	Calculates $\sin x$, with x in radians.
sqrt(x)	Calculates the square root of x.
tan(x)	Calculates $\tan x$, with x in radians.
Rounding Functions	
ceil(x)	Rounds x to the nearest integer towards positive infinity: ceil(3.1) = 4 and ceil(-3.1) = -3.
fix(x)	Rounds x to the nearest integer towards zero: fix(3.1) = 3 and fix(-3.1) = -3.
floor(x)	Rounds x to the nearest integer towards minus infinity: floor(3.1) = 3 and floor(-3.1) = -4.
round(x)	Rounds x to the nearest integer.
String Conversion Functions	
char(x)	Converts a matrix of numbers into a character string. For ASCII characters the matrix should contain numbers ≤ 127.
double(x)	Converts a character string into a matrix of numbers.
int2str(x)	Converts x into an integer character string.
num2str(x)	Converts x into a character string.
str2num(s)	Converts character string s into a numeric array.

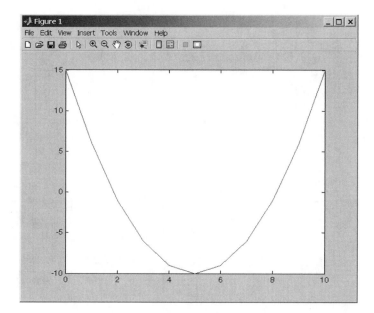

Figure 2.5   Plot of $y = x^2 - 10x + 15$ from 0 to 10.

2.11.1   Using Simple *xy* Plots

As we have seen previously, plotting is *very* easy in MATLAB. Any pair of vectors can be plotted versus each other as long as both vectors have the same length. However, the result is not a finished product, since there are no titles, axis labels, or grid lines on the plot.

Titles and axis labels can be added to a plot with the `title`, `xlabel`, and `ylabel` functions. Each function is called with a string containing the title or label to be applied to the plot. Grid lines can be added or removed from the plot with the `grid` command: `grid on` turns on grid lines, and `grid off` turns off grid lines. For example, the statements that follow generate a plot of the function $y = x^2 - 10x + 15$ with titles, labels, and gridlines. The resulting plot is shown in Figure 2.6.

```
x = 0:1:10;
y = x.^2 - 10.*x + 15;
plot(x,y);
title ('Plot of y = x.^2 - 10.*x + 15');
xlabel ('x');
ylabel ('y');
grid on;
```

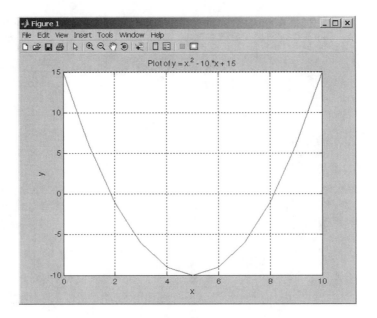

Figure 2.6 Plot of $y = x^2 - 10x + 15$ with a title, axis labels, and gridlines.

2.11.2 Printing a Plot

Once created, a plot may be printed on a printer with the `print` command by clicking on the "print" icon in the Figure Window, or by selecting the "File/Print" menu option in the Figure Window.

The `print` command is especially useful because it can be included in a MATLAB program, allowing the program to automatically print graphical images. The form of the `print` command is:

```
print <options> <filename>
```

If no filename is included, this command prints a copy of the current figure on the system printer. If a filename is specified, the command prints a copy of the current figure to the specified file.

2.11.3 Exporting a Plot as a Graphical Image

The `print` command can be used to save a plot as a graphical image by specifying appropriate options and a file name.

```
print <options> <filename>
```

There are many different options that specify the format of the output sent to a file. One very important option is `-dtiff`. This option specifies that the output

Table 2.9   `print` Options to Create Graphics Files

Option	Description
-deps	Creates a monochrome encapsulated postcript image.
-depsc	Creates a color encapsulated postcript image.
-djpeg	Creates a JPEG image.
-dpng	Creates a Portable Network Graphic color image.
-dtiff	Creates a compressed TIFF image.

will be to a file in Tagged Image File Format (TIFF). Since this format can be imported into all of the important word processors on PC, Mac, Unix, and Linux platforms, it is a great way to include MATLAB plots in a document. The following command will create a TIFF image of the current figure and store it in a file called my_image.tif:

```
print -dtiff my_image.tif
```

Other options allow image files to be created in other formats. Some of the most important image file formats are given in Table 2.9.

In addition, the "File/Save As" menu option on the Figure Window can be used to save a plot as a graphical image. In this case, the user selects the file name and the type of image from a standard dialog box (see Figure 2.7).

2.11.4   Multiple Plots

It is possible to plot multiple functions on the same graph by simply including more than one set of (x, y) values in the plot function. For example, suppose that we wanted to plot the function $f(x) = \sin 2x$ and its derivative on the same plot. The derivative of $f(x) = \sin 2x$ is:

$$\frac{d}{dt} \sin 2x = 2 \cos 2x \qquad (2\text{-}4)$$

To plot both functions on the same axes, we must generate a set of x values and the corresponding y values for each function. Then to plot the functions, we would simply list both sets of (x, y) values in the plot function as shown in the following statements:

```
x = 0:pi/100:2*pi;
y1 = sin(2*x);
y2 = 2*cos(2*x);
plot(x,y1,x,y2);
```

The resulting plot is shown in Figure 2.8.

Figure 2.7 Exporting a plot as an image file using the "File/Save As" menu item.

Figure 2.8 Plot of $f(x) = \sin 2x$ and $f(x) = 2 \cos 2x$ on the same axes.

2.11.5   Line Color, Line Style, Marker Style, and Legends

MATLAB allows a programmer to select the color of a line to be plotted, the style of the line to be plotted, and the type of marker to be used for data points on the line. These traits may be selected using an attribute character string after the x and y vectors in the plot function.

The attribute character string can have up to three characters, with the first character specifying the color of the line, the second character specifying the style of the marker, and the last character specifying the style of the line. The characters for various colors, markers, and line styles are shown in Table 2.10.

The attribute characters may be mixed in any combination, and more than one attribute string may be specified if more than one pair of (x, y) vectors is included in a single `plot` function call. For example, the following statements will plot the function $y = x^2 - 10x + 15$ with a dashed red line and will include the actual data points as blue circles (see Figure 2.9).

```
x = 0:1:10;
y = x.^2 - 10.*x + 15;
plot(x,y,'r--',x,y,'bo');
```

Legends may be created with the `legend` function. The basic form of this function is

```
legend('string1', 'string2',..., pos)
```

Table 2.10   Table of Plot Colors, Marker Styles, and Line Styles

Color		Marker Style		Line Style	
y	yellow	.	point	–	solid
m	magenta	o	circle	:	dotted
c	cyan	x	x-mark	-.	dash-dot
r	red	+	plus	--	dashed
g	green	*	star	\<none\>	no line
b	blue	s	square		
w	white	d	diamond		
k	black	v	triangle (down)		
		^	triangle (up)		
		<	triangle (left)		
		>	triangle (right)		
		p	pentagram		
		h	hexagram		
		\<none\>	no marker		

Figure 2.9 Plot of the function $y = x2 - 10x + 15$ with a dashed red line, showing the actual data points as blue circles.

where `string1`, `string2`, etc. are the labels associated with the lines plotted, and `pos` is an string specifying where to place the legend. The possible values for `pos` are given in Table 2.11[2], and are shown graphically in Figure 2.10.

The command `legend off` will remove an existing legend.

An example of a complete plot is shown in Figure 2.11, and the statements to produce that plot are shown below. They plot the function $f(x) = \sin 2x$ and its derivative $f'(x) = 2\cos 2x$ on the same axes, with a solid black line for $f(x)$ and a dashed red line for its derivative. The plot includes a title, axis labels, a legend in the top left corner of the plot, and grid lines.

```
x = 0:pi/100:2*pi;
y1 = sin(2*x);
y2 = 2*cos(2*x);
plot(x,y1,'k-',x,y2,'b--');
title ('Plot of f(x) = sin(2x) and its derivative');
xlabel ('x');
ylabel ('y');
legend ('f(x)','d/dx f(x)','tl')
grid on;
```

---

[2]Before MATLAB 7.0, the `pos` parameter took a number in the range 0–4 to specify the location of a legend. This usage is now obsolete, but is still supported for backwards compatibility.

Table 2.11   Values of `pos` in the `legend` Command

Value	Legend Location
`'NW'`	Above and to the left
`'NL'`	Above top left corner
`'NC'`	Above center of top edge
`'NR'`	Above top right corner
`'NE'`	Above and to right
`'TW'`	At top and to left
`'TL'`	Top left corner
`'TC'`	At top center
`'TR'`	Top right corner
`'TE'`	At top and to right
`'MW'`	At middle and to left
`'ML'`	Middle left edge
`'MC'`	Middle and center
`'MR'`	Middle right edge
`'ME'`	At middle and to right
`'BW'`	At bottom and to left
`'BL'`	Bottom left corner
`'BC'`	At bottom center
`'BR'`	Bottom right corner
`'BE'`	At bottom and to right
`'SW'`	Below and to left
`'SL'`	Below bottom left corner
`'SC'`	Below center of bottom edge
`'SR'`	Below bottom right corner
`'SE'`	Below and to right

Figure 2.10   Possible locations for a plot legend.

Figure 2.11 A complete plot with title, axis labels, legend, grid, and multiple line styles.

2.11.6 Logarithmic Scales

It is possible to plot data on logarithmic scales as well as linear scales. There are four possible combinations of linear and logarithmic scales on the x and y axes, and each combination is produced by a separate function.

1. The plot function plots both x and y data on linear axes.
2. The semilogx function plots x data on logarithmic axes and y data on linear axes.
3. The semilogy function plots x data on linear axes and y data on logarithmic axes.
4. The loglog function plots both x and y data on logarithmic axes.

All of these functions have identical calling sequences: the only difference is the type of axis used to plot the data. Examples of each plot are shown in Figure 2.12.

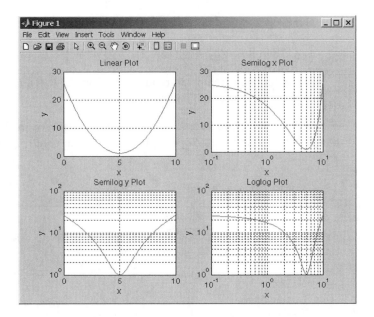

Figure 2.12 Comparison of linear, semilog x, semilog y, and log-log plots.

2.12 Examples

The following examples illustrate problem-solving with MATLAB.

▶

Example 2.3—Temperature Conversion

Design a MATLAB program that reads an input temperature in degrees Fahrenheit, converts it to an absolute temperature in kelvin, and writes out the result.

SOLUTION The relationship between temperature in degrees Fahrenheit (°F) and temperature in kelvin (K) can be found in any physics textbook. It is

$$T(\text{in kelvin}) = \left[\frac{5}{9} \, T(\text{in °F}) - 32.0\right] + 273.15 \qquad (2\text{-}5)$$

The physics books also give us sample values on both temperature scales, which we can use to check the operation of our program. Two such values are

The boiling point of water	212° F	373.15 K
The sublimation point of dry ice	–110° F	194.26 K

Our program must perform the following steps:

1. Prompt the user to enter an input temperature in °F.
2. Read the input temperature.
3. Calculate the temperature in kelvin from Equation (2-5).
4. Write out the result and stop.

We will use function input to get the temperature in degrees Fahrenheit and function fprintf to print the answer. The resulting program is shown as follows:

```
% Script file: temp_conversion
%
% Purpose:
%   To convert an input temperature from degrees
%   Fahrenheit to an output temperature in kelvin.
%
% Record of revisions:
%   Date          Programmer        Description of change
%   ====          ==========        =====================
%   01/03/07    S. J. Chapman     Original code
%
% Define variables:
%   temp_f    -- Temperature in degrees Fahrenheit
%   temp_k    -- Temperature in kelvin

% Prompt the user for the input temperature.
temp_f = input('Enter the temperature in degrees Fahrenheit:');

% Convert to kelvin.
temp_k = (5/9) * (temp_f - 32) + 273.15;

% Write out the result.
fprintf('%6.2f degrees Fahrenheit = %6.2f kelvin.\n', ... temp_f,temp_k);
```

To test the completed program, we will run it with the known input values that have been given. Note that user inputs appear in boldface in the following:

```
» temp_conversion
Enter the temperature in degrees Fahrenheit: 212
212.00 degrees Fahrenheit = 373.15 kelvin.
» temp_conversion
Enter the temperature in degrees Fahrenheit: -110
-110.00 degrees Fahrenheit = 194.26 kelvin.
```

The results of the program match the values from the physics book. ◀

In the foregoing program, we echoed the input values and printed the output values together with their units. The results of this program make sense only if the units (degrees Fahrenheit and kelvin) are included together with their values. As a general rule, the units associated with any input value should always be printed along with the prompt that requests the value, and the units associated with any output value should always be printed along with that value.

✳ Good Programming Practice

Always include the appropriate units with any values that you read or write in a program.

The foregoing program exhibits many of the good programming practices that we have described in this chapter. It includes a data dictionary defining the meanings of all of the variables in the program. It also uses descriptive variable names, and appropriate units are attached to all printed values.

▶

Example 2.4—Electrical Engineering: Maximum Power Transfer to a Load

Figure 2.13 shows a voltage source $V = 120$ V with an internal resistance R_S of 50 Ω supplying a load of resistance R_L. Find the value of load resistance R_L that will result in the maximum possible power being supplied by the source to the load. How much power be supplied in this case? Also, plot the power supplied to the load as a function of the load resistance R_L.

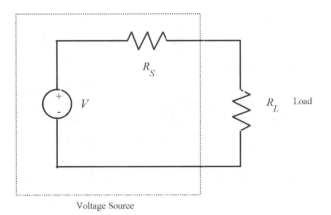

Voltage Source

Figure 2.13   A voltage source with a voltage V and an internal resistance R_S supplying a load of resistance R_L.

SOLUTION

In this program, we need to vary the load resistance R_L and compute the power supplied to the load at each value of R_L. The power supplied to the load resistance is given by the equation

$$P_L = I^2 R_L \tag{2-6}$$

where I is the current supplied to the load. The current supplied to the load can be calculated by Ohm's Law:

$$I = \frac{V}{R_{\text{TOT}}} = \frac{V}{R_S + R_L} \tag{2-7}$$

The program must perform the following steps:

1. Create an array of possible values for the load resistance R_L. The array will vary R_L from 1 Ω to 100 Ω in 1 Ω steps.
2. Calculate the current for each value of R_L.
3. Calculate the power supplied to the load for each value of R_L.
4. Plot the power supplied to the load for each value of R_L, and determine the value of load resistance resulting in the maximum power.

The final MATLAB program is shown here.

```
% Script file: calc_power.m
%
% Purpose:
%   To calculate and plot the power supplied to a load as
%   as a function of the load resistance.
%
% Record of revisions:
%     Date          Programmer          Description of change
%     ====          ==========          =====================
%   01/03/07      S. J. Chapman          Original code
%
% Define variables:
%   amps   -- Current flow to load (amps)
%   pl     -- Power supplied to load (watts)
%   rl     -- Resistance of the load (ohms)
%   rs     -- Internal resistance of the power source (ohms)
%   volts  -- Voltage of the power source (volts)

% Set the values of source voltage and internal resistance
volts = 120;
rs = 50;

% Create an array of load resistances
rl = 1:1:100;
```

```
% Calculate the current flow for each resistance
amps = volts ./ ( rs + rl );

% Calculate the power supplied to the load
pl = (amps .^ 2) .* rl;

% Plot the power versus load resistance
plot(rl,pl);
title('Plot of power versus load resistance');
xlabel('Load resistance (ohms)');
ylabel('Power (watts)');
grid on;
```

When this program is executed, the resulting plot is shown in Figure 2.14. From this plot, we can see that the maximum power is supplied to the load when the load's resistance is 50 Ω. The power supplied to the load at this resistance is 72 watts.

Figure 2.14   Plot of power supplied to load versus load resistance.

◄

Note the use of the array operators . *, . ^, and . / in the preceding program. These operators cause the arrays amps and pl to be calculated on an element-by-element basis.

▶

Example 2.5—Carbon 14 Dating

A radioactive isotope of an element is a form of the element that is not stable. Instead, it spontaneously decays into another element over a period of time. Radioactive decay is an exponential process. If Q_0 is the initial quantity of a radioactive substance at time $t = 0$, then the amount of that substance which will be present at any time t in the future is given by

$$Q(t) = Q_0 e^{-\lambda t} \qquad (2\text{-}8)$$

where λ is the radioactive decay constant.

Because radioactive decay occurs at a known rate, it can be used as a clock to measure the time that has elapsed since the decay started. If we know the initial amount of the radioactive material Q_0 present in a sample and the amount of the material Q left at the current time, we can solve for t in Equation (2-8) to determine how long the decay has been going on. The resulting equation is

$$t_{\text{decay}} = -\frac{1}{\lambda} \log_e \frac{Q}{Q_0} \qquad (2\text{-}9)$$

Equation (2-9) has practical applications in many areas of science. For example, archaeologists use a radioactive clock based on carbon 14 to determine the time that has passed since a once-living thing died. Carbon 14 is continually taken into the body while a plant or animal is living, so the amount of it present in the body at the time of death is assumed to be known. The decay constant λ of carbon 14 is well known to be 0.00012097/year. Therefore, the amount of carbon 14 remaining now can be accurately measured, then Equation (2-9) can be used to determine how long ago the living thing died. The amount of carbon 14 remaining as a function of time is shown in Figure 2.15.

Write a program that reads the percentage of carbon 14 remaining in a sample, calculates the age of the sample from it, and prints out the result with proper units.

SOLUTION   Our program must perform the following steps:

1. Prompt the user to enter the percentage of carbon 14 remaining in the sample.

2. Read in the percentage.

3. Convert the percentage into the fraction $\dfrac{Q}{Q_0}$.

4. Calculate the age of the sample in years using Equation (2-9).

5. Write out the result, and stop.

The resulting code is shown here.

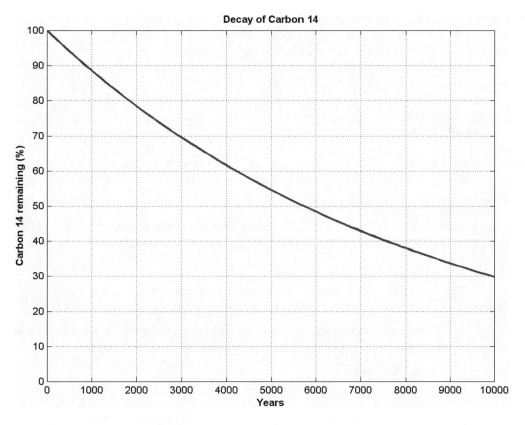

Figure 2.15 The radioactive decay of carbon 14 as a function of time. Notice that 50 percent of the original carbon 14 is left after about 5730 years have elapsed.

```
% Script file: c14_date.m
%
% Purpose:
%   To calculate the age of an organic sample from the
%   percentage of the original carbon 14 remaining in
%   the sample.
%
% Record of revisions:
%     Date         Programmer          Description of change
%     ====         ==========          =====================
%   01/05/07     S. J. Chapman         Original code
%
% Define variables:
%   age        -- The age of the sample in years
%   lambda     -- The radioactive decay constant for
%                  carbon-14, in units of 1/years.
```

```
%  percent    -- The percentage of carbon 14 remaining
%                at the time of the measurement
%  ratio      -- The ratio of the carbon 14 remaining at
%                the time of the measurement to the
%                original amount of carbon 14.

% Set decay constant for carbon-14
lambda = 0.00012097;

% Prompt the user for the percentage of C-14 remaining.
percent = input('Enter the percentage of carbon 14 remaining:\n');

% Perform calculations
ratio = percent / 100;              % Convert to fractional ratio
age = (-1.0 / lambda) * log(ratio);  % Get age in years

% Tell the user about the age of the sample.
string = ['The age of the sample is' num2str(age) ' years.'];
disp(string);
```

To test the completed program, we will calculate the time it takes for half of the carbon 14 to disappear. This time is known as the *half-life* of carbon 14.

```
» c14_date
Enter the percentage of carbon 14 remaining:
50
The age of the sample is 5729.9097 years.
```

The *CRC Handbook of Chemistry and Physics* states that the half-life of carbon 14 is 5730 years, so output of the program agrees with the reference book.

◀

2.13   Debugging MATLAB Programs

There is an old saying that the only sure things in life are death and taxes. We can add one more certainty to that list: if you write a program of any significant size, it won't work the first time you try it! Errors in programs are known as **bugs**, and the process of locating and eliminating them is known as **debugging**. Given that we have written a program and it is not working, how do we debug it?

Three types of error are found in MATLAB programs. The first type of error is a **syntax error**. Syntax errors are errors in the MATLAB statement itself, such as spelling errors or punctuation errors. These errors are detected by the MATLAB compiler the first time that an M-file is executed. For example, the statement

```
x = (y + 3) / 2);
```

contains a syntax error because it has unbalanced parentheses. If this statement appears in an M-file named `test.m`, the following message appears when `test` is executed.

```
» test
??? x = (y + 3) / 2)
                   |
Missing operator, comma, or semi-colon.

Error in ==> d:\book\matlab\chap1\test.m
On line 2 ==>
```

The second type of error is the **run-time error**. A run-time error occurs when an illegal mathematical operation is attempted during program execution (for example, attempting to divide by 0). These errors cause the program to return `Inf` or `NaN`, which is then used in further calculations. The results of a program that contains calculations using `Inf` or `NaN` are usually invalid.

The third type of error is a **logical error**. Logical errors occur when the program compiles and runs successfully but produces the wrong answer.

The most common mistakes made during programming are *typographical errors*. Some typographical errors create invalid MATLAB statements. These errors produce syntax errors that are caught by the compiler. Other typographical errors occur in variable names. For example, the letters in some variable names might have been transposed, or an incorrect letter might be typed. The result will be a new variable, and MATLAB simply creates the new variable the first time that it is referenced. MATLAB cannot detect this type of error. Typographical errors can also produce logical errors. For example, if variables `vel1` and `vel2` are both used for velocities in the program, then one of them might be inadvertently used instead of the other one at some point. You must check for that sort of error by manually inspecting the code.

Sometimes a program will start to execute, but run-time errors or logical errors occur during execution. In this case, there is either something wrong with the input data or something wrong with the logical structure of the program. The first step in locating this sort of bug should be to *check the input data to the program*. Either remove semicolons from input statements or add extra output statements to verify that the input values are what you expect them to be.

If the variable names seem to be correct and the input data is correct, then you are probably dealing with a logical error. You should check each of your assignment statements.

1. If an assignment statement is very long, break it into several smaller assignment statements. Smaller statements are easier to verify.
2. Check the placement of parentheses in your assignment statements. It is a very common error to have the operations in an assignment statement evaluated in the wrong order. If you have any doubts as to the order in which the variables are being evaluated, add extra sets of parentheses to make your intentions clear.

3. Make sure that you have initialized all of your variables properly.
4. Be sure that any functions you use are in the correct units. For example, the input to trigonometric functions must be in units of radians, not degrees.

If you are still getting the wrong answer, add output statements at various points in your program to see the results of intermediate calculations. If you can locate the point where the calculations go bad, then you know just where to look for the problem, which is 95 percent of the battle.

If you still cannot find the problem after following all of these steps, explain what you are doing to another student or to your instructor, and let him or her look at the code. It is very common for people to see just what they expect to see when they look at their own code. Another person can often quickly spot an error that you have overlooked time after time.

✳ **Good Programming Practice**

To reduce your debugging effort, make sure that during your program design you

1. Initialize all variables.
2. Use parentheses to make the functions of assignment statements clear.

MATLAB includes a special debugging tool called a *symbolic debugger*, which is embedded into the Edit/Debug Window. A symbolic debugger is a tool that allows you to walk through the execution of your program one statement at a time, and to examine the values of any variables at each step along the way. Symbolic debuggers allow you to see all of the intermediate results without having to insert a lot of output statements into your code. We will learn how to use MATLAB's symbolic debugger in Chapter 3.

2.14 Summary

In this chapter, we have presented many of the fundamental concepts required to write functional MATLAB programs. We learned about the basic types of MATLAB windows, the workspace, and how to get on-line help.

We introduced two data types: `double` and `char`. We also introduced assignment statements, arithmetic calculations, intrinsic functions, input/output statements, and data files.

The order in which MATLAB expressions are evaluated follows a fixed hierarchy, with operations at a higher level evaluated before operations at lower levels. The hierarchy of operations is summarized in Table 2.12.

The MATLAB language includes an extremely large number of built-in functions to help us solve problems. This list of functions is *much* richer than the list of functions found in other languages such as Fortran or C++, and it includes

Table 2.12   Hierarchy of Operations

Precedence	Operation
1	The contents of all parentheses are evaluated, starting from the innermost parentheses and working outward.
2	All exponentials are evaluated, working from left to right.
3	All multiplications and divisions are evaluated, working from left to right.
4	All additions and subtractions are evaluated, working from left to right.

device-independent plotting capabilities. A few of the common intrinsic functions are summarized in Table 2.8, and many others will be introduced throughout the remainder of the book. A complete list of all MATLAB functions is available through MATLAB Help, which can be accessed by clicking the ? icon, or by typing `helpdesk` in the Command Window.

2.14.1   Summary of Good Programming Practice

Every MATLAB program should be designed so that another person who is familiar with MATLAB can easily understand it. This is very important, given that a good program may be used for a long period of time. Over that time, conditions will change, and the program will need to be modified to reflect the changes. The program modifications may be done by someone other than the original programmer. The programmer making the modifications must understand the original program well before attempting to change it.

It is much harder to design clear, understandable, and maintainable programs than it is to simply write programs. To do so, a programmer must develop the discipline to properly document his or her work. In addition, the programmer must be careful to avoid known pitfalls along the path to good programs. The following guidelines will help you to develop good programs:

1. Use meaningful variable names whenever possible. Use names that can be understood at a glance, like `day`, `month`, and `year`.
2. Create a data dictionary for each program to make program maintenance easier.
3. Use only lower-case letters in variable names, so that there won't be errors due to capitalization differences in different occurrences of a variable name.
4. Use a semicolon at the end of all MATLAB assignment statements to suppress echoing of assigned values in the Command Window. If you need to examine the results of a statement during program debugging, you may remove the semicolon from that statement only.

5. If data must be exchanged between MATLAB and other programs, save the MATLAB data in ASCII format. If the data will only be used in MATLAB, save the data in MAT-file format.
6. Save ASCII data files with a "dat" file extent to distinguish them from MAT-files, which have a "mat" file extent.
7. Use parentheses as necessary to make your equations clear and easy to understand.
8. Always include the appropriate units with any values that you read or write in a program.

2.14.2   MATLAB Summary

The following summary lists all of the MATLAB special symbols, commands, and functions described in this chapter, along with a brief description of each one.

Special Symbols

[ ]	Array constructor
( )	Forms subscripts
' '	Marks the limits of a character string.
,	1. Separates subscripts or matrix elements. 2. Separates assignment statements on a line.
,	Separates subscripts or matrix elements.
;	1. Suppresses echoing in Command Window. 2. Separates matrix rows. 3. Separates assignment statements on a line.
%	Marks the beginning of a comment.
:	Colon operator, used to create shorthand lists
+	Array and matrix addition
–	Array and matrix subtraction
.*	Array multiplication
*	Matrix multiplication
./	Array right division
.\	Array left division
/	Matrix right division
\	Matrix left division
.^	Array exponentiation
'	Transpose operator

Commands and Functions

`...`	Continues a MATLAB statement on the following line.
`abs(x)`	Calculates the absolute value of x.
`ans`	Default variable used to store the result of expressions not assigned to another variable
`acos(x)`	Calculates the inverse cosine of x. The resulting angle is in radians between 0 and π.
`asin(x)`	Calculates the inverse sine of x. The resulting angle is in radians between $-\pi/2$ and $\pi/2$.
`atan(x)`	Calculates the inverse tangent of x. The resulting angle is in radians between $-\pi/2$ and $\pi/2$.
`atan2(y,x)`	Calculates the inverse tangent of y/x, valid over the entire circle. The resulting angle is in radians between $-\pi$ and π.
`ceil(x)`	Rounds x to the nearest integer towards positive infinity: `floor(3.1) = 4` and `floor(-3.1) = -3`.
`char`	Converts a matrix of numbers into a character string. For ASCII characters the matrix should contain numbers ≤ 127.
`clock`	Current time
`cos(x)`	Calculates cosine of x, where x is in radians.
`date`	Current date
`disp`	Displays data in Command Window.
`doc`	Open HTML Help Desk directly at a particular function description.
`double`	Converts a character string into a matrix of numbers.
`eps`	Represents machine precision.
`exp(x)`	Calculates e^x.
`eye(m,n)`	Generates an identity matrix.
`fix(x)`	Rounds x to the nearest integer towards zero: `fix(3.1) = 3` and `fix(-3.1) = -3`.
`floor(x)`	Rounds x to the nearest integer towards minus infinity: `floor(3.1) = 3` and `floor(-3.1) = -4`.
`format +`	Print + and – signs only.
`format bank`	Print in "dollars and cents" format.
`format compact`	Suppress extra linefeeds in output.
`format hex`	Print hexadecimal display of bits.
`format long`	Print with 14 digits after the decimal.
`format long e`	Print with 15 digits plus exponent.
`format long g`	Print with 15 digits with or without exponent.
`format loose`	Print with extra linefeeds in output.
`format rat`	Print as an approximate ratio of small integers.
`format short`	Print with 4 digits after the decimal.

(continued)

Commands and Functions *(Continued)*

`format short e`	Print with 5 digits plus exponent.
`format short g`	Print with 5 digits with or without exponent.
`fprintf`	Print formatted information.
`grid`	Add or remove a grid from a plot.
`i`	$\sqrt{-1}$
`inf`	Represents machine infinity (∞).
`input`	Writes a prompt and reads a value from the keyboard.
`int2str`	Converts x into an integer character string.
`j`	$\sqrt{-1}$
`legend`	Adds a legend to a plot.
`length(arr)`	Returns the length of a vector, or the longest dimension of a 2-D array.
`load`	Load data from a file.
`log(x)`	Calculates the natural logarithm of x.
`loglog`	Generates a log-log plot.
`lookfor`	Look for a matching term in the one-line MATLAB function descriptions.
`max(x)`	Returns the maximum value in vector x, and optionally the location of that value.
`min(x)`	Returns the minimum value in vector x, and optionally the location of that value.
`mod(m,n)`	Remainder or modulo function.
`nan`	Represents not-a-number.
`num2str(x)`	Converts x into a character string.
`ones(m,n)`	Generates an array of ones.
`pi`	Represents the number π.
`plot`	Generates a linear xy plot.
`print`	Prints a Figure Window.
`round(x)`	Rounds x to the nearest integer.
`save`	Saves data from workspace into a file.
`semilogx`	Generates a log-linear plot.
`semilogy`	Generates a linear-log plot.
`sin(x)`	Calculates sine of x, where x is in radians.
`size`	Get number of rows and columns in an array.
`sqrt`	Calculates the square root of a number.
`str2num`	Converts a character string into a number.
`tan(x)`	Calculates tangent of x, where x is in radians.
`title`	Adds a title to a plot.
`zeros`	Generate an array of zeros.

2.15   Exercises

2.1   Answer the following questions for the array shown here.

$$
array1 = \begin{bmatrix}
0.0 & 0.5 & 2.1 & -3.5 & 6.0 \\
0.0 & -1.1 & -6.6 & 2.8 & 3.4 \\
2.1 & 0.1 & 0.3 & -0.4 & 1.3 \\
1.1 & 5.1 & 0.0 & 1.1 & -2.0
\end{bmatrix}
$$

(a) What is the size of `array1`?
(b) What is the value of `array1(1,4)`?
(c) What is the size and value of `array1(:,1:2:5)`?
(d) What is the size and value of `array1([1 3],end)`?

2.2   Are the following MATLAB variable names legal or illegal? Why?

(a) `dog1`
(b) `1dog`
(c) `Do_you_know_the_way_to_san_jose`
(d) `_help`
(e) `What's_up?`

2.3   Determine the size and contents of the following arrays. Note that the later arrays may depend on the definitions of arrays defined earlier in this exercise.

(a) `a = 2:3:8;`
(b) `b = [a' a' a'];`
(c) `c = b(1:2:3,1:2:3);`
(d) `d = a + b(2,:);`
(e) `w = [zeros(1,3) ones(3,1)' 3:5'];`
(f) `b([1 3],2) = b([3 1],2);`
(g) `e = 1:-1:5;`

2.4   Assume that array `array1` is defined as shown, and determine the contents of the following subarrays:

$$
array1 = \begin{bmatrix}
1.1 & 0.0 & -2.1 & -3.5 & 6.0 \\
0.0 & -3.0 & -5.6 & 2.8 & 4.3 \\
2.1 & 0.3 & 0.1 & -0.4 & 1.3 \\
-1.4 & 5.1 & 0.0 & 1.1 & -3.0
\end{bmatrix}
$$

(a) `array1(3,:)`
(b) `array1(:,3)`
(c) `array1(1:2:3,[3 3 4])`
(d) `array1([1 1],:)`

2.5 Assume that `value` has been initialized to 10π, and determine what is printed out by each of the following statements.

```
disp (['value = ' num2str(value)]);
disp (['value = ' int2str(value)]);
fprintf('value = %e\n',value);
fprintf('value = %f\n',value);
fprintf('value = %g\n',value);
fprintf('value = %12.4f\n',value);
```

2.6 Assume that a, b, c, and d are defined as follows, and calculate the results of the following operations if they are legal. If an operation is illegal, explain why it is illegal.

$$a = \begin{bmatrix} 2 & 1 \\ -1 & 4 \end{bmatrix} \qquad b = \begin{bmatrix} -1 & 3 \\ 0 & 2 \end{bmatrix}$$

$$c = \begin{bmatrix} 2 \\ 1 \end{bmatrix} \qquad d = eye(2)$$

(*a*) `result = a + b;`
(*b*) `result = a * d;`
(*c*) `result = a .* d;`
(*d*) `result = a * c;`
(*e*) `result = a .* c;`
(*f*) `result = a \ b;`
(*g*) `result = a .\ b;`
(*h*) `result = a .^ b;`

2.7 Evaluate each of the following expressions:

(*a*) `11 / 5 + 6`
(*b*) `(11 / 5) + 6`
(*c*) `11 / (5 + 6)`
(*d*) `3 ^ 2 ^ 3`
(*e*) `3 ^ (2 ^ 3)`
(*f*) `(3 ^ 2) ^ 3`
(*g*) `round(-11/5) + 6`
(*h*) `ceil(-11/5) + 6`
(*i*) `floor(-11/5) + 6`

2.8 Use MATLAB to evaluate each of the following expressions:

(*a*) $(3 - 4i)(-4 + 3i)$
(*b*) $\cos^{-1}(1.2)$

2.9   Solve the following system of simultaneous equations for x:

$$
\begin{aligned}
-2.0\,x_1 + 5.0\,x_2 + 1.0\,x_3 + 3.0\,x_4 + 4.0\,x_5 - 1.0\,x_6 &= 0.0 \\
2.0\,x_1 - 1.0\,x_2 - 5.0\,x_3 - 2.0\,x_4 + 6.0\,x_5 + 4.0\,x_6 &= 1.0 \\
-1.0\,x_1 + 6.0\,x_2 - 4.0\,x_3 - 5.0\,x_4 + 3.0\,x_5 - 1.0\,x_6 &= -6.0 \\
4.0\,x_1 + 3.0\,x_2 - 6.0\,x_3 - 5.0\,x_4 - 2.0\,x_5 - 2.0\,x_6 &= 10.0 \\
-3.0\,x_1 + 6.0\,x_2 + 4.0\,x_3 + 2.0\,x_4 - 6.0\,x_5 + 4.0\,x_6 &= -6.0 \\
2.0\,x_1 + 4.0\,x_2 + 4.0\,x_3 + 4.0\,x_4 + 5.0\,x_5 - 4.0\,x_6 &= -2.0
\end{aligned}
$$

2.10   **Position and Velocity of a Ball** If a stationary ball is released at a height h_0 above the surface of the Earth with a vertical velocity v_0, the position and velocity of the ball as a function of time will be given by the equations

$$h(t) = \frac{1}{2}gt^2 + v_0t + h_0 \tag{2-10}$$

$$v(t) = gt + v_0 \tag{2-11}$$

where g is the acceleration due to gravity (-9.81 m/s²), h is the height above the surface of the Earth (assuming no air friction), and v is the vertical component of velocity. Write a MATLAB program that prompts a user for the initial height of the ball in meters and velocity of the ball in meters per second, and plots the height and velocity as a function of time. Be sure to include proper labels in your plots.

2.11   The distance between two points (x_1, y_1) and (x_2, y_2) on a Cartesian coordinate plane is given by the equation

$$d = \sqrt{(x_1 - x_2)^2 + (y_1 - y_2)^2} \tag{2-12}$$

(See Figure 2.16.) Write a program to calculate the distance between any two points (x_1, y_1) and (x_2, y_2) specified by the user. Use good programming practices in your program. Use the program to calculate the distance between the points $(-3, 2)$ and $(3, -6)$.

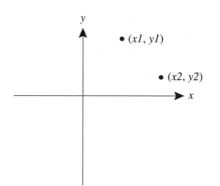

Figure 2.16   Distance between two points on a Cartesian plane.

2.12 The distance between two points (x_1, y_1, z_1) and (x_2, y_2, z_2) in a three-dimensional Cartesian coordinate system is given by the equation

$$d = \sqrt{(x_1 - x_2)^2 + (y_1 - y_2)^2 + (z_1 - z_2)^2} \qquad (2\text{-}13)$$

Write a program to calculate the distance between any two points (x_1, y_1, z_1) and (x_2, y_2, z_2) specified by the user. Use good programming practices in your program. Use the program to calculate the distance between the points $(-3, 2, 5)$ and $(3, -6, -5)$.

2.13 **Decibels** Engineers often measure the ratio of two power measurements in *decibels*, or dB. The equation for the ratio of two power measurements in decibels is

$$dB = 10 \log_{10} \frac{P_2}{P_1} \qquad (2\text{-}14)$$

where P_2 is the power level being measured, and P_1 is some reference power level.

(*a*) Assume that the reference power level P_1 is 1 milliwatt, and write a program that accepts an input power P_2 and converts it into dB with respect to the 1 mW reference level. (Engineers have a special unit for dB power levels with respect to a 1 mW reference: dBm.) Use good programming practices in your program.

(*b*) Write a program that creates a plot of power in watts versus power in dBm with respect to a 1 mW reference level. Create both a linear *xy* plot and a log-linear *xy* plot.

2.14 **Power in a Resistor** The voltage across a resistor is related to the current flowing through it by Ohm's law

$$V = IR \qquad (2\text{-}15)$$

and the power consumed in the resistor is given by the equation

$$P = IV \qquad (2\text{-}16)$$

Write a program that creates a plot of the power consumed by a 1000 Ω resistor as the voltage across it is varied from 1 V to 200 V. Create two plots, one showing power in watts, and one showing power in dBW (dB power levels with respect to a 1 W reference).

Figure 2.17 Voltage and current in a resistor.

2.15 Hyperbolic cosine The hyperbolic cosine function is defined by the equation

$$\cosh x = \frac{e^x + e^{-x}}{2} \tag{2-17}$$

Write a program to calculate the hyperbolic cosine of a user-supplied value x. Use the program to calculate the hyperbolic cosine of 3.0. Compare the answer that your program produces to the answer produced by the MATLAB intrinsic function cosh(x). Also, use MATLAB to plot the function cosh(x). What is the smallest value that this function can have? At what value of x does it occur?

2.16 Energy Stored in a Spring The force required to compress a linear spring is given by the equation

$$F = kx \tag{2-18}$$

where F is the force in newtons and k is the spring constant in newtons per meter. The potential energy stored in the compressed spring is given by the equation

$$E = \frac{1}{2} kx^2 \tag{2-19}$$

where E is the energy in joules. The following information is available for four springs:

	Spring 1	Spring 2	Spring 3	Spring 4
Force (N)	20	30	25	20
Spring constant k (N/m)	200	250	300	400

Determine the compression of each spring, and the potential energy stored in each spring. Which spring has the most energy stored in it?

2.17 Radio Receiver A simplified version of the front end of an AM radio receiver is shown in Figure 2.18. This receiver consists of an *RLC* tuned circuit containing a resistor, capacitor, and an inductor connected in series. The *RLC* circuit is connected to an external antenna and ground as shown in the picture.

The tuned circuit allows the radio to select a specific station out of all the stations transmitting on the AM band. At the resonant frequency of the circuit, essentially all of the signal V_0 appearing at the antenna appears across the resistor, which represents the rest of the radio. In other words, the radio receives its strongest signal at the resonant frequency. The resonant frequency of the LC circuit is given by the equation

$$f_0 = \frac{1}{2\pi\sqrt{LC}} \tag{2-20}$$

Figure 2.18 A simplified version of the front end of an AM radio receiver.

where L is inductance in henrys (H) and C is capacitance in farads (F). Write a program that calculates the resonant frequency of this radio set given specific values of L and C. Test your program by calculating the frequency of the radio when $L = 0.25$ mH and $C = 0.10$ nF.

2.18 **Radio Receiver** The average (rms) voltage across the resistive load in Figure 2.18 varies as a function of frequency according to Equation (2-18).

$$V_R = \frac{R}{\sqrt{R^2 + \left(\omega L - \dfrac{1}{\omega C} \right)^2}} V_0 \qquad (2\text{-}21)$$

where $\omega = 2\pi f$ and f is the frequency in hertz. Assume that $L = 0.25$ mH, $C = 0.10$ nF, $R = 50\ \Omega$, and $V_0 = 10$ mV.

(a) Plot the rms voltage on the resistive load as a function of frequency. At what frequency does the voltage on the resitive load peak? What is the voltage on the load at this frequency? This frequency is called the resonant frequency f_0 of the circuit.

(b) If the frequency is changed to 10% greater than the resonant frequency, what is the voltage on the load? How selective is this radio receiver?

(c) At what frequencies will the voltage on the load drop to half of the voltage at the resonant frequency?

2.19 Suppose two signals were received at the antenna of the radio receiver described in the previous problem. One signal has a strength of 1 V at a frequency of 1000 kHz, and the other signal has a strength of 1 V at 950 kHz. Calculate the voltage V_R that will be received for each of these signals. How much power will the first signal supply to the resistive load R? How much power will the second signal supply to the resistive load R? Express the ratio of the power supplied by signal 1 to the power supplied by signal 2 in decibels (see Problem 2.12 above for the definition of a decibel). How much is the second signal enhanced or suppressed compared to the first signal? (*Note:* The power supplied to the resistive load can be calculated from the equation $P = V_R^2 / R$).

2.20   **Aircraft Turning Radius** An object moving in a circular path at a constant tangential velocity v is shown in Figure 2.19. The radial acceleration required for the object to move in the circular path is given by the Equation (2-22)

$$a = \frac{v^2}{r} \qquad (2\text{-}22)$$

where a is the centripetal acceleration of the object in m/s², v is the tangential velocity of the object in m/s, and r is the turning radius in meters. Suppose that the object is an aircraft, and answer the following questions about it:

(*a*)   Suppose that the aircraft is moving at Mach 0.85, or 85% of the speed of sound. If the centripetal acceleration is 2 g, what is the turning radius of the aircraft? (*Note:* For this problem, you may assume that Mach 1 is equal to 340 m/s, and that 1 g = 9.81 m/s².)

(*b*)   Suppose that the speed of the aircraft increases to Mach 1.5. What is the turning radius of the aircraft now?

(*c*)   Plot the turning radius as a function of aircraft speed for speeds between Mach 0.5 and Mach 2.0, assuming that the acceleration remains 2 g.

(*d*)   Suppose that the maximum acceleration that the pilot can stand is 7 g. What is the minimum possible turning radius of the aircraft at Mach 1.5?

(*e*)   Plot the turning radius as a function of centripetal acceleration for accelerations between 2 g and 8 g, assuming a constant speed of Mach 0.85.

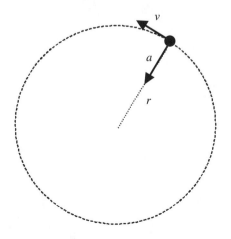

Figure 2.19   An object moving in uniform circular motion due to the centripetal acceleration a.

CHAPTER 3

Branching Statements and Program Design

In the previous chapter, we developed several complete working MATLAB programs. However, all of the programs were very simple, consisting of a series of MATLAB statements that were executed one after another in a fixed order. Such programs are called *sequential* programs. They read input data, process it to produce a desired answer, print out the answer, and quit. There is no way to repeat sections of the program more than once, and there is no way to selectively execute only certain portions of the program depending on values of the input data.

In the next two chapters, we will introduce a number of MATLAB statements that allow us to control the order in which statements are executed in a program. There are two broad categories of control statements: **branches**, which select specific sections of the code to execute, and **loops**, which cause specific sections of the code to be repeated. Branches will be discussed in this chapter, and loops will be discussed in Chapter 4.

With the introduction of branches and loops, our programs are going to become more complex, and it will get easier to make mistakes. To help avoid programming errors, we will introduce a formal program design procedure based on the technique known as top-down design. We will also introduce a common algorithm development tool known as pseudocode.

We will also study the MATLAB logical data type before discussing branches, because branches are controlled by logical values and expressions.

3.1 Introduction to Top-Down Design Techniques

Suppose that you are an engineer working in industry and that you need to write a program to solve some problem. How do you begin?

87

When given a new problem, there is a natural tendency to sit down at a keyboard and start programming without "wasting" a lot of time thinking about the problem first. It is often possible to get away with this "on-the-fly" approach to programming for very small problems, such as many of the examples in this book. In the real world, however, problems are larger, and a programmer attempting this approach will become hopelessly bogged down. For larger problems, it pays to completely think out the problem and the approach you are going to take to it before writing a single line of code.

We will introduce a formal program design process in this section and then apply that process to every major application developed in the remainder of the book. For some of the simple examples that we will be doing, the design process will seem like overkill. However, as the problems that we solve get larger and larger, the process becomes more and more essential to successful programming.

When I was an undergraduate, one of my professors was fond of saying, "Programming is easy. It's knowing what to program that's hard." His point was forcefully driven home to me after I left university and began working in industry on larger-scale software projects. I found that the most difficult part of my job was to *understand the problem* I was trying to solve. Once I really understood the problem, it became easy to break the problem apart into smaller, more easily manageable pieces with well-defined functions, and then to tackle those pieces one at a time.

Top-down design is the process of starting with a large task and breaking it down into smaller, more easily understandable pieces (subtasks) which perform a portion of the desired task. Each subtask may in turn be subdivided into smaller subtasks if necessary. Once the program is divided into small pieces, each piece can be coded and tested independently. We do not attempt to combine the subtasks into a complete task until each of the subtasks has been verified to work properly by itself.

The concept of top-down design is the basis of our formal program design process. We will now introduce the details of the process, which is illustrated in Figure 3.1. The steps involved are as follows:

1. *Clearly state the problem that you are trying to solve.*
   Programs are usually written to fill some perceived need, but that need may not be articulated clearly by the person requesting the program. For example, a user may ask for a program to solve a system of simultaneous linear equations. This request is not clear enough to allow a programmer to design a program to meet the need; he or she must first know much more about the problem to be solved. Is the system of equations to be solved real or complex? What is the maximum number of equations and unknowns that the program must handle? Are there any symmetries in the equations which might be exploited to make the task easier? The program designer will have to talk with the user requesting the program, and the two of them will have to come up with a clear statement of exactly what they are trying to accomplish. A clear statement of the problem will prevent misunderstandings, and it will also help the program designer to properly organize his or her thoughts. In the example we were describing, a proper statement of the problem might have been:

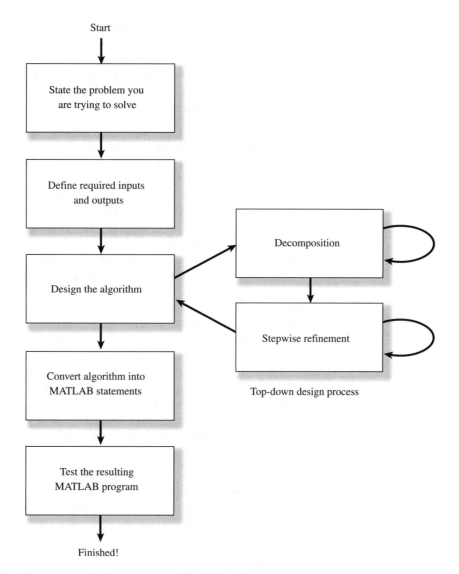

Figure 3.1   The program design process used in this book.

Design and write a program to solve a system of simultaneous linear equations having real coefficients and with up to 20 equations in 20 unknowns.

2. *Define the inputs required by the program and the outputs to bepro- duced by the program.*

   The inputs to the program and the outputs to be produced by the program must be specified so that the new program will properly fit into the overall processing scheme. In the preceding example, the coefficients of the equations to be solved are probably in some preexisting order, and our new program needs to be able to read them in that order. Similarly, it needs to produce the answers required by the programs that may follow it in the overall processing scheme, and to write out those answers in the format needed by the programs following it.

3. *Design the algorithm that you intend to implement in the program.*
An **algorithm** is a step-by-step procedure for finding the solution to a problem. It is at this stage in the process that top-down design techniques come into play. The designer looks for logical divisions within the problem and divides it up into sub-tasks along those lines. This process is called *decomposition*. If the sub-tasks are themselves large, the designer can break them up into even smaller sub–sub-tasks. This process continues until the problem has been divided into many small pieces, each of which does a simple, clearly understandable job.

After the problem has been decomposed into small pieces, each piece is further refined through a process called *stepwise refinement*. In stepwise refinement, a designer starts with a general description of what the piece of code should do and then defines the functions of the piece in greater and greater detail until they are specific enough to be turned into MATLAB statements. Stepwise refinement is usually done with **pseudocode**, which will be described in the next section.

It is often helpful to solve a simple example of the problem by hand during the algorithm development process. If the designer understands the steps that he or she went through in solving the problem by hand, then he or she will be in better able to apply decomposition and stepwise refinement to the problem.

4. *Turn the algorithm into MATLAB statements.*
If the decomposition and refinement process was carried out properly, this step will be very simple. All the programmer will have to do is to replace pseudocode with the corresponding MATLAB statements on a one-for-one basis.

5. *Test the resulting MATLAB program.*
This step is the real killer. The components of the program must first be tested individually, if possible, and then the program as a whole must be tested. When testing a program, we must verify that it works correctly for *all legal input data sets*. It is very common for a program to be written, tested with some standard data set, and released for use, only to find that it produces the wrong answers (or crashes) with a different input data set. If the algorithm implemented in a program includes different branches, we must test all of the possible branches to confirm that the program operates correctly under every possible circumstance.

Large programs typically go through a series of tests before they are released for general use (see Figure 3.2). The first stage of testing is sometimes called **unit testing**. During unit testing, the individual sub-tasks of the program are tested separately to confirm that they work correctly. After the unit testing is completed, the program goes through a series of *builds* during which the individual sub-tasks are combined to produce the final program. The first build of the program

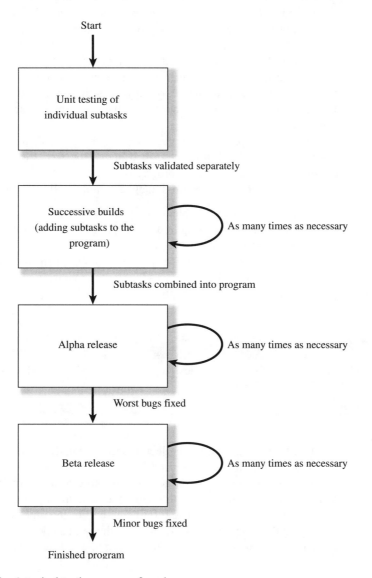

Figure 3.2   A typical testing process for a large program.

typically includes only a few of the sub-tasks. It is used to check the interactions among those sub-tasks and the functions performed by the combinations of the sub-tasks. In successive builds, more and more sub-tasks are added, until the entire program is complete. Testing is performed on each build, and any errors (bugs) that are detected are corrected before moving on to the next build.

Testing continues even after the program is complete. The first complete version of the program is usually called the **alpha release**. It is exercised by the programmers and others very close to them in as many different ways as possible, and the bugs discovered during the testing are corrected. When the most serious bugs have been removed from the program, a new version called the **beta release** is prepared. The beta release is normally given to "friendly" outside users who have a need for the program in their normal day-to-day jobs. These users put the program through its paces under many different conditions and with many different input data sets, and they report any bugs that they find to the programmers. When those bugs have been corrected, the program is ready to be released for general use.

Because the programs in this book are fairly small, we will not go through the sort of extensive testing described here. However, we will follow the basic principles in testing all of our programs.

The program design process may be summarized as follows:

1. Clearly state the problem that you are trying to solve.
2. Define the inputs required by the program and the outputs to be produced by the program.
3. Design the algorithm that you intend to implement in the program.
4. Turn the algorithm into MATLAB statements.
5. Test the MATLAB program.

✴ Good Programming Practice

Follow the steps of the program design process to produce reliable, understandable MATLAB programs.

In a large programming project, the time actually spent programming is surprisingly small. In his book *The Mythical Man-Month*[1], Frederick P. Brooks, Jr. suggests that in a typical large software project, one-third of the time is spent planning what to do (steps 1 through 3), one-sixth of the time is spent actually writing the program (step 4), and fully one-half of the time is spent in testing and debugging the program! Clearly, anything that we can do to reduce the testing and debugging time will be very helpful. We can best reduce the testing and debugging

---

[1] *The Mythical Man-Month, Anniversary Edition*, by Frederick P. Brooks, Jr., Addison-Wesley, 1995.

time by doing a very careful job in the planning phase, and by using good programming practices. Good programming practices will reduce the number of bugs in the program and will make the ones that do creep in easier to find.

3.2   Use of Pseudocode

As a part of the design process, it is necessary to describe the algorithm that you intend to implement. The description of the algorithm should be in a standard form that is easy for both you and other people to understand, and the description should aid you in turning your concept into MATLAB code. The standard forms that we use to describe algorithms are called **constructs** (or sometimes structures), and an algorithm described using these constructs is called a structured algorithm. When the algorithm is implemented in a MATLAB program, the resulting program is called a **structured program**.

The constructs used to build algorithms can be described in a special way called pseudocode. **Pseudocode** is a hybrid mixture of MATLAB and English. It is structured like MATLAB, with a separate line for each distinct idea or segment of code, but the descriptions on each line are in English. Each line of the pseudocode should describe its idea in plain, easily understandable English. Pseudocode is very useful for developing algorithms, since it is flexible and easy to modify. It is especially useful since pseudocode can be written and modified with the same editor or word processor used to write the MATLAB program—no special graphical capabilities are required.

For example, the pseudocode for the algorithm in Example 1.3 is

```
Prompt user to enter temperature in degrees Fahrenheit
Read temperature in degrees Fahrenheit (temp_f)
temp_k in kelvins <- (5/9) * (temp_f - 32) + 273.15
Write temperature in kelvins
```

Notice that a left arrow ($\leftarrow$) is used instead of an equal sign ($=$) to indicate that a value is stored in a variable, since this avoids any confusion between assignment and equality. Pseudocode is intended to aid you in organizing your thoughts before converting them into MATLAB code.

3.3   The Logical Data Type

The `logical` data type[2] is a special type of data that can have one of only two possible values: `true` or `false`. These values are produced by the two special functions `true` and `false`. They are also produced by two types of MATLAB operators: relational operators and logic operators.

---

[2]The logical data type was introduced in MATLAB 6.5.

Logical values are stored in a single byte of memory, so they take up much less space than numbers, which usually occupy 8 bytes.

The operation of many MATLAB branching constructs is controlled by logical variables or expressions. If the result of a variable or expression is true, then one section of code is executed. If not, then a different section of code is executed.

To create a logical variable, just assign a logical value it to in an assignment statement. For example, the statement

```
a1 = true;
```

creates a logical variable a1 containing the logical value true. If this variable is examined with the whos command, we can see that it has the logical data type:

```
» whos a1
Name    Size    Bytes    Class
a1      1x1         1    logical array
```

Unlike programming languages such as Java, C++, and Fortran, it is legal in MATLAB to mix numerical and logical data in expressions. If a logical value is used in a place where a numerical value is expected, true values are converted to 1, and false values are converted to 0 and then used as numbers. If a numerical value is used in a place where a logical value is expected, nonzero values are converted to true, and 0 values are converted to false and then used as logical values.

It is also possible to explicitly convert numerical values to logical values, and vice versa. The logical function converts numerical data to logical data, and the real function converts logical data to numerical data.

3.3.1 Relational Operators

Relational operators are operators with two numerical or string operands that yield a logical result, depending on the relationship between the two operands. The general form of a relational operator is

$$a_1 \text{ op } a_2$$

where a_1 and a_2 are arithmetic expressions, variables, or strings, and op is one of the following relational operators:

Table 3.1   Relational Operators

Operator	Operation
==	Equal to
~=	Not equal to
>	Greater than
>=	Greater than or equal to
<	Less than
<=	Less than or equal to

If the relationship between a_1 and a_2 expressed by the operator is true, then the operation returns a true value; otherwise, the operation returns false.

Some relational operations and their results are given here:

Operation	Result
3 < 4	true (1)
3 <= 4	true (1)
3 == 4	false (0)
3 > 4	false (0)
4 <= 4	true (1)
'A' < 'B'	true (1)

The last relational operation is true, because characters are evaluated in alphabetical order.

Note that both `true` and `1` are shown as the result of true operations, and both `false` and `0` are shown as the result of false operations. MATLAB is a bit schizophrenic about how the results of logical operations are displayed. When a relational operator is evaluated in the Command Window, the result of the operation will be displayed as a 0 or a 1. When it is displayed in the Workspace Browser, the same value will be show as `false` or `true` (see Figure 3.3).

Relational operators may be used to compare a scalar value with an array. For example, if a $= \begin{bmatrix} 1 & 0 \\ -2 & 1 \end{bmatrix}$ and b $= 0$, then the expression a > b will yield the logical array $\begin{bmatrix} \text{true} & \text{false} \\ \text{false} & \text{true} \end{bmatrix}$ (shown as $\begin{bmatrix} 1 & 0 \\ 0 & 1 \end{bmatrix}$ in the Command Window). Relational operators may also be used to compare two arrays, as long as both arrays have the same size. For example, if a $= \begin{bmatrix} 1 & 0 \\ -2 & 1 \end{bmatrix}$ and b $= \begin{bmatrix} 0 & 2 \\ -2 & -1 \end{bmatrix}$, then the expression a >= b will yield the logical array $\begin{bmatrix} \text{true} & \text{false} \\ \text{true} & \text{true} \end{bmatrix}$ (shown as $\begin{bmatrix} 1 & 0 \\ 1 & 1 \end{bmatrix}$ in the Command Window). If the arrays have different sizes, a run-time error will result.

Note that since strings are really arrays of characters, *relational operators can only compare two strings if they are of equal lengths*. If they are of unequal lengths, the comparison operation will produce an error. We will learn of a more general way to compare strings in Chapter 6.

The equivalence relational operator is written with two equal signs, while the assignment operator is written with a single equal sign. These are very different operators that beginning programmers often confuse. The == symbol is a *comparison* operation that returns a logical result, while the = symbol *assigns* the value of the

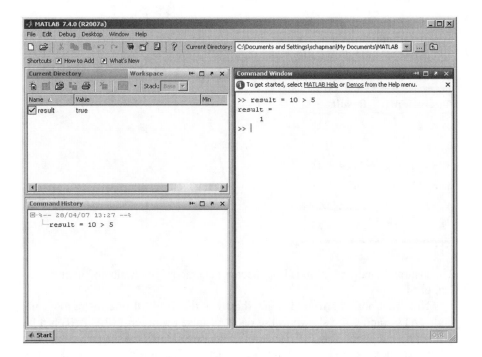

Figure 3.3 The result of a relational operator is a true or false value that can be stored in a `logical` variable. In the example shown here, the result of the operator $10 > 5$ is displayed as a 1 on the Command Window and as a `true` in the Workspace Browser.

expression to the right of the equal sign to the variable on the left of the equal sign. It is a very common mistake for beginning programmers to use a single equal sign when trying to do a comparison.

⬢ Programming Pitfalls

Be careful not to confuse the equivalence relational operator (==) with the assignment operator (=).

In the hierarchy of operations, relational operators are evaluated after all arithmetic operators have been evaluated. Therefore, the following two expressions are equivalent (both are true).

```
7 + 3 < 2 + 11
(7 + 3) < (2 + 11)
```

3.3.2   A Caution About the == and ~= Operators

The equivalence operator (==) returns a `true` value (1) when the two values being compared are equal, and a `false` (0) when the two values being compared are different. Similarly, non-equivalence operator (~=) returns a `false` (0) when the two values being compared are equal, and a `true` (1) when the two values being compared are different. These operators are generally safe to use for comparing strings, but they can sometimes produce surprising results when two numeric values are compared. Due to **roundoff errors** during computer calculations, two theoretically equal numbers can differ slightly, causing an equality or inequality test to fail.

For example, consider the following two numbers, both of which should be equal to 0.0.

```
a = 0;
b = sin(pi);
```

Since these numbers are theoretically the same, the relational operation a == b *should* produce a 1. In fact, the results of this MATLAB calculation are

```
» a = 0;
» b = sin(pi);
» a == b
ans =
          0
```

MATLAB reports that a and b are different because a slight roundoff error in the calculation of `sin(pi)` makes the result be 1.2246×10^{-16} instead of exactly zero. The two theoretically equal values differ slightly due to roundoff error!

Instead of comparing two numbers for *exact* equality, you should set up your tests to determine if the two numbers are *nearly* equal to each other within some accuracy that takes into account the roundoff error expected for the numbers being compared. The test

```
» abs(a - b) < 1.0E-14
ans =
          1
```

produces the correct answer, despite the roundoff error in calculating b.

✳ **Good Programming Practice**

Be cautious about testing for equality with numeric values, since roundoff errors may cause two variables that should be equal to fail a test for equality. Instead, test to see if the variables are *nearly* equal within the roundoff error to be expected on the computer you are working with.

Table 3.2   Logic Operators

Operator	Operation
&	Logical AND
&&	Logical AND with shortcut evaluation
|	Logical Inclusive OR
||	Logical Inclusive OR with shortcut evaluation
xor	Logical Exclusive OR
~	Logical NOT

3.3.3   Logic Operators

Logic operators are operators with one or two logical operands that yield a logical result. There are five binary logic operators: AND (& and &&), inclusive OR (| and ||), and exclusive OR (xor), and one unary operator: NOT (~). The general form of a binary logic operation is

$$l_1 \text{ op } l_2$$

and the general form of a unary logic operation is

$$\text{op } l_1$$

where l_1 and l_2 are expressions or variables, and op is one of the following logic operators shown in Table 3.2.

If the relationship between l_1 and l_2 expressed by the operator is true, then the operation returns a value of `true` (displayed as 1 in the Command Window); otherwise, the operation returns a value of `false` (0 in the Command Window).

The results of the operators are summarized in **truth tables**, which show the result of each operation for all possible combinations of l_1 and l_2. Table 3.3 shows the truth tables for all logic operators.

Logical ANDs

The result of an AND operator is `true` if and only if both input operands are `true`. If either or both operands are `false`, the result is `false`, as shown in Table 3.3.

Note that there are two logical AND operators: && and &. Why are there two AND operators, and what is the difference between them? The basic difference between && and & is that && supports *short-circuit evaluations* (or *partial evaluations*), while & doesn't. That is, && will evaluate expression l_1 and immediately return a `false` value if l_1 is `false`. If l_1 is `false`, the operator never evaluates l_2, because the result of the operator will be `false` regardless of the value of l_2. In contrast, the & operator always evaluates both l_1 and l_2 before returning an answer.

A second difference between && and & is that && works only between scalar values, whereas & works with either scalar or array values, as long as the sizes of the arrays are compatible.

Table 3.3   Truth Tables for Logic Operators

Inputs		and		or		xor	not
l_1	l_2	$l_1 \& l_2$	$l_1 \&\& l_2$	$l_1 \mid l_2$	$l_1 \mid\mid l_2$	$\text{xor}(l_1, l_2)$	$\sim l_1$
false	false	false	false	false	false	false	true
false	true	false	false	true	true	true	true
true	false	false	false	true	true	true	false
true	true	true	true	true	true	false	false

When should you use && and when should you use & in a program? Most of the time, it doesn't matter which AND operation is used. If you are comparing scalars and it is not necessary to always evaluate l_2, then use the && operator. The partial evaluation will make the operation faster in the cases where the first operand is false.

Sometimes it is important to use shortcut expressions. For example, suppose that we wanted to test for the situation where the ratio of two variables a and b is greater than 10. The code to perform this test is:

```
x = a / b > 10.0
```

This code normally works fine, but what about the case in which b is zero? In that case, we would be dividing by zero, which produces an Inf instead of a number. The test could be modified to avoid this problem as follows:

```
x = (b ~= 0) && (a/b > 10.0)
```

This expression uses partial evaluation, so if b = 0, the expression a/b > 10.0 will never be evaluated, and no Inf will occur.

✳ Good Programming Practice

Use the & AND operator if it is necessary to ensure that both operands are evaluated in an expression, or if the comparison is between arrays. Otherwise, use the && AND operator, since the partial evaluation will make the operation faster in the cases where the first operand is false. The & operator is preferred in most practical cases.

Logical Inclusive ORs

The result of an inclusive OR operator is true if either of the input operands are true. If both operands are false, the result is false, as shown in Table 3.3.

Note that there are two inclusive OR operators: || and |. Why are there two inclusive OR operators, and what is the difference between them? The basic difference between | | and | is that | | supports partial evaluations, while | doesn't. That is, | | will evaluate expression l_1 and immediately return a true value if l_1 is true. If l_1 is true, the operator never evaluates l_2, because the result of the operator will be true regardless of the value of l_2. In contrast, the | operator always evaluates both l_1 and l_2 before returning an answer.

A second difference between | | and | is that | | only works between scalar values, while | works with either scalar or array values, as long as the sizes of the arrays are compatible.

When should you use | | and when should you use | in a program? Most of the time, it doesn't matter which OR operation is used. If you are comparing scalars, and it is not necessary to always evaluate l_2, use the | | operator. The partial evaluation will make the operation faster in the cases where the first operand is true.

✳ Good Programming Practice

Use the | inclusive OR operator if it is necessary to ensure that both operands are evaluated in an expression, or if the comparison is between arrays. Otherwise, use the | | operator, since the partial evaluation will make the operation faster in the cases where the first operand is true. The | operator is preferred in most practical cases.

Logical Exclusive OR

The result of an exclusive OR operator is true if and only if one operand is true and the other one is false. If both operands are true or both operands are false, then the result is false, as shown in Table 3.3. Note that both operands must always be evaluated in order to calculate the result of an exclusive OR.

The logical exclusive OR operation is implemented as a function. For example:

```
a = 10;
b = 0;
x = xor(a, b);
```

This result is true. The value of a is nonzero, so it will be converted to true. The value of b is zero, so it will be converted to false. Therefore, the result of the xor operation will be true.

Logical NOT

The NOT operator is a unary operator, having only one operand. The result of a NOT operator is true if its operand is false, and false if its operand is true, as shown in Table 3.3.

Using Numeric Data with Logic Operators

Real numeric data can also be use with logic operators. Since logic operators expect logical input values, MATLAB converts nonzero values to true and zero values to false before performing the operation. Thus, the result of ~5 is false (0 in the Command Window) and the result of ~0 is true (1 in the Command Window).

Logic operators may be used to compare a scalar value with an array. For example, if a = $\begin{bmatrix} \text{true} & \text{false} \\ \text{false} & \text{true} \end{bmatrix}$ and b = false, then the expression a & b will yield the result $\begin{bmatrix} \text{false} & \text{false} \\ \text{false} & \text{false} \end{bmatrix}$ (displayed as $\begin{bmatrix} 0 & 0 \\ 0 & 0 \end{bmatrix}$ in the Command Window). Logic operators may also be used to compare two arrays, as long as both arrays have the same size. For example, if a = $\begin{bmatrix} \text{true} & \text{false} \\ \text{false} & \text{true} \end{bmatrix}$ and b = $\begin{bmatrix} \text{true} & \text{true} \\ \text{false} & \text{false} \end{bmatrix}$, then the expression a | b will yield the result $\begin{bmatrix} \text{true} & \text{true} \\ \text{false} & \text{true} \end{bmatrix}$ (displayed as $\begin{bmatrix} 1 & 1 \\ 0 & 1 \end{bmatrix}$ in the Command Window). If the arrays have different sizes, a run-time error will result.

Logic operators may *not* be used with complex or imaginary numeric data. For example, an expression such as "2i & 2i" will produce an error when it is evaluated.

Hierarchy of Operations

In the hierarchy of operations, logic operators are evaluated *after all arithmetic operations and all relational operators have been evaluated*. The order in which the operators in an expression are evaluated is as follows:

1. All arithmetic operators are evaluated first in the order previously described.
2. All relational operators (==, ~=, >, >=, <, <=) are evaluated, working from left to right.
3. All ~ operators are evaluated.
4. All & and && operators are evaluated, working from left to right.
5. All |, ||, and xor operators are evaluated, working from left to right.

As with arithmetic operations, parentheses can be used to change the default order of evaluation. Examples of some logic operators and their results follow.

▶
Example 3.1

Assume that the following variables are initialized with the values shown, and calculate the result of the specified expressions:

```
value1 = true
value2 = false
value3 = 1
```

```
value4 = -10
value5 = 0
value6 = [1  2; 0  1]
```

	Expression	Result	Comment
(a)	~value1	false	
(b)	~value3	false	The number 1 is converted to `true` before operation is applied.
(c)	value1 | value2	true	
(d)	value1 & value2	false	
(e)	value4 & value5	false	-10 is converted to `true` and 0 is converted to `false` before the operation is applied.
(f)	~(value4 & value5)	true	-10 is converted to `true` and 0 is converted to `false` before the operation is applied.
(g)	value1 + value4	-9	value1 is converted to the number 1 before the addition is performed.
(h)	value1 + (~value4)	1	The logical value1 is converted to the number 1 before the addition is performed. The number value4 is converted to `true` before the NOT is performed. Then ~value4 is evaluated to be `false`. This `false` value is converted to 0 before the addition, so the final result is $1 + 0 = 1$.
(i)	value3 && value6	Illegal	The && operator must be used with scalar operands.
(j)	value3 & value6	$\begin{bmatrix} \text{true} & \text{true} \\ \text{false} & \text{true} \end{bmatrix}$	AND between a scalar and an array operand.

The ~ operator is evaluated before other logic operators. Therefore, the parentheses in part (f) of the above example were required. If they had been absent, the expression in part (f) would have been evaluated in the order (~value4) & value5.

3.3.4 Logical Functions

MATLAB includes a number of logical functions that return `true` whenever the condition they test for is true, and `false` whenever the condition they test for is false. These functions can be used with relational and logic operator to control the operation of branches and loops.

A few of the more important logical functions are given in Table 3.4.

Table 3.4 Selected MATLAB Logical Functions

Function	Purpose
ischar(a)	Returns true if a is a character array and false otherwise.
isempty(a)	Returns true if a is an empty array and false otherwise.
isinf(a)	Returns true if the value of a is infinite (Inf) and false otherwise.
isnan(a)	Returns true if the value of a is NaN (not a number) and false otherwise.
isnumeric(a)	Returns true if a is a numeric array and false otherwise.
logical	Converts numerical values to logical values: if a value is non-zero, it is converted to true. If it is zero, it is converted to false.

Quiz 3.1

This quiz provides a quick check to see if you have understood the concepts introduced in Section 3.3. If you have trouble with the quiz, reread the sections, ask your instructor, or discuss the material with a fellow student. The answers to this quiz are found in the back of the book.

Assume that a, b, c, and d are as defined, and evaluate the following expressions:

$$a = 20; \qquad b = -2;$$
$$c = 0; \qquad d = 1;$$

1. a > b
2. b > d
3. a > b && c > d
4. a == b
5. a && b > c
6. ~~b

Assume that a, b, c, and d are as defined, and evaluate the following expressions:

$$a = 2; \qquad b = \begin{bmatrix} 1 & -2 \\ 0 & 10 \end{bmatrix};$$

$$c = \begin{bmatrix} 0 & 1 \\ 2 & 0 \end{bmatrix}; \qquad d = \begin{bmatrix} -2 & 1 & 2 \\ 0 & 1 & 0 \end{bmatrix};$$

7. ~(a > b)
8. a > c && b > c
9. c <= d

10. `logical(d)`
11. `a * b > c`
12. `a * (b > c)`

Assume that a, b, c, and d are as defined. Explain the order in which each of the following expressions are evaluated, and specify the results in each case:

$$a = 2; \qquad b = 3;$$
$$c = 10; \qquad d = 0;$$

13. `a*b^2 > a*c`
14. `d || b > a`
15. `(d | b) > a`

Assume that a, b, c, and d are as defined, and evaluate the following expressions:

$$a = 20; \qquad b = -2;$$
$$c = 0; \qquad d = 'Test';$$

16. `isinf(a/b)`
17. `isinf(a/c)`
18. `a > b && ischar(d)`
19. `isempty(c)`
20. `(~a) & b`
21. `(~a) + b`

3.4 Branches

Branches are MATLAB statements that permit us to select and execute specific sections of code (called *blocks*) while skipping other sections of code. They are variations of the `if` construct, the `switch` construct, and the `try/catch` construct.

3.4.1 The `if` Construct

The `if` construct has the form

```
if control_expr_1
    Statement 1
    Statement 2      } Block 1
    ...
elseif control_expr_2
    Statement 1
    Statement 2      } Block 2
    ...
```

```
    else
            Statement 1  ⎫
            Statement 2  ⎬  Block 3
               . . .     ⎭
    end
```

where the control expressions are logical expressions that control the operation of the `if` construct. If *control_expr_1* is true (nonzero), then the program executes the statements in Block 1 and skips to the first executable statement following the `end`. Otherwise, the program checks for the status of *control_expr_2*. If *control_expr_2* is true (nonzero), then the program executes the statements in Block 2, and skips to the first executable statement following the `end`. If all control expressions are false (zero), then the program executes the statements in the block associated with the `else` clause.

There can be any number of `elseif` clauses (0 or more) in an `if` construct, but there can be at most one `else` clause. The control expression in each clause will be tested only if the control expressions in every clause above are false (0). Once one of the expressions proves to be true and the corresponding code block is executed, the program skips to the first executable statement following the `end`. If all control expressions are false, then the program executes the statements in the block associated with the `else` clause. If there is no `else` clause, then execution continues after the `end` statement without executing any part of the `if` construct.

Note that the MATLAB keyword `end` in this construct is *completely different* from the MATLAB function `end` that we used in Chapter 2 to return the highest value of a given subscript. MATLAB tells the difference between these two uses of `end` from the context in which the word appears within an M-file.

In most circumstances, *the control expressions will be some combination of relational and logic operators.* As we learned earlier in this chapter, relational and logic operators produce a true (1) when the corresponding condition is true and a false (0) when the corresponding condition is false. When an operator is true, its result is nonzero, and the corresponding block of code will be executed.

As an example of an `if` construct, consider the solution of a quadratic equation of the form

$$ax^2 + bx + c = 0 \tag{3-1}$$

The solution to this equation is

$$x = \frac{-b \pm \sqrt{b^2 - 4ac}}{2a} \tag{3-2}$$

The term $b^2 - 4ac$ is known as the *discriminant* of the equation. If $b^2 - 4ac > 0$, then there are two distinct real roots to the quadratic equation. If $b^2 - 4ac = 0$, then there is a single repeated root to the equation, and if $b^2 - 4ac < 0$, then there are two complex roots to the quadratic equation.

Suppose that we wanted to examine the discriminant of a quadratic equation and to tell a user whether the equation has two complex roots, two identical real roots, or two distinct real roots. In pseudocode, this construct would take the form

```
if (b^2 - 4*a*c) < 0
   Write msg that equation has two complex roots.
elseif (b**2 - 4.*a*c) == 0
   Write msg that equation has two identical real roots.
else
   Write msg that equation has two distinct real roots.
end
```

The MATLAB statements to do this are

```
if (b^2 - 4*a*c) < 0
   disp('This equation has two complex roots.');
elseif (b^2 - 4*a*c) == 0
   disp('This equation has two identical real roots.');
else
   disp('This equation has two distinct real roots.');
end
```

For readability, the blocks of code within an if construct are usually indented by two or three spaces, but this is not actually required.

✳ **Good Programming Practice**

Always indent the body of an if construct by two or more spaces to improve the readability of the code. Note that indentation is automatic if you use the MATLAB editor to write your programs.

It is possible to write a complete if construct on a single line by separating the parts of the construct by commas or semicolons. Thus the following two constructs are identical:

```
if x < 0
   y = abs(x);
end
```

and

```
if x < 0; y = abs(x); end
```

However, this should be done only for very simple constructs.

3.4.2 Examples Using if Constructs

We will now look at two examples that illustrate the use of if constructs.

▶

Example 3.2—The Quadratic Equation

Write a program to solve for the roots of a quadratic equation, regardless of type.

SOLUTION   We will follow the design steps outlined earlier in the chapter.

1. **State the problem.**

   The problem statement for this example is very simple. We want to write a program that will solve for the roots of a quadratic equation, whether they are distinct real roots, repeated real roots, or complex roots.

2. **Define the inputs and outputs.**

   The inputs required by this program are the coefficients a, b, and c of the quadratic equation

   $$ax^2 + bx + c = 0 \qquad (3\text{-}1)$$

   The output from the program will be the roots of the quadratic equation, whether they are distinct real roots, repeated real roots, or complex roots.

3. **Design the algorithm.**

   This task can be broken down into three major sections, whose functions are input, processing, and output:

   ```
   Read the input data
   Calculate the roots
   Write out the roots
   ```

   We will now break each of the foregoing major sections into smaller, more detailed pieces. There are three possible ways to calculate the roots, depending on the value of the discriminant, so it is logical to implement this algorithm with a three-branched `if` construct. The resulting pseudocode is

```
Prompt the user for the coefficients a, b,and c.
Read a, b,and c
discriminant <- b^2 - 4 * a * c
if discriminant > 0
   x1 <- ( -b + sqrt(discriminant) ) / ( 2 * a )
   x2 <- ( -b - sqrt(discriminant) ) / ( 2 * a )
   Write msg that equation has two distinct real roots.
   Write out the two roots.
elseif discriminant == 0
   x1 <- -b / ( 2 * a )
   Write msg that equation has two identical real roots.
   Write out the repeated root.
else
   real_part <- -b / ( 2 * a )
```

```
      imag_part <- sqrt ( abs ( discriminant ) ) / ( 2 * a )
      Write msg that equation has two complex roots.
      Write out the two roots.
   end
```

4. **Turn the algorithm into MATLAB statements.**
   The final MATLAB code is shown here:

```
%   Script file: calc_roots.m
%
%   Purpose:
%      This program solves for the roots of a quadratic equation
%      of the form a*x**2 + b*x + c = 0. It calculates the answers
%      regardless of the type of roots that the equation possesses.
%
%   Record of revisions:
%        Date          Programmer           Description of change
%        ====          ==========           =====================
%      01/02/07     S. J. Chapman            Original code
%
%   Define variables:
%      a               -- Coefficient of x^2 term of equation
%      b               -- Coefficient of x term of equation
%      c               -- Constant term of equation
%      discriminant    -- Discriminant of the equation
%      imag_part       -- Imag part of equation (for complex roots)
%      real_part       -- Real part of equation (for complex roots)
%      x1              -- First solution of equation (for real roots)
%      x2              -- Second solution of equation (for real roots)

%   Prompt the user for the coefficients of the equation
disp ('This program solves for the roots of a quadratic ');
disp ('equation of the form A*X^2 + B*X + C = 0. ');
a = input ('Enter the coefficient A: ');
b = input ('Enter the coefficient B: ');
c = input ('Enter the coefficient C: ');
% Calculate discriminant
discriminant = b^2 - 4 * a * c;

%   Solve for the roots, depending on the value of the discriminant
if discriminant > 0 % there are two real roots, so...

      x1 = ( -b + sqrt(discriminant) ) / ( 2 * a );
      x2 = ( -b - sqrt(discriminant) ) / ( 2 * a );
      disp ('This equation has two real roots:');
      fprintf ('x1 = %f\n', x1);
      fprintf ('x2 = %f\n', x2);
```

```
elseif discriminant == 0 % there is one repeated root, so...

   x1 = ( -b ) / ( 2 * a );
   disp ('This equation has two identical real roots:');
   fprintf ('x1 = x2 = %f\n', x1);

else % there are complex roots, so ...

   real_part = ( -b ) / ( 2 * a );
   imag_part = sqrt ( abs ( discriminant ) ) / ( 2 * a );
   disp ('This equation has complex roots:');
   fprintf('x1 = %f +i %f\n', real_part, imag_part );
   fprintf('x1 = %f -i %f\n', real_part, imag_part );

end
```

5. **Test the program.**

Next, we must test the program using real input data. Since there are three possible paths through the program, we must test all three paths before we can be certain that the program is working properly. From Equation (3-2), it is possible to verify the solutions to the following equations:

$$x^2 + 5x + 6 = 0 \qquad x = -2 \text{ and } x = -3$$
$$x^2 + 4x + 4 = 0 \qquad x = -2$$
$$x^2 + 2x + 5 = 0 \qquad x = -1 \pm i2$$

If this program is executed three times with these coefficients, the results are as shown in the following (user inputs are shown in boldface):

```
» calc_roots
This program solves for the roots of a quadratic
equation of the form A*X^2 + B*X + C = 0.
Enter the coefficient A: 1
Enter the coefficient B: 5
Enter the coefficient C: 6
This equation has two real roots:
x1 = -2.000000
x2 = -3.000000
» calc_roots
This program solves for the roots of a quadratic
equation of the form A*X^2 + B*X + C = 0.
Enter the coefficient A: 1
Enter the coefficient B: 4
Enter the coefficient C: 4
This equation has two identical real roots:
x1 = x2 = -2.000000
» calc_roots
This program solves for the roots of a quadratic
```

```
equation of the form A*X^2 + B*X + C = 0.
Enter the coefficient A: 1
Enter the coefficient B: 2
Enter the coefficient C: 5
This equation has complex roots:
x1 = -1.000000 +i 2.000000
x1 = -1.000000 -i 2.000000
```

The program gives the correct answers for our test data in all three possible cases.

◀

▶

Example 3.3—Evaluating a Function of Two Variables

Write a MATLAB program to evaluate a function $f(x, y)$ for any two user-specified values x and y. The function $f(x, y)$ is defined as follows:

$$f(x, y) = \begin{cases} x + y & x \geq 0 \text{ and } y \geq 0 \\ x + y^2 & x \geq 0 \text{ and } y < 0 \\ x^2 + y & x < 0 \text{ and } y \geq 0 \\ x^2 + y^2 & x < 0 \text{ and } y < 0 \end{cases}$$

SOLUTION   The function $f(x, y)$ is evaluated differently depending on the signs of the two independent variables x and y. To determine the proper equation to apply, it will be necessary to check for the signs of the x and y values supplied by the user.

1. **State the problem.**
   This problem statement is very simple: Evaluate the function $f(x, y)$ for any user-supplied values of x and y.

2. **Define the inputs and outputs.**
   The inputs required by this program are the values of the independent variables x and y. The output from the program will be the value of the function $f(x, y)$.

3. **Design the algorithm.**
   This task can be broken down into three major sections, whose functions are input, processing, and output.

   ```
   Read the input values x and y
   Calculate f(x,y)
   Write out f(x,y)
   ```

   We will now break each of the foregoing major sections into smaller, more detailed pieces. There are four possible ways to calculate the function $f(x, y)$, depending upon the values of x and y, so it is logical to implement this algorithm with a four-branched if statement. The resulting pseudocode is as follows:

```
        Prompt the user for the values x and y.
        Read x and y
        if x ≥ 0 and y ≥ 0
            fun <- x + y
        elseif x ≥ 0 and y < 0
            fun <- x + y^2
        elseif x < 0 and y ≥ 0
            fun <- x^2 + y
        else
            fun <- x^2 + y^2
        end
        Write out f(x,y)
```

4. **Turn the algorithm into MATLAB statements.**
   The final MATLAB code is shown here.

```
%   Script file: funxy.m
%
%   Purpose:
%       This program solves the function f(x,y) for a
%       user-specified x and y, where f(x,y) is defined as:
%
%
%
%                    ⎡x + y                   x >= 0 and y >= 0
%                    ⎪x + y^2                 x >= 0 and y < 0
%         f(x,y)  =  ⎪x^2 + y                 x < 0  and y >= 0
%                    ⎣x^2 + y^2               x < 0  and y < 0
%
%
%   Record of revisions:
%       Date          Programmer            Description of change
%       ====          ==========            =====================
%       01/03/04      S. J. Chapman         Original code
%
% Define variables:
%     x      -- First independent variable
%     y      -- Second independent variable
%     fun    -- Resulting function

% Prompt the user for the values x and y
x = input ('Enter the x coefficient: ');
y = input ('Enter the y coefficient: ');

% Calculate the function f(x,y) based upon
% the signs of x and y.
if x >= 0 && y >= 0
   fun = x + y;
```

```
elseif x >= 0 && y < 0
   fun = x + y^2;
elseif x < 0 && y >= 0
   fun = x^2 + y;
else % x < 0 and y < 0, so
   fun = x^2 + y^2;
end

% Write the value of the function.
disp (['The value of the function is ' num2str(fun)]);
```

5. **Test the program.**

Next, we must test the program using real input data. Since there are four possible paths through the program, we must test all four paths before we can be certain that the program is working properly. To test all four possible paths, we will execute the program with the four sets of input values $(x, y) = (2, 3)$, $(2 -3)$, $(-2, 3)$, and $(-2, -3)$. Calculating by hand, we see that

$$f(2, 3) = 2 + 3 = 5$$
$$f(2, -3) = 2 + (-3)^2 = 11$$
$$f(-2, 3) = (-2)^2 + 3 = 7$$
$$f(-2, -3) = (-2)^2 + (-3)^2 = 13$$

If this program is compiled and then run four times with the foregoing values, the results are

```
» funxy
Enter the x coefficient: 2
Enter the y coefficient: 3
The value of the function is 5
» funxy
Enter the x coefficient: 2
Enter the y coefficient: -3
The value of the function is 11
» funxy
Enter the x coefficient: -2
Enter the y coefficient: 3
The value of the function is 7
» funxy
Enter the x coefficient: -2
Enter the y coefficient: -3
The value of the function is 13
```

The program gives the correct answers for our test values in all four possible cases.

3.4.3 Notes Concerning the Use of `if` Constructs

The `if` construct is very flexible. It must have one `if` statement and one `end` statement. In between, it can have any number of `elseif` clauses, and it may also have one `else` clause. With this combination of features, it is possible to implement any desired branching construct.

In addition, `if` constructs may be **nested**. Two `if` constructs are said to be nested if one of them lies entirely within a single code block of the other one. The following two `if` constructs are properly nested.

```
if x > 0
   ...
   if y < 0
      ...
   end
   ...
end
```

The MATLAB interpreter always associates a given `end` statement with the most recent `if` statement, so the first `end` above closes the `if y < 0` statement, while the second `end` closes the `if x > 0` statement. This works well for a properly written program, but can cause the interpreter to produce confusing error messages in cases where the programmer makes a coding error. For example, suppose that we have a large program containing a construct like the one shown here.

```
   ...
if (test1)
   ...
   if (test2)
      ...
      if (test3)
         ...
      end
      ...
   end
   ...
end
```

This program contains three nested `if` constructs that may span hundreds of lines of code. Now suppose that the first `end` statement is accidentally deleted during an editing session. When that happens, the MATLAB interpreter will automatically associate the second `end` with the innermost `if (test3)` construct, and the third `end` with the middle `if (test2)`. When the interpreter reaches the end of the file, it will notice that the first `if (test1)` construct was never ended, and it will generate an error message saying that there is a missing `end`. Unfortunately, it can't tell *where* the problem occurred, so we will have to go back and manually search the entire program to locate the problem.

It is sometimes possible to implement an algorithm using either multiple
`elseif` clauses or nested `if` statements. In that case, a programmer may choose
whichever style he or she prefers.

▶

Example 3.4—Assigning Letter Grades

Suppose that we are writing a program that reads in a numerical grade and assigns
a letter grade to it according to the following table:

```
95 < grade            A
86 < grade ≤ 95       B
76 < grade ≤ 86       C
66 < grade ≤ 76       D
 0 < grade ≤ 66       F
```

Write an `if` construct that will assign the grades as described herein using
(*a*) multiple `elseif` clauses and (*b*) nested `if` constructs.

SOLUTION

(*a*) One possible structure using `elseif` clauses is

```
if grade > 95.0
    disp('The grade is A.');
elseif grade > 86.0
    disp('The grade is B.');
elseif grade > 76.0
    disp('The grade is C.');
elseif grade > 66.0
    disp('The grade is D.');
else
    disp('The grade is F.');
end
```

(*b*) One possible structure using nested `if` constructs is

```
if grade > 95.0
    disp('The grade is A.');
else
    if grade > 86.0
        disp('The grade is B.');
    else
        if grade > 76.0
            disp('The grade is C.');
        else
            if grade > 66.0
                disp('The grade is D.');
            else
                disp('The grade is F.');
```

```
                              end
                        end
                  end
            end
```

It should be clear from the preceding example that if there are a lot of mutually exclusive options, a single `if` construct with multiple `elseif` clauses will be simpler than a nested `if` construct.

✳ Good Programming Practice

For branches in which there are many mutually exclusive options, use a single `if` construct with multiple `elseif` clauses in preference to nested `if` constructs.

3.4.4 The `switch` Construct

The `switch` construct is another form of branching construct. It permits a programmer to select a particular code block to execute based on the value of a single integer, character, or logical expression. The general form of a `switch` construct is

```
switch (switch_expr)
case case_expr_1
   Statement 1
   Statement 2          Block 1
   ...
case case_expr_2
   Statement 1
   Statement 2          Block 2
   ...
...
otherwise
   Statement 1
   Statement 2          Block n
   ...
end
```

If the value of *switch_expr* is equal to *case_expr_1*, then the first code block will be executed and the program will jump to the first statement following the end of the `switch` construct. Similarly, if the value of *switch_expr* is equal to *case_expr_2*, then the second code block will be executed, and the program will jump to the first statement following the end of the `switch` construct. The same idea applies for any other cases in the construct. The `otherwise` code block is optional. If it is present, it will be executed whenever the value of *switch_expr* is outside the range of all of the case selectors. If it is not present and the value of *switch_expr* is outside the range of all of the case selectors, then none of the code blocks will be executed. The pseudocode for the case construct looks just like its MATLAB implementation.

If many values of the *switch_expr* should cause the same code to execute, all of those values may be included in a single block by enclosing them in brackets, as shown in the following statements. If the switch expression matches any of the case expressions in the list, then the block will be executed.

```
switch (switch_expr)
case {case_expr_1, case_expr_2, case_expr_3}
    Statement 1          ⎫
    Statement 2          ⎬  Block 1
    ...                  ⎭
otherwise
    Statement 1          ⎫
    Statement 2          ⎬  Block n
    ...                  ⎭
end
```

The *switch_expr* and each *case_expr* may be either numerical or string values.

Note that at most one code block can be executed. After a code block is executed, execution skips to the first executable statement after the end statement. Thus, if the switch expression matches more than one case expression, *only the first one of them will be executed.*

Let's look at a simple example of a switch construct. The following statements determine whether an integer between 1 and 10 is even or odd and print out an appropriate message. It illustrates the use of a list of values as case selectors, as well as the use of the otherwise block.

```
switch (value)
case {1,3,5,7,9}
    disp('The value is odd.');
case {2,4,6,8,10}
    disp('The value is even.');
otherwise
    disp('The value is out of range.');
end
```

3.4.5 The try/catch Construct

The try/catch construct is a special form branching construct designed to trap errors. Ordinarily, when a MATLAB program encounters an error while running, the program aborts. The try/catch construct modifies this default behavior. If an error occurs in a statement in the try block of this construct, then instead of aborting, the code in the catch block is executed and the program keeps running. This allows a programmer to handle errors within the program without causing the program to stop.

The general form of a try/catch construct is as follows:

```
try
    Statement 1
    Statement 2          ⎫
    ...                  ⎬  Try Block
                         ⎭
catch
    Statement 1
    Statement 2          ⎫
    ...                  ⎬  Catch Block
                         ⎭
end
```

When a try/catch construct is reached, the statements in the try block of a will be executed. If no error occurs, the statements in the catch block will be skipped, and execution will continue at the first statement following the end of the construct. On the other hand, if an error *does* occur in the try block, the program will stop executing the statements in the try block and immediately execute the statements in the catch block.

An example program containing a try/catch construct follows. This program creates an array and asks the user to specify an element of the array to display. The user will supply a subscript number, and the program displays the corresponding array element. The statements in the try block will always be executed in this program, while the statements in the catch block will be executed only if an error occurs in the try block.

```
% Initialize array
a = [ 1 -3 2 5];
try
    % Try to display an element
    index = input('Enter subscript of element to display: ');
    disp( ['a(' int2str(index) ') = ' num2str(a(index))] );
catch
    % If we get here an error occurred
    disp( ['Illegal subscript: ' int2str(index)] );
end
```

When this program is executed, the results are as follows:

```
» try_catch
Enter subscript of element to display: 3
a(3) = 2
» try_catch
Enter subscript of element to display: 8
Illegal subscript: 8
```

Quiz 3.2

This quiz provides a quick check to see if you have understood the concepts introduced in Section 3.4. If you have trouble with the quiz, reread the section, ask your instructor, or discuss the material with a fellow student. The answers to this quiz are found in the back of the book.

Write MATLAB statements that perform the functions described below.

1. If x is greater than or equal to zero, then assign the square root of x to variable `sqrt_x` and print out the result. Otherwise, print out an error message about the argument of the square root function and set `sqrt_x` to zero.

2. A variable `fun` is calculated as `numerator/denominator`. If the absolute value of `denominator` is less than $1.0E-300$, write "Divide by 0 error." Otherwise, calculate and print out `fun`.

3. The cost per mile for a rented vehicle is $1.00 for the first 100 miles, $0.80 for the next 200 miles, and $0.70 for all miles in excess of 300 miles. Write MATLAB statements that determine the total cost and the average cost per mile for a given number of miles (stored in variable `distance`).

Examine the following MATLAB statements. Are they correct or incorrect? If they are correct, what do they output? If they are incorrect, what is wrong with them?

4.
```
if volts > 125
    disp('WARNING: High voltage on line.');

if volts < 105
    disp('WARNING: Low voltage on line.');
else
    disp('Line voltage is within tolerances.');
end
```

5.
```
color = 'yellow';
switch ( color )
case 'red',
   disp('Stop now!');
case 'yellow',
   disp('Prepare to stop.');
case 'green',
   disp('Proceed through intersection.');
otherwise,
   disp('Illegal color encountered.');
end
```

```
6. if temperature > 37
      disp('Human body temperature exceeded.');
   elseif temperature > 100
      disp('Boiling point of water exceeded.');
   end
```

3.5 Additional Plotting Features

This section describes additional features of the simple two-dimensional plots introduced in Chapter 2. These features permit us to control the range of *x* and *y* values displayed on a plot, to lay multiple plots on top of each other, to create multiple figures, to create multiple subplots within a figure, and to provide greater control of the plotted lines and text strings. In addition, we will learn how to create polar plots.

3.5.1 Controlling *x*- and *y*-Axis Plotting Limits

By default, a plot is displayed with *x*- and *y*-axis ranges wide enough to show every point in an input data set. However, it is sometimes useful to display only the subset of the data that is of particular interest. This can be done using the **axis** command/function (see the Sidebar about the relationship between MATLAB commands and functions).

Some of the forms of the axis command/function are shown in Table 3.5. The two most important forms are shown in bold type—they let a programmer get

Table 3.5   Forms of the axis Function/Command

Command	Description
v = axis;	This function returns a 4-element row vector containing [xmin xmax ymin ymax], where xmin, xmax, ymin, and ymax are the current limits of the plot.
axis ([xmin xmax ymin ymax]);	This function sets the *x* and *y* limits of the plot to the specified values.
axis equal	This command sets the axis increments to be equal on both axes.
axis square	This command makes the current axis box square.
axis normal	This command cancels the effect of axis equal and axis square.
axis off	This command turns off all axis labeling, tick marks, and background.
axis on	This command turns on all axis labeling, tick marks, and background (default case).

Command/Function Duality

Some items in MATLAB seem to be unable to make up their minds whether they are commands (words typed out on the command line) or functions (with arguments in parentheses). For example, sometimes `axis` seems to behave like a command, and sometimes it seems tobehave like a function. Sometimes we treat it as a command: `axis on`, and other times we might treat it as a function: `axis([0 20 0 35])`. How is this possible?

The short answer is that MATLAB commands are really implemented by functions, and the MATLAB interpreter is smart enough to substitute the function call whenever it encounters the command. It is always possible to call the command directly as a function instead of using the command syntax. Thus the following two statements are identical:

```
axis on;
axis ('on');
```

Whenever MATLAB encounters a command, it forms a function from the command by treating each command argument as a character string and calling the equivalent function with those character strings as arguments. Thus MATLAB interprets the command

```
garbage 1 2 3
```

as the following function call:

```
garbage('1','2','3')
```

Note that *only functions with character arguments can be treated as commands*. Functions with numerical arguments must be used in function form only. This fact explains why `axis` is sometimes treated as a command and sometimes treated as a function.

the current limits of a plot and modify them. A complete list of all options can be found in the MATLAB on-line documentation.

To illustrate the use of `axis`, we will plot the function $f(x) = \sin x$ from -2π to 2π, and then restrict the axes to the region to $0 \le x \le \pi$ and $0 \le y \le 1$. The statements to create this plot are shown as follows, and the resulting plot is shown in Figure 3.4*a*.

```
x = -2*pi:pi/20:2*pi;
y = sin(x);
plot(x,y);
title ('Plot of sin(x) vs x');
grid on;
```

The current limits of this plot can be determined from the basic `axis` function.

(a)

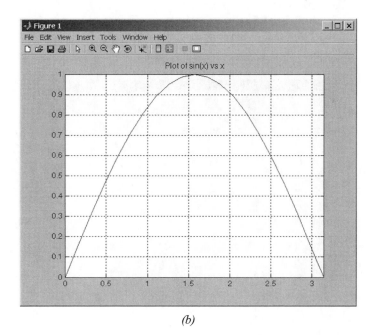

(b)

Figure 3.4 *(a)* Plot of sin *x* versus *x*. *(b)* Closeup of the region [0 π 0 1].

```
» limits=axis
limits =
    -8   8  -1   1
```

These limits can be modified with the function call axis([0 pi 0 1]). After that function is executed, the resulting plot is shown in Figure 3.4*b*.

3.5.2 Plotting Multiple Plots on the Same Axes

Normally, a new plot is created each time that a plot command is issued, and the previous data are lost. This behavior can be modified with the **hold command**. After a hold on command is issued, all additional plots will be laid on top of the previously existing plots. A hold off command switches plotting behavior back to the default situation, in which a new plot replaces the previous one.

For example, the following commands plot sin *x* and cos *x* on the same axes. The resulting plot is shown in Figure 3.5.

```
x = -pi:pi/20:pi;
y1 = sin(x);
y2 = cos(x);
plot(x,y1,'b-');
hold on;
plot(x,y2,'k--');
hold off;
legend ('sin x','cos x');
```

Figure 3.5 Multiple curves plotted on a single set of axes using the hold command.

3.5.3   Creating Multiple Figures

MATLAB can create multiple Figure Windows, with different data displayed in each window. Each Figure Window is identified by a *figure number*, which is a small positive integer. The first Figure Window is Figure 1, the second is Figure 2, *etc*. One of the Figure Windows will be the **current figure**, and all new plotting commands will be displayed in that window.

The current figure is selected with the **figure function**. This function takes the form "figure(n)", where n is a figure number. When this command is executed, Figure n becomes the current figure and is used for all plotting commands. The figure is automatically created if it does not already exist. The current figure may also be selected by clicking on it with the mouse.

The function gcf returns the number of the current figure. This function can be used by an M-file if it needs to know the current figure.

The following commands illustrate the use of the figure function. They create two figures, displaying e^x in the first figure and e^{-x} in the second one.

```
figure(1)
x = 0:0.05:2;
y1 = exp(x);
plot(x,y1);
figure(2)
y2 = exp(-x);
plot(x,y2);
```

3.5.4   Subplots

It is possible to place more than one set of axes on a single figure, creating multiple **subplots**. Subplots are created with a subplot command of the form

```
subplot(m,n,p)
```

This command divides the current figure into m × n equal-sized regions, arranged in m rows and n columns, and creates a set of axes at position p to receive all current plotting commands. The subplots are numbered from left to right and from top to bottom. For example, the command subplot(2,3,4) would divide the current figure into six regions arranged in two rows and three columns, and create an axis in position 4 (the lower left one) to accept new plot data (see Figure 3.6).

If a subplot command creates a new set of axes that conflict with a previously existing set, then the older axes are automatically deleted.

The commands that follow create two subplots within a single window and display the separate graphs in each subplot. The resulting figure is shown in Figure 3.7.

```
figure(1)
subplot(2,1,1)
x = -pi:pi/20:pi;
y = sin(x);
```

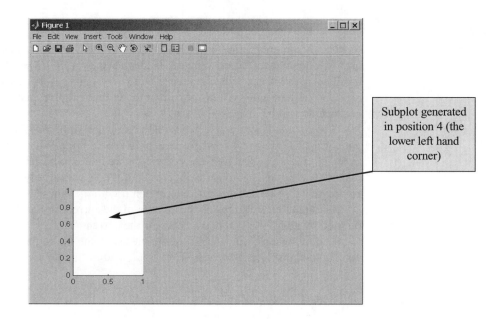

Figure 3.6 The axis created by the subplot(2,3,4) command.

Figure 3.7 A figure containing two subplots.

```
plot(x,y);
title('Subplot 1 title');
subplot(2,1,2)
x = -pi:pi/20:pi;
y = cos(x);
plot(x,y);
title('Subplot 2 title');
```

3.5.5 Enhanced Control of Plotted Lines

In Chapter 1 we learned how to set the color, style, and marker type for a line. It is also possible to set four additional properties associated with each line:

- `LineWidth`—specifies the width of each line in points.
- `MarkerEdgeColor`—specifies the color of the marker or the edge color for filled markers.
- `MarkerFaceColor`—specifies the color of the face of filled markers.
- `MarkerSize`—specifies the size of the marker in points.

These properties are specified in the `plot` command after the data to be plotted in the following fashion:

```
plot(x,y,'PropertyName',value,...)
```

For example, the following command plots a 3-point-wide solid black line with 6-point-wide circular markers at the data points. Each marker has a red edge and a green center, as shown in Figure 3.8.

```
x = 0:pi/15:4*pi;
y = exp(2*sin(x));
plot(x,y,'-ko','LineWidth',3.0,'MarkerSize',6,...
    'MarkerEdgeColor','r','MarkerFaceColor','g')
```

3.5.6 Enhanced Control of Text Strings

It is possible to enhance plotted text strings (titles, axis labels, etc.) with formatting such as bold face, italics, and so forth, and with special characters such as Greek and mathematical symbols.

The font used to display the text can be modified by **stream modifiers**. A stream modifier is a special sequence of characters that tells the MATLAB interpreter to change its behavior. The most common stream modifiers are

- `\bf`—Boldface.
- `\it`—Italics.
- `\rm`—Remove modifiers, restoring normal font.
- `\fontname{fontname}`—Specify the font name to use.
- `\fontsize{fontsize}`—Specify font size.
- `_{xxx}`—The characters inside the braces are subscripts.
- `^{xxx}`—The characters inside the braces are superscripts.

Figure 3.8 A plot illustrating the use of the `LineWidth` and `Marker` properties.

Once a stream modifier has been inserted into a text string, it will remain in effect until the end of the string or until canceled. Any stream modifier can be followed by braces {}. If a modifier is followed by braces, only the text within the braces is affected.

Special Greek and mathematical symbols may also be used in text strings. They are created by embedding *escape sequences* into the text string. These escape sequences are the same as those defined in the TeX language. A sample of the possible escape sequences is shown in Table 3.6; the full set of possibilities is included in the MATLAB on-line documentation.

If one of the special escape characters \, {, }, _, or ^ must be printed, precede it by a backslash character.

The following examples illustrate the use of stream modifiers and special characters.

String	Result
`\tau_{ind} versus \omega_{\itm}`	τ_{ind} versus ω_m
`\theta varies from 0\circ to 90\circ`	θ varies from 0° to 90°
`\bf{B}_{\itS}`	$\mathbf{B}_S$

Table 3.6   Selected Greek and Mathematical Symbols

Character Sequence	Symbol	Character Sequence	Symbol	Character Sequence	Symbol
\alpha	α			\int	∫
\beta	β			\cong	≅
\gamma	γ	\Gamma	Γ	\sim	~
\delta	δ	\Delta	Δ	\infty	∞
\epsilon	ε			\pm	±
\eta	η			\leq	≤
\theta	θ			\geq	≥
\lambda	λ	\Lambda	Λ	\neq	≠
\mu	μ			\propto	∝
\nu	ν			\div	÷
\pi	π	\Pi	Π	\circ	°
\phi	φ			\leftrightarrow	<->
\rho	ρ			\leftarrow	<-
\sigma	σ	\Sigma	Σ	\rightarrow	->
\tau	τ			\uparrow	↑
\omega	ω	\Omega	Ω	\downarrow	↓

3.5.7   Polar Plots

MATLAB includes a special function called `polar`, which plots data in polar coordinates. The basic form of this function is

```
polar(theta,r)
```

where `theta` is an array of angles in radians and r is an array of distances. It is useful for plotting data that is intrinsically a function of angle.

▶

Example 3.5—Cardioid Microphone

Most microphones designed for use on a stage are directional microphones, which are specifically built to enhance the signals received from the singer in the front of the microphone while suppressing the audience noise from behind the microphone. The gain of such a microphone varies as a function of angle according to the equation

$$Gain = 2g(1 + \cos\theta) \tag{3-3}$$

where g is a constant associated with a particular microphone, and θ is the angle from the axis of the microphone to the sound source. Assume that g is 0.5 for a particular microphone, and make a polar plot the gain of the microphone as a function of the direction of the sound source.

SOLUTION   We must calculate the gain of the microphone versus angle and then plot it with a polar plot. The MATLAB code to do this is as follows:

```
%   Script file: microphone.m
%
%   Purpose:
%      This program plots the gain pattern of a cardioid
%      microphone.
%
%   Record of revisions:
%        Date           Programmer            Description of change
%        ====           ==========            =====================
%      01/05/07     S. J. Chapman          Original code
%
% Define variables:
%     g              -- Microphone gain constant
%     gain       -- Gain as a function of angle
%     theta      -- Angle from microphone axis  (radians)

% Calculate gain versus angle
g = 0.5;
theta = 0:pi/20:2*pi;
gain = 2*g*(1+cos(theta));

% Plot gain
polar (theta,gain,'r-');
title ('\bfGain versus angle \theta');
```

The resulting plot is shown in Figure 3.9. Note that this type of microphone is called a "cardioid microphone" because its gain pattern is heart-shaped.

◄

►

Example 3.6—Electrical Engineering: Frequency Response of a Low-Pass Filter

A simple low-pass filter circuit is shown in Figure 3.10. This circuit consists of a resistor and capacitor in series, and the ratio of the output voltage V_o to the input voltage V_i is given by the equation

$$\frac{V_o}{V_i} = \frac{1}{1 + j2\pi fRC} \tag{3-4}$$

Figure 3.9 Gain of a cardioid microphone.

Figure 3.10 A simple low-pass filter circuit.

where V_i is a sinusoidal input voltage of frequency f, R is the resistance in ohms, C is the capacitance in farads, and j is $\sqrt{-1}$ (electrical engineers use j instead of i for $\sqrt{-1}$, because the letter i is traditionally reserved for the current in a circuit).

Assume that the resistance $R = 16\ \text{k}\Omega$, and capacitance $C = 1\ \mu\text{F}$, and plot the amplitude and frequency response of this filter.

SOLUTION The amplitude response of a filter is the ratio of the amplitude of the output voltage to the amplitude of the input voltage, and the phase response of the filter is the difference between the phase of the output voltage and the phase of the input voltage. The simplest way to calculate the amplitude and phase response of the filter is to evaluate Equation (3-4) at many different frequencies. The plot of the magnitude of Equation (3-4) versus frequency is the amplitude response of the filter, and the plot of the angle of Equation (3-4) versus frequency is the phase response of the filter.

Because the frequency and amplitude response of a filter can vary over a wide range, it is customary to plot both of these values on logarithmic scales. On the other hand, the phase varies over a very limited range, so it is customary to plot the phase of the filter on a linear scale. Therefore, we will use a `loglog` plot for the amplitude response, and a `semilogx` plot for the phase response of the filter. We will display both responses as two subplots within a figure.

The MATLAB code required to create and plot the responses is shown here.

```
%   Script file: plot_filter.m
%
%   Purpose:
%     This program plots the amplitude and phase responses
%     of a low-padd RC filter.
%
%   Record of revisions:
%       Date          Programmer          Description of change
%       ====          ==========          =====================
%     01/05/07      S. J. Chapman         Original code
%
% Define variables:
%     amp        -- Amplitude response
%     C          -- Capacitiance (farads)
%     f          -- Frequency of input signal (Hz)
%     phase      -- Phase response
%     R          -- Resistance (ohms)
%     res        -- Vo/Vi

% Initialize R & C
R = 16000;              % 16 k ohms
C = 1.0E-6;             % 1 uF

% Create array of input frequencies
f = 1:2:1000;

% Calculate response
res = 1 ./ ( 1 + j*2*pi*f*R*C );

% Calculate amplitude response
amp = abs(res);
```

```
% Calculate phase response
phase = angle(res);

% Create plots
subplot(2,1,1);
loglog( f, amp );
title('Amplitude Response');
xlabel('Frequency (Hz)');
ylabel('Output/Input Ratio');
grid on;

subplot(2,1,2);
semilogx( f, phase );
title('Phase Response');
xlabel('Frequency (Hz)');
ylabel('Output-Input Phase (rad)');
grid on;
```

The resulting amplitude and phase responses are shown in Figure 3.11. Note that this circuit is called a low-pass filter because low frequencies are passed through with little attenuation, while high frequencies are strongly attenuated. ◀

Example 3.7—Thermodynamics: The Ideal Gas Law

An ideal gas is one in which all collisions between molecules are perfectly elastic. It is possible to think of the molecules in an ideal gas as perfectly hard billiard balls that collide and bounce off of each other without losing kinetic energy.

Such a gas can be characterized by three quantities: absolute pressure (P), volume (V), and absolute temperature (T). The relationship among these quantities in an ideal gas is known as the Ideal Gas Law:

$$PV = nRT \tag{3-5}$$

where P is the pressure of the gas in kilopascals (kPa), V is the volume of the gas in liters (L), n is the number of molecules of the gas in units of moles (mol), R is the universal gas constant (8.314 L·kPa/mol·K), and T is the absolute temperature in kelvins (K). (Note: 1 mol = 6.02×10^{23} molecules)

Assume that a sample of an ideal gas contains 1 mole of molecules at a temperature of 273 K, and answer the following questions.

(a) How does the volume of this gas vary as its pressure varies from 1 to 1000 kPa? Plot pressure versus volume for this gas on an appropriate set of axes. Use a solid red line with a width of 2 pixels.

Figure 3.11 The amplitude and phase response of the low-pass filter circuit.

(b) Suppose that the temperature of the gas is increased to 373 K. How does the volume of this gas vary with pressure now? Plot pressure versus volume for this gas on an the same set of axes as in part (a). Use a dashed blue line with a width of 2 pixels.

Include a bold-face title and *x*- and *y*-axis labels on the plot, as well as legends for each line.

SOLUTION The values that we wish to plot both vary by a factor of 1000, so an ordinary linear plot will not produce a useful plot. Therefore, we will plot the data on a log-log scale.

Note that we must plot two curves on the same set of axes, so we must issue the command `hold on` after the first one is plotted, and `hold off` after the plot is complete. It will also be necessary to specify the color, style, and width of each line, and to specify that labels be in bold face.

A program that calculates the volume of the gas as a function of pressure and creates the appropriate plot is shown below. Note that the special features controlling the style of the plot are shown in boldface.

```
%   Script file: ideal_gas.m
%
%   Purpose:
%     This program plots the pressure versus volume of an
%     ideal gas.
%
%   Record of revisions:
%       Date          Programmer          Description of change
%       ====          ==========          =====================
%     01/16/07      S. J. Chapman         Original code
%
% Define variables:
%     n             -- Number of atoms (mol)
%     P             -- Pressure (kPa)
%     R             -- Ideal gas constant (L kPa/mol K)
%     T             -- Temperature (K)
%     V             -- volume (L)

% Initialize nRT
n = 1;                % Moles of atoms
R = 8.314;            % Ideal gas constant
T = 273;              % Temperature (K)

% Create array of input pressures. Note that this
% array must be quite dense to catch the major
% changes in volume at low pressures.
P = 1:0.1:1000;

% Calculate volumes
V = (n * R * T) ./ P;

% Create first plot
figure(1);
loglog( P, V, 'r-', 'LineWidth', 2 );
title('\bfVolume vs Pressure in an Ideal Gas');
xlabel('\bfPressure (kPa)');
ylabel('\bfVolume (L)');
grid on;
hold on;

% Now increase temperature
T = 373;              % Temperature (K)

% Calculate volumes
V = (n * R * T) ./ P;
```

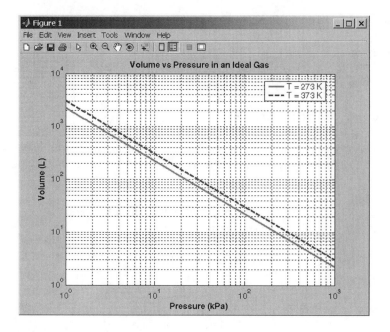

Figure 3.12 Pressure versus volume for an ideal gas.

```
% Add second line to plot
figure(1);
loglog( P, V, 'b--', 'LineWidth', 2 );
hold off;

% Add legend
legend('T = 273 K','T = 373 k');
```

The resulting volume versus pressure plot is shown in Figure 3.12. ◄

3.5.8 Annotating and Saving Plots

Once a plot has been created by a MATLAB program, a user can edit and annotate the plot using the GUI-based tools available from the plot toolbar. Figure 3.13
shows the tools available, which allow the user to edit the properties of any
objects on the plot, or to add annotations to the plot. When the Editing Button
( ) is selected from the toolbar, the editing tools become available for use.
When the button is depressed, clicking any line or text on the figure will cause it
to be selected for editing, and double-clicking the line or text will open a Property
Editor Window that allows you to modify any or all of the characteristics of that
object. Figure 3.14 shows Figure 3.12 after a user has clicked on the blue line to
change it to a 3-pixel-wide dashed line.

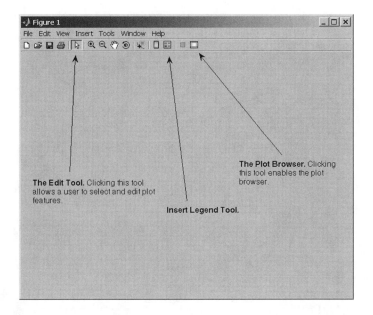

Figure 3.13 The editing tools on the figure toolbar.

The figure toolbar also includes a Plot Browser Button (▢). When this button is depressed, the Plot Browser is displayed. This tool gives the user complete control over the figure. He or she can add axes, edit object properties, modify data values, and add annotations such as lines and text boxes.

If it is not otherwise displayed, the user can enable a Plot Edit Toolbar by selecting the "View/Plot Edit Toolbar" menu item. This toolbar allows a user to add lines, arrows, text, rectangles, and ellipses to annotate and explain a plot. Figure 3.15 shows a Figure Window with the Plot Edit Toolbar enabled.

Figure 3.16 the plot in Figure 3.12 after the Plot Browser and the Plot Edit Toolbar have been enabled. In this figure, the user has used the controls on the Plot Edit Toolbar to add an arrow and a comment to the plot.

When the plot has been edited and annotated, you can save the entire plot in a modifiable form using the "File/Save As" menu item from the Figure Window. The resulting figure file (`*.fig`) contains all the information required to re-create the figure plus annotations at any time in the future.

Quiz 3.3

This quiz provides a quick check to see if you have understood the concepts introduced in Section 3.5. If you have trouble with the quiz, reread the section, ask your instructor, or discuss the material with a fellow student. The answers to this quiz are found in the back of the book.

1. Write the MATLAB statements required to plot sin x versus cos $2x$ from 0 to 2π in steps of $\pi/10$. The points should be connected by a

Figure 3.14 Figure 3.12 after the blue line has been modified using the editing tools built into the figure toolbar.

2-pixel-wide red line, and each point should marked with a 6-pixel-wide blue circular marker.

2. Use the Figure Editing tools to change the markers on the previous plot into black squares. Add an arrow and annotation pointing to the location $x = \pi$ on the plot.

Write the MATLAB text string that will produce the following expressions:

3. $f(x) = \sin \theta \cos 2\phi$

4. **Plot of $\sum x^2$ versus x**

Write the expression produced by the following text strings:

5. `'\tau\it_{m}'`

6. `'\bf\itx_{1}^{2} + x_{2}^{2} \rm(units: \bfm^{2}`
   `                              \rm)'`

7. How do you display the backslash (\) character in a text string?

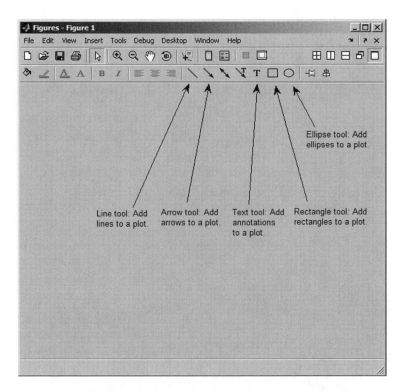

Figure 3.15   A figure window showing the Plot Edit Toolbar.

3.6   More on Debugging MATLAB Programs

It is much easier to make a mistake when writing a program containing branches and loops than it is when writing simple sequential programs. Even after a user has gone through the full design process, a program of any size is almost guaranteed not to be completely correct the first time it is used. Suppose that we have built the program and tested it, only to find that the output values are in error. How do we go about finding the bugs and fixing them?

Once programs start to include loops and branches, the best way to locate an error is to use the symbolic debugger supplied with MATLAB. This debugger is integrated with the MATLAB editor.

To use the debugger, first open the file that you would like to debug using the "File/Open" menu selection in the MATLAB Command Window. When the file is opened, it is loaded into the editor and the syntax is automatically color-coded. Comments in the file appear in green, variables and numbers appear in black, character strings appear in red, and language keywords appear in blue. Figure 3.17 shows an example Edit/Debug window containing the file `calc_roots.m`.

Figure 3.16 Figure 3.12 after the Plot Browser has been used to add an arrow and annotation.

Let's say that we would like to determine what happens when the program is executed. To do this, we can set one or more **breakpoints** by right-clicking the mouse on the lines of interest and choosing the "Set/Clear Breakpoint" option. When a breakpoint is set, a red dot appears to the left of that line containing the breakpoint, as shown in Figure 3.18.

Once the breakpoints have been set, execute the program as usual by typing `calc_roots` in the Command Window. The program will run until it reaches the first breakpoint and stop there. A green arrow will appear by the current line during the debugging process, as shown in Figure 3.19. When the breakpoint is reached, the programmer can examine and/or modify any variable in the workspace by typing its name in the Command Window. When the programmer is satisfied with the program at that point, he or she can either step through the program a line at a time by repeatedly pressing F10 or run to the next breakpoint by pressing F5. It is always possible to examine the values of any variable at any point in the program.

When a bug is found, the programmer can use the Editor to correct the MATLAB program and save the modified version to disk. Note that all breakpoints may

Figure 3.17   An Edit/Debug window with a MATLAB program loaded.

be lost when the program is saved to disk, so they may have to be set again before debugging can continue. This process is repeated until the program appears to be bug-free.

Two other very important features of the debugger are found in the "Debug" menu. The first feature is "Set/Modify Conditional Breakpoint." A **conditional breakpoint** is a breakpoint where the code stops only if some condition is true. For example, a conditional breakpoint can be used to stop execution inside a `for` loop on its 200th execution. This can be very important if a bug appears only after a loop has been executed many times. The condition that causes the breakpoint to stop execution can be modified, and the breakpoint can be enabled or disabled during debugging.

The second feature is "Set Error Breakpoints for All Files." If an error is occurring in a program that causes it to crash or generate warning messages, the

```
 1     % Script file: calc_roots.m
 2     %
 3     % Purpose:
 4     %   This program solves for the roots of a quadratic equation
 5     %   of the form a*x**2 + b*x + c = 0.  It calculates the answers
 6     %   regardless of the type of roots that the equation possesses.
 7     %
 8     % Record of revisions:
 9     %     Date         Programmer          Description of change
10     %     ====         ==========          =====================
11     %   01/12/07     S. J. Chapman           Original code
12     %
13     % Define variables:
14     %   a             -- Coefficient of x^2 term of equation
15     %   b             -- Coefficient of x term of equation
16     %   c             -- Constant term of equation
17     %   discriminant  -- Discriminant of the equation
18     %   imag_part     -- Imag part of equation (for complex roots)
19     %   real_part     -- Real part of equation (for complex roots)
20     %   x1            -- First solution of equation (for real roots)
21     %   x2            -- Second solution of equation (for real roots)
22
23     % Prompt the user for the coefficients of the equation
24 -   disp ('This program solves for the roots of a quadratic ');
25 -   disp ('equation of the form A*X^2 + B*X + C = 0. ');
26 -   a = input ('Enter the coefficient A: ');
27 -   b = input ('Enter the coefficient B: ');
28 -   c = input ('Enter the coefficient C: ');
29
30     % Calculate discriminant
31 O   discriminant = b^2 - 4 * a * c;
32
33     % Solve for the roots, depending on the value of the discriminant
34 -   if discriminant > 0 % there are two real roots, so...
35
36 -      x1 = ( -b + sqrt(discriminant) ) / ( 2 * a );
37 -      x2 = ( -b - sqrt(discriminant) ) / ( 2 * a );
38 -      disp ('This equation has two real roots:');
39 -      fprintf ('x1 = %f\n', x1);
40 -      fprintf ('x2 = %f\n', x2);
```

Figure 3.18 The window after a breakpoint has been set. Note the red dot to the left of the line with the breakpoint.

programmer can turn this item on and execute the program. It will run to the point of the error and stop there, allowing the programmer to examine the values of variables and determine exactly what is causing the problem.

A final critical feature is found on the "Tools" menu. It is "Show M-Lint Report." M-Lint is a program that examines one or more M-files and reports any examples of improper or questionable usage. It is a *great* tool for locating errors, poor usage, or obsolete features in MATLAB code, including such things as variables that are defined but never used. You should always run M-Lint over your programs when they are finished as a final check that everything has been done properly.

Take some time now to become familiar with the Editor/Debugger and its supporting tools—it is a very worthwhile investment.

```
 1    %  Script file: calc_roots.m
 2    %
 3    %  Purpose:
 4    %    This program solves for the roots of a quadratic equation
 5    %    of the form a*x**2 + b*x + c = 0.  It calculates the answers
 6    %    regardless of the type of roots that the equation possesses.
 7    %
 8    %  Record of revisions:
 9    %      Date        Programmer          Description of change
10    %      ====        ==========          =====================
11    %    01/12/07   S. J. Chapman          Original code
12    %
13    %  Define variables:
14    %    a            -- Coefficient of x^2 term of equation
15    %    b            -- Coefficient of x term of equation
16    %    c            -- Constant term of equation
17    %    discriminant -- Discriminant of the equation
18    %    imag_part    -- Imag part of equation (for complex roots)
19    %    real_part    -- Real part of equation (for complex roots)
20    %    x1           -- First solution of equation (for real roots)
21    %    x2           -- Second solution of equation (for real roots)
22
23    % Prompt the user for the coefficients of the equation
24    disp ('This program solves for the roots of a quadratic ');
25    disp ('equation of the form A*X^2 + B*X + C = 0. ');
26    a = input ('Enter the coefficient A: ');
27    b = input ('Enter the coefficient B: ');
28    c = input ('Enter the coefficient C: ');
29
30    % Calculate discriminant
31    discriminant = b^2 - 4 * a * c;
32
33    % Solve for the roots, depending on the value of the discriminant
34    if discriminant > 0 % there are two real roots, so...
35
36       x1 = ( -b + sqrt(discriminant) ) / ( 2 * a );
37       x2 = ( -b - sqrt(discriminant) ) / ( 2 * a );
38       disp ('This equation has two real roots:');
39       fprintf ('x1 = %f\n', x1);
40       fprintf ('x2 = %f\n', x2);
```

Figure 3.19 A green arrow will appear by the current line during the debugging process.

3.7 Summary

In Chapter 3 we have presented the basic types of MATLAB branches and the relational and logic operations used to control them. The principal type of branch is the `if` construct. This construct is very flexible. It can have as many `elseif` clauses as needed to construct any desired test. Furthermore, `if` constructs can be nested to produce more complex tests. A second type of branch is the `switch` construct. It may be used to select among mutually exclusive alternatives speci-fied by a control expression. A third type of branch is the `try/catch` construct. It is used to trap errors that might occur during execution.

Chapter 3 also provided additional information about plots. The `axis` com-mand allows a programmer to select the specific range of *x* and *y* data to be plotted. The `hold` command allows later plots to be plotted on top of earlier ones, so that elements can be added to a graph a piece at a time. The `figure` command allows the programmer to create and select among multiple Figure Windows, so that a program can create multiple plots in separate windows. The `subplot` command allows the programmer to create and select among multiple plots within a single Figure Window.

In addition, we learned how to control additional characteristics of our plots, such as the line width and marker color. These properties may be controlled by specifying `'PropertyName'`,`value` pairs in the plot command after the data to be plotted.

Text strings in plots may be enhanced with stream modifiers and escape sequences. Stream modifiers allow a programmer to specify features such as boldface, italic, superscripts, subscripts, font size, and font name. Escape sequences allow the programmer to include special characters such as Greek and mathematical symbols in the text string.

The MATLAB symbolic debugger and related tools such as M-Lint make debugging MATLAB code much easier. You should invest some time to become familiar with these tools.

3.7.1   Summary of Good Programming Practice

The following guidelines should be adhered to when programming with branch or loop constructs. If you follow them consistently, your code will contain fewer bugs, it will be easier to debug, and it will be more understandable to others who may need to work with it in the future.

1. Follow the steps of the program design process to produce reliable, understandable MATLAB programs.
2. Be cautious about testing for equality with numeric values, since roundoff errors may cause two variables that should be equal to fail a test for equality. Instead, test to see if the variables are *nearly* equal within the roundoff error to be expected on the computer you are working with.
3. Use the `&` AND operator if it is necessary to ensure that both operands are evaluated in an expression, or if the comparison is between arrays. Otherwise, use the `&&` AND operator, since the partial evaluation will make the operation faster in the cases where the first operand is `false`. The `&` operator is preferred in most practical cases.
4. Use the `|` inclusive OR operator if it is necessary to ensure that both operands are evaluated in an expression or if the comparison is between arrays. Otherwise, use the `||` operator, since the partial evaluation will make the operation faster in the cases where the first operand is `true`. The `|` operator is preferred in most practical cases.
5. Always indent code blocks in `if`, `switch`, and `try/catch` constructs to make them more readable.
6. For branches in which there are many mutually exclusive options, use a single `if` construct with multiple `elseif` clauses in preference to nested `if` constructs.

3.7.2 MATLAB Summary

The following summary lists all of the MATLAB commands and functions described in this chapter, along with a brief description of each one.

Commands and Functions

axis	(a) Set the x and y limits of the data to be plotted.
	(b) Get the x and y limits of the data to be plotted.
	(c) Set other axis-related properties.
figure	Select a Figure Window to be the current Figure Window. If the selected Figure Window does not exist, it is automatically created.
hold	Allows multiple plot commands to write on top of each other.
if construct	Selects a block of statements to execute if a specified condition is satisfied.
ischar(a)	Returns a 1 if a is a character array and a 0 otherwise.
isempty(a)	Returns a 1 if a is an empty array and a 0 otherwise.
isinf(a)	Returns a 1 if the value of a is infinite (Inf) and a 0 otherwise.
isnan(a)	Returns a 1 if the value of a is NaN (not a number) and a 0 otherwise.
isnumeric(a)	Returns a 1 if the a is a numeric array and a 0 otherwise.
logical	Converts numeric data to logical data, with nonzero values becoming true and zero values becoming false.
polar	Create a polar plot.
subplot	Select a subplot in the current Figure Window. If the selected subplot does not exist, it is automatically created. If the new subplot conflicts with a previously existing set of axes, they are automatically deleted.
switch construct	Selects a block of statements to execute from a set of mutually-exclusive choices based on the result of a single expression.
try/catch construct	A special construct used to trap errors. It executes construct the code in the try block. If an error occurs, execution stops immediately and transfers to the code in the catch construct.

3.8 Exercises

3.1 Evaluate the following MATLAB expressions:

(a) 5 >= 5.5
(b) 20 > 20
(c) xor(17 - pi < 15, pi < 3)
(d) true > false
(e) ~~(35 / 17) == (35 / 17)
(f) (7 <= 8) == (3 / 2 == 1)
(g) 17.5 && (3.3 > 2.)

3.2 The tangent function is defined as tan θ = sin θ/cos θ. This expression can be evaluated to solve for the tangent as long as the magnitude of cos θ is not too near to 0. (If cos θ is 0, evaluating the equation for tan θ will produce the nonnumerical value Inf.) Assume that θ is given in *degrees*, and write the MATLAB statements to evaluate tan θ as long as the magnitude of cos θ is greater than or equal to 10^{-20}. If the magnitude of cos θ is less than 10^{-20}, write out an error message instead.

3.3 The following statements are intended to alert a user to dangerously high oral thermometer readings (values are in degrees Fahrenheit). Are they correct or incorrect? If they are incorrect, explain why and correct them.

```
if temp < 97.5
    disp('Temperature below normal');
elseif temp > 97.5
    disp('Temperature normal');
    elseif temp > 99.5
    disp('Temperature slightly high');
elseif temp > 103.0
    disp('Temperature dangerously high');
end
```

3.4 The cost of sending a package by an express delivery service is $15.00 for the first two pounds, and $5.00 for each pound or fraction thereof over two pounds. If the package weighs more than 70 pounds, a $15.00 excess weight surcharge is added to the cost. No package over 100 pounds will be accepted. Write a program that accepts the weight of a package in pounds and computes the cost of mailing the package. Be sure to handle the case of overweight packages.

3.5 In Example 3.3, we wrote a program to evaluate the function $f(x, y)$ for any two user-specified values x and y, where the function $f(x, y)$ was defined as follows:

$$f(x, y) = \begin{cases} x + y & x \geq 0 \text{ and } y \geq 0 \\ x + y^2 & x \geq 0 \text{ and } y < 0 \\ x^2 + y & x < 0 \text{ and } y \geq 0 \\ x^2 + y^2 & x < 0 \text{ and } y < 0 \end{cases}$$

The problem was solved by using a single if construct with four code blocks to calculate $f(x, y)$ for all possible combinations of x and y. Rewrite program funxy to use nested if constructs, where the outer construct evaluates the value of x and the inner constructs evaluate the value of y.

3.6 Write a MATLAB program to evaluate the function:

$$y(x) = \ln \frac{1}{1 - x}$$

for any user-specified value of x, where x is a number < 1.0 (note that ln is the natural logarithm, the logarithm to the base e). Use an if structure to verify that the value passed to the program is legal. If the value of x is legal, calculate $y(x)$. If not, write a suitable error message and quit.

3.7 Write a program that allows a user to enter a string containing a day of the week ("Sunday," "Monday," "Tuesday," etc.) and uses a `switch` construct to convert the day to its corresponding number, where Sunday is considered the first day of the week and Saturday is considered the last day of the week. Print out the resulting day number. Also, be sure to handle the case of an illegal day name! (*Note:* Be sure to use the `'s'` option on function `input` so that the input is treated as a string.)

3.8 Suppose that a student has the option of enrolling for a single elective during a term. The student must select a course from a limited list of options: "English," "History," "Astronomy," or "Literature." Construct a fragment of MATLAB code that will prompt the student for his or her choice, read in the choice, and use the answer as the case expression for a `switch` construct. Be sure to include a default case to handle invalid inputs.

3.9 **Ideal Gas Law** The Ideal Gas Law was defined in Example 3.7. Assume that the volume of 1 mole of this gas is 10 L, and plot the pressure of the gas as a function of temperature as the temperature is changed from 250 to 400 kelvins. What sort of plot (linear, semilogx, *etc.*) is most appropriate for this data?

3.10 **Antenna Gain Pattern** The gain G of a certain microwave dish antenna can be expressed as a function of angle by the equation

$$G(\theta) = |\text{sinc } 4\theta| \quad \text{for} -\frac{\pi}{2} \leq \theta \leq \frac{\pi}{2} \qquad (3\text{-}5)$$

where θ is measured in radians from the boresite of the dish, and sinc $x = \sin x/x$. Plot this gain function on a polar plot, with the title "**Antenna Gain vs θ**" in boldface.

3.11 The author of this book now lives in Australia. Australia is a great place to live, but it is also a land of high taxes. In 2002, individual citizens and residents of Australia paid the following income taxes:

Taxable Income (in A$)	Tax on This Income
$0–$6,000	None
$6,001–$20,000	17¢ for each $1 over $6,000
$20,001–$50,000	$2,380 plus 30¢ for each $1 over $20,000
$50,001–$60,000	$11,380 plus 42¢ for each $1 over $50,000
Over $60,000	$15,580 plus 47¢ for each $1 over $60,000

In addition, a flat 1.5% Medicare levy is charged on all income. Write a program to calculate how much income tax a person will owe based on this information. The program should accept a total income figure from the user and calculate the income tax, Medicare levy, and total tax payable by the individual.

3.12   Refraction   When a ray of light passes from a region with an index of refraction n_1 into a region with a different index of refraction n_2, the light ray is bent (see Figure 3.20). The angle at which the light is bent is given by *Snell's Law*

$$n_1 \sin \theta_1 = n_2 \sin \theta_2 \tag{3-6}$$

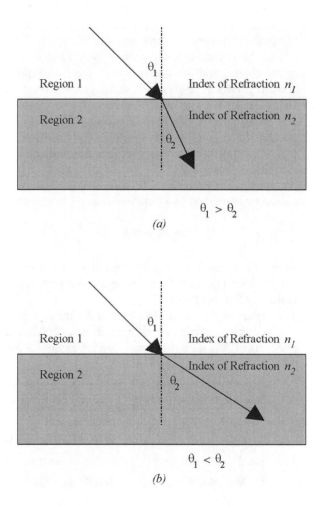

(a)

(b)

Figure 3.20   A ray of light bends as it passes from one medium into another one.   *(a)* If the ray of light passes from a region with a low index of refraction into a region with a higher index of refraction, the ray of light bends more towards the vertical. *(b)* If the ray of light passes from a region with a high index of refraction into a region with a lower index of refraction, the ray of light bends away from the vertical.

where θ_1 is the angle of incidence of the light in the first region and θ_2 is the angle of incidence of the light in the second region. Using Snell's Law, it is possible to predict the angle of incidence of a light ray in Region 2 if the angle of incidence θ_1 in Region 1 and the indices of refraction n_1 and n_2 are known. The equation to perform this calculation is

$$\theta_2 = \sin^{-1}\left(\frac{n_1}{n_2}\sin\theta_1\right) \qquad (3\text{-}7)$$

Write a program to calculate the angle of incidence (in degrees) of a light ray in Region 2 given the angle of incidence θ_1 in Region 1 and the indices of refraction n_1 and n_2. (*Note:* If $n_1 > n_2$, then for some angles θ_1, Equation (3-7) will have no real solution because the absolute value of the quantity $\left(\frac{n_1}{n_2}\sin\theta_1\right)$ will be greater than 1.0. When this occurs, all light is reflected back into Region 1, and no light passes into Region 2 at all. Your program must be able to recognize and properly handle this condition.)

The program should also create a plot showing the incident ray, the boundary between the two regions, and the refracted ray on the other side of the boundary.

Test your program by running it for the following two cases: *(a)* $n_1 = 1.0$, $n_2 = 1.7$, and $\theta_1 = 45°$. *(b)* $n_1 = 1.7$, $n_2 = 1.0$; and $\theta_1 = 45°$.

3.13 Assume that the complex function $f(t)$ is defined by the equation

$$f(t) = (1 + 0.25i)\,t - 2.0$$

Plot the amplitude and phase of function f for $0 \le t \le 4$.

3.14 High-Pass Filter Figure 3.21 shows a simple high-pass filter consisting of a resistor and a capacitor. The ratio of the output voltage V_o to the input voltage V_i is given by the equation

$$\frac{V_o}{V_i} = \frac{j2\pi fRC}{1 + j2\pi fRC} \qquad (3\text{-}8)$$

Figure 3.21 A simple high-pass filter circuit.

Assume that $R = 16$ kΩ and $C = 1$ μF. Calculate and plot the amplitude and phase response of this filter as a function of frequency.

3.15 The Spiral of Archimedes The spiral of Archimedes is a curve described in polar coordinates by the equation

$$r = k\theta \qquad (3\text{-}9)$$

where r is the distance of a point from the origin and θ is the angle of that point in radians with respect to the origin. Plot the spiral of Archimedes for $0 \le \theta \le 6\pi$ when $k = 0.5$. Be sure to label your plot properly.

3.16 Output Power from a Motor The output power produced by a rotating motor is given by the equation

$$P = \tau_{\text{IND}}\, \omega_m \qquad (3\text{-}10)$$

where τ_{IND} is the induced torque on the shaft in newton-meters, ω_m is the rotational speed of the shaft in radians per second, and P is in watts. Assume that the rotational speed of a particular motor shaft is given by the equation

$$\omega_m = 188.5(1 - e^{-0.2t})\ \text{rad/s}$$

and the induced torque on the shaft is given by

$$\tau_{\text{IND}} = 10e^{-0.2t}\ \text{N} \cdot \text{m}$$

Plot the torque, speed, and power supplied by this shaft versus time for $0 \le t \le 10$ s. Be sure to label your plot properly with the symbols τ_{IND} and ω_m where appropriate. Create two plots, one with the power displayed on a linear scale, and one with the output power displayed on a logarithmic scale. Time should always be displayed on a linear scale.

3.17 Plotting Orbits When a satellite orbits the Earth, the satellite's orbit will form an ellipse with the Earth located at one of the focal points of the ellipse. The satellite's orbit can be expressed in polar coordinates as

$$r = \frac{p}{1 - \varepsilon \cos\theta} \qquad (3\text{-}11)$$

where r and θ are the distance and angle of the satellite from the center of the Earth, p is a parameter specifying the size of the size of the orbit, and ε is a parameter representing the eccentricity of the orbit. A circular orbit has an eccentricity ε of 0. An elliptical orbit has an eccentricity of $0 \le \varepsilon \le 1$. If $\varepsilon > 1$, the satellite follows a hyperbolic path and escapes from the Earth's gravitational field.

Consider a satellite with a size parameter $p = 1000$ km. Plot the orbit of this satellite if (a) $\varepsilon = 0$; (b) $\varepsilon = 0.25$; (c) $\varepsilon = 0.5$. How close does each orbit come to the Earth? How far away does each orbit get from the Earth? Compare the three plots you created. Can you determine what the parameter p means from looking at the plots?

CHAPTER 4

Loops

Loops are MATLAB constructs that permit us to execute a sequence of statements more than once. There are two basic forms of loop constructs: **while loops** and **for loops**. The major difference between these two types of loop is in how the repetition is controlled. The code in a `while` loop is repeated an indefinite number of times until some user-specified condition is satisfied. By contrast, the code in a `for` loop is repeated a specified number of times, and the number of repetitions is known before the loops starts.

4.1 The `while` Loop

A **while loop** is a block of statements that are repeated indefinitely as long as some condition is satisfied. The general form of a while loop is

```
while expression
    ...
    ...          } Code block
    ...
end
```

The controlling expression produces a logical value. If the *expression* is `true`, the code block will be executed, and then control will return to the `while` statement. If the *expression* is still `true`, the statements will be executed again. This process will be repeated until the *expression* becomes `false`. When control returns to the `while` statement and the expression is `false`, the program will execute the first statement after the `end`.

The pseudocode corresponding to a `while` loop is

```
while expr
   ...
   ...
   ...
end
```

We will now show an example statistical analysis program that is implemented using a `while` loop.

►

Example 4.1—Statistical Analysis

It is very common in science and engineering to work with large sets of numbers, each of which is a measurement of some particular property that we are interested in. A simple example would be the grades on the first test in this course. Each grade would be a measurement of how much a particular student has learned in the course to date.

Much of the time, we are not interested in looking closely at every single measurement that we make. Instead, we want to summarize the results of a set of measurements with a few numbers that tell us a lot about the overall data set. Two such numbers are the *average* (or *arithmetic mean*) and the *standard deviation* of the set of measurements. The average or arithmetic mean $\bar{x}$ of a set of numbers is defined as

$$\bar{x} = \frac{1}{N}\sum_{i=1}^{N} x_i \tag{4-1}$$

where x_i is sample i out of N samples. If all of the input values are available in an array, the average of a set of number may be calculated by the MATLAB function `mean`. The standard deviation of a set of numbers is defined as

$$s = \sqrt{\frac{N\sum\limits_{i=1}^{N} x_i^2 - \left(\sum\limits_{i=1}^{N} x_i\right)^2}{N(N-1)}} \tag{4-2}$$

Standard deviation is a measure of the amount of scatter on the measurements; the greater the standard deviation, the more scattered the points in the data set are.

Implement an algorithm that reads in a set of measurements and calculates the mean and the standard deviation of the input data set[1].

SOLUTION   This program must be able to read in an arbitrary number of measurements and then calculate the mean and standard deviation of those measure-

---

[1]MATLAB has a built-in function called `std` to calculate the standard deviation of the data in an array. In this exercise, we are creating our on program to calculate the standard deviation. In a real problem, we would normally just use the built-in function.

ments. We will use a while loop to accumulate the input measurements before performing the calculations.

When all of the measurements have been read, we must have some way of telling the program that there is no more data to enter. For now, we will assume that all the input measurements are either positive or zero, and we will use a negative input value as a *flag* to indicate that there is no more data to read. If a negative value is entered, then the program will stop reading input values and will calculate the mean and standard deviation of the data set.

1. **State the problem.**

   Since we assume that the input numbers must be positive or zero, a proper statement of this problem would be: *calculate the average and the standard deviation of a set of measurements, assuming that all of the measurements are either positive or zero and assuming that we do not know in advance how many measurements are included in the data set. A negative input value will mark the end of the set of measurements.*

2. **Define the inputs and outputs.**

   The inputs required by this program are an unknown number of positive or zero numbers. The outputs from this program are a printout of the mean and the standard deviation of the input data set. In addition, we will print out the number of data points input to the program, since this is a useful check that the input data was read correctly.

3. **Design the algorithm.**

   This program can be broken down into three major steps:

   ```
   Accumulate the input data
   Calculate the mean and standard deviation
   Write out the mean, standard deviation, and
     number of points
   ```

   The first major step of the program is to accumulate the input data. To do this, we will have to prompt the user to enter the desired numbers. When the numbers are entered, we will have to keep track of the number of values entered, plus the sum and the sum of the squares of those values. The pseudocode for these steps is

   ```
   Initialize n, sum_x, and sum_x2 to 0
   Prompt user for first number
   Read in first x
   while x >= 0
      n <- n + 1
      sum_x <- sum_x + x
      sum_x2 <- sum_x2 + x^2
      Prompt user for next number
      Read in next x
   end
   ```

Note that we have to read in the first value before the while loop starts so that the while loop can have a value to test the first time it executes.

Next, we must calculate the mean and standard deviation. The pseudocode for this step is just the MATLAB versions of Equations (4-1) and (4-2).

```
x_bar <- sum_x / n
std_dev <- sqrt((n*sum_x2 - sum_x^2) / (n*(n-1)))
```

Finally, we must write out the results.

```
Write out the mean value x_bar
Write out the standard deviation std_dev
Write out the number of input data points n
```

4. **Turn the algorithm into MATLAB statements.**

The final MATLAB program is shown below:

```
%  Script file: stats_1.m
%
%  Purpose:
%    To calculate mean and the standard deviation of
%    an input data set containing an arbitrary number
%    of input values.
%
%  Record of revisions:
%      Date         Programmer        Description of change
%      ====         ==========        =====================
%    01/24/07      S. J. Chapman      Original code
%
%  Define variables:
%    n        -- The number of input samples
%    std_dev  -- The standard deviation of the input samples
%    sum_x    -- The sum of the input values
%    sum_x2   -- The sum of the squares of the input values
%    x        -- An input data value
%    xbar     -- The average of the input samples

% Initialize sums.
n = 0; sum_x = 0; sum_x2 = 0;

% Read in first value
x = input('Enter first value: ');

% While Loop to read input values.
while x >= 0
```

```
% Accumulate sums.
n       = n + 1;
sum_x   = sum_x + x;
sum_x2  = sum_x2 + x^2;

% Read in next value
x = input('Enter next value:  ');

end

% Calculate the mean and standard deviation
x_bar = sum_x / n;
std_dev = sqrt( (n * sum_x2 - sum_x^2) / (n * (n-1)) );

% Tell user.
fprintf('The mean of this data set is:   %f\n', x_bar);
fprintf('The standard deviation is:      %f\n', std_dev);
fprintf('The number of data points is:   %f\n', n);
```

5. **Test the program.**
   To test this program, we will calculate the answers by hand for a simple data set, and then compare the answers to the results of the program. If we used three input values: 3, 4, and 5, then the mean and standard deviation would be

$$\bar{x} = \frac{1}{N}\sum_{i=1}^{N}x_i = \frac{1}{3}(12) = 4$$

$$s = \sqrt{\frac{N\sum_{i=1}^{N}x_i^2 - \left(\sum_{i=1}^{N}x_i\right)^2}{N(N-1)}} = 1$$

   When these values are fed into the program, the results are as follows:

```
» stats_1
Enter first value: 3
Enter next value:  4
Enter next value:  5
Enter next value:  -1
The mean of this data set is:  4.000000
The standard deviation is:     1.000000
The number of data points is:  3.000000
```

   The program gives the correct answers for our test data set.   ◄

In the preceding example, we failed to follow the design process completely. This failure has left the program with a fatal flaw! Did you spot it?

We have failed because *we did not completely test the program for all possible types of input*. Look at the example once again. If we enter either no numbers or only one number, then we will be dividing by zero in the foregoing equations! The division-by-zero error will cause divide-by-zero warnings to be printed, and the output values will be NaN. We need to modify the program to detect this problem, tell the user what the problem is, and stop gracefully.

A modified version of the program called stats_2 follows. Here, we check to see if there are enough input values before performing the calculations. If not, the program will print out an intelligent error message and quit. Test the modified program for yourself.

```
%  Script file: stats_2.m
%
%  Purpose:
%    To calculate mean and the standard deviation of
%    an input data set containing an arbitrary number
%    of input values.
%
%  Record of revisions:
%      Date        Programmer         Description of change
%      ====        ==========         =====================
%    01/24/07   S. J. Chapman      Original code
%  1. 01/24/07   S. J. Chapman      Correct divide-by-0 error if
%                                   0 or 1 input values given.
%
%  Define variables:
%    n          -- The number of input samples
%    std_dev    -- The standard deviation of the input samples
%    sum_x      -- The sum of the input values
%    sum_x2     -- The sum of the squares of the input values
%    x          -- An input data value
%    xbar       -- The average of the input samples

%  Initialize sums.
n = 0; sum_x = 0; sum_x2 = 0;

%  Read in first value
x = input('Enter first value: ');

%  While Loop to read input values.
while x >= 0

    %  Accumulate sums.
    n = n + 1;
```

```
        sum_x  = sum_x + x;
        sum_x2 = sum_x2 + x^2;

        % Read in next value
        x = input('Enter next value: ');

end


% Check to see if we have enough input data.
if n < 2    % Insufficient information

    disp('At least 2 values must be entered!');

else  % There is enough information, so
      % calculate the mean and standard deviation

        x_bar = sum_x / n;
        std_dev = sqrt( (n * sum_x2 - sum_x^2) / (n * (n-1)) );

        % Tell user.
        fprintf('The mean of this data set is:   %f\n', x_bar);
        fprintf('The standard deviation is:      %f\n', std_dev);
        fprintf('The number of data points is:   %f\n', n);

end
```

Note that the average and standard deviation could have been calculated with the built-in MATLAB functions mean and std if all of the input values are saved in a vector and that vector is passed to these functions. You will be asked to create a version of the program that uses the standard MATLAB functions in an exercise at the end of this chapter.

4.2   The for Loop

The for **loop** is a loop that executes a block of statements a specified number of times. The for loop has the form

```
        for index = expr
            . . .
            . . .            } Body
            . . .
        end
```

where index is the loop variable (also known as the **loop index**) and *expr* is the loop control expression, whose result is an array. The columns in the array produced by *expr* are stored one at a time in the variable index and then the loop body is executed, so that the loop is executed once for each column in the array produced by *expr*. The expression usually takes the form of a vector in shortcut notation first:incr:last.

The statements between the for statement and the end statement are known as the *body* of the loop. They are executed repeatedly during each pass of the for loop. The for loop construct functions as follows:

1. At the beginning of the loop, MATLAB generates an array by evaluating the control expression.
2. The first time through the loop, the program assigns the first column of the array to the loop variable index, and the program executes the statements within the body of the loop.
3. After the statements in the body of the loop have been executed, the program assigns the next column of the array to the loop variable index, and the program executes the statements within the body of the loop again.
4. Step 3 is repeated over and over as long as there are additional columns in the array.

Let's look at a number of specific examples to make the operation of the for loop clearer. First, consider the following example:

```
for ii = 1:10
    Statement 1
    ...
    Statement n
end
```

In this case, the control expression generates a 1×10 array, so statements 1 through n will be executed 10 times. The loop index ii will be 1 the first time, 2 the second time, and so on. The loop index will be 10 on the last pass through the statements. When control is returned to the for statement after the tenth pass, there are no more columns in the control expression, so execution transfers to the first statement after the end statement. Note that the loop index ii is still set to 10 after the loop finishes executing.

Second, consider the following example:

```
for ii = 1:2:10
    Statement 1
    ...
    Statement n
end
```

In this case, the control expression generates a 1×5 array, so statements 1 through n will be executed 5 times. The loop index ii will be 1 the first time, 3 the second time, and so on. The loop index will be 9 on the fifth and last pass through the statements. When control is returned to the for statement after the fifth pass, there are no more columns in the control expression, so execution transfers to the first statement after the end statement. Note that the loop index ii is still set to 9 after the loop finishes executing.

Third, consider the following example:

```
for ii = [5 9 7]
    Statement 1
    ...
    Statement n
end
```

Here, the control expression is an explicitly written 1×3 array, so statements 1 through n will be executed three times with the loop index set to 5 the first time, 9 the second time, and 7 the final time. The loop index ii is still set to 7 after the loop finishes executing.

Finally, consider the example:

```
for ii = [1 2 3;4 5 6]
    Statement 1
    ...
    Statement n
end
```

In this case, the control expression is a 2×3 array, so statements 1 through n will be executed three times. The loop index ii will be the column vector $\begin{bmatrix} 1 \\ 4 \end{bmatrix}$ the first time, $\begin{bmatrix} 2 \\ 5 \end{bmatrix}$ the second time, and $\begin{bmatrix} 3 \\ 6 \end{bmatrix}$ the third time. The loop index ii is still set to $\begin{bmatrix} 3 \\ 6 \end{bmatrix}$ after the loop finishes executing. This example illustrates the fact that a loop index can be a vector.

The pseudocode corresponding to a for loop looks like the loop itself:

```
for index = expression
    Statement 1
    ...
    Statement n
end
```

Example 4.2—The Factorial Function

To illustrate the operation of a for loop, we will use a for loop to calculate the factorial function. The factorial function is defined as

```
N! = 1                                            N = 0
N! = N * (N-1) * (N-2) * ... * 3 * 2 * 1          N > 0
```

The MATLAB code to calculate N factorial for positive value of N would be

```
n_factorial = 1
for ii = 1:n
    n_factorial = n_factorial * ii;
end
```

Suppose that we wish to calculate the value of 5!. If n is 5, the `for` loop control expression would be the row vector [1 2 3 4 5]. This loop will be executed 5 times, with the variable `ii` taking on values of 1, 2, 3, 4, and 5 in the successive loops. The resulting value of `n_factorial` will be $1 \times 2 \times 3 \times 4 \times 5 = 120$. ◀

▶

Example 4.3—Calculating the Day of Year

The *day of year* is the number of days (including the current day) that have elapsed since the beginning of a given year. It is a number in the range 1 to 365 for ordinary years, and 1 to 366 for leap years. Write a MATLAB program that accepts a day, month, and year, and calculates the day of year corresponding to that date.

SOLUTION   To determine the day of year, this program will need to sum up the number of days in each month preceding the current month, plus the number of elapsed days in the current month. A `for` loop will be used to perform this sum. Since the number of days in each month varies, it is necessary to determine the correct number of days to add for each month. A `switch` construct will be used to determine the proper number of days to add for each month.

During a leap year, an extra day must be added to the day of year for any month after February. This extra day accounts for the presence of February 29 in the leap year. Therefore, to perform the day of year calculation correctly, we must determine which years are leap years. In the Gregorian calendar, leap years are determined by the following rules:

1. Years evenly divisible by 400 are leap years.
2. Years evenly divisible by 100 but *not* by 400 are not leap years.
3. All years divisible by 4 but *not* by 100 are leap years.
4. All other years are not leap years.

We will use the `mod` (for modulus) function to determine whether or not a year is evenly divisible by a given number. The `mod` function returns the remainder after the division of two numbers. For example, the remainder of 9/4 is 1, since 4 goes into 9 twice with a remainder of 1. If the result of the function mod(year, 4) is zero, then we know that the year was evenly divisible by 4. Similarly, if the result of the function mod(year, 400) is zero, then we know that the year was evenly divisible by 400.

A program to calculate the day of year follows. Note that the program sums up the number of days in each month before the current month, and that it uses a `switch` construct to determine the number of days in each month.

```
%   Script file: doy.m
%
%   Purpose:
%     This program calculates the day of year corresponding
%     to a specified date. It illustrates the use switch and
%     for constructs.
%
%   Record of revisions:
%       Date        Programmer            Description of change
%       ====        ==========            =====================
%     01/27/07    S. J. Chapman           Original code
%
%   Define variables:
%     day              -- Day (dd)
%     day_of_year      -- Day of year
%     ii               -- Loop index
%     leap_day         -- Extra day for leap year
%     month            -- Month (mm)
%     year             -- Year (yyyy)


% Get day, month, and year to convert
disp('This program calculates the day of year given the ');
disp('specified date.');
month = input('Enter specified month (1-12): ');
day   = input('Enter specified day(1-31):    ');
year  = input('Enter specified year(yyyy):   ');


% Check for leap year, and add extra day if necessary
if mod(year,400) == 0
   leap_day = 1;     % Years divisible by 400 are leap years
elseif mod(year,100) == 0
   leap_day = 0;     % Other centuries are not leap years
elseif mod(year,4) == 0
   leap_day = 1;     % Otherwise every 4th year is a leap year
else
   leap_day = 0;     % Other years are not leap years
end


% Calculate day of year by adding current day to the
% days in previous months.
day_of_year = day;
for ii = 1:month-1
```

```
% Add days in months from January to last month
switch (ii)
case {1,3,5,7,8,10,12},
    day_of_year = day_of_year + 31;
case {4,6,9,11},
    day_of_year = day_of_year + 30;
case 2,
    day_of_year = day_of_year + 28 + leap_day;
end

end

% Tell user
fprintf('The date %2d/%2d/%4d is day of year %d.\n', ...
        month, day, year, day_of_year);
```

We will use the following known results to test the program:

1. Year 1999 is not a leap year. January 1 must be day of year 1, and December 31 must be day of year 365.
2. Year 2000 is a leap year. January 1 must be day of year 1, and December 31 must be day of year 366.
3. Year 2001 is not a leap year. March 1 must be day of year 60, since January has 31 days, February has 28 days, and this is the first day of March.

If this program is executed five times with the above dates, the results are

```
» doy
This program calculates the day of year given the
specified date.
Enter specified month (1-12):  1
Enter specified day(1-31):     1
Enter specified year(yyyy):    1999
The date  1/ 1/1999 is day of year 1.
» doy
This program calculates the day of year given the
specified date.
Enter specified month (1-12):  12
Enter specified day(1-31):     31
Enter specified year(yyyy):    1999
The date 12/31/1999 is day of year 365.
» doy
This program calculates the day of year given the
specified date.
Enter specified month (1-12):  1
Enter specified day(1-31):     1
Enter specified year(yyyy):    2000
The date  1/ 1/2000 is day of year 1.
```

» **doy**
This program calculates the day of year given the
specified date.
Enter specified month (1-12):   **12**
Enter specified day(1-31):   **31**
Enter specified year(yyyy):   **2000**
The date 12/31/2000 is day of year 366.
» **doy**
This program calculates the day of year given the
specified date.
Enter specified month (1-12):   **3**
Enter specified day(1-31):   **1**
Enter specified year(yyyy):   **2001**
The date 3/ 1/2001 is day of year 60.

The program gives the correct answers for our test dates in all five test cases. ◀

▶

Example 4.4—Statistical Analysis

Implement an algorithm that reads in a set of measurements and calculates the mean and the standard deviation of the input data set, when any value in the data set can be positive, negative, or zero.

SOLUTION   This program must be able to read in an arbitrary number of measurements, and then calculate the mean and standard deviation of those measurements. Each measurement can be positive, negative, or zero.

Since we cannot use a data value as a flag this time, we will ask the user for the number of input values and then use a for loop to read in those values. The modified program that permits the use of any input value is shown below. Verify its operation for yourself by finding the mean and standard deviation of the following five input values: 3., $-1.$, 0., 1., and -2.

```
%  Script file: stats_3.m
%
%  Purpose:
%    To calculate mean and the standard deviation of
%    an input data set, where each input value can be
%    positive, negative, or zero.
%
%  Record of revisions:
%     Date          Programmer          Description of change
%     ====          ==========          =====================
%   01/27/07     S. J. Chapman          Original code
%
```

```
% Define variables:
%    ii       -- Loop index
%    n        -- The number of input samples
%    std_dev  -- The standard deviation of the input samples
%    sum_x    -- The sum of the input values
%    sum_x2   -- The sum of the squares of the input values
%    x        -- An input data value
%    xbar     -- The average of the input samples

% Initialize sums.
sum_x = 0; sum_x2 = 0;

% Get the number of points to input.
n = input('Enter number of points: ');

% Check to see if we have enough input data.
if n < 2          % Insufficient data

   disp ('At least 2 values must be entered.');

else % we will have enough data, so let's get it.

   % Loop to read input values.
   for ii = 1:n

      % Read in next value
      x = input('Enter value:  ');

      % Accumulate sums.
      sum_x  = sum_x + x;
      sum_x2 = sum_x2 + x^2;

   end

   % Now calculate statistics.
   x_bar = sum_x / n;
   std_dev = sqrt( (n * sum_x2 - sum_x^2) / (n * (n-1)) );

   % Tell user.
   fprintf('The mean of this data set is:    %f\n', x_bar);
   fprintf('The standard deviation is:       %f\n', std_dev);
   fprintf('The number of data points is:    %f\n', n);

end
```

◄

4.2.1   Details of Operation

Now that we have seen examples of a for loop in operation, we must examine some important details required to use for loops properly.

1. **Indent the bodies of loops.** It is not necessary to indent the body of a for loop as we have shown above. MATLAB will recognize the loop even if every statement in it starts in column 1. However, the code is much more readable if the body of the for loop is indented, so you should always indent the bodies of loops.

✳ **Good Programming Practice**

Always indent the body of a for loop by two or more spaces to improve the readability of the code.

---

2. **Don't modify the loop index within the body of a loop.** The loop index of a for loop *should not be modified anywhere within the body of the loop*. The index variable is often used as a counter within the loop, and modifying its value can cause strange and hard-to-find errors. The example shown below is intended to initialize the elements of an array, but the statement "ii = 5" has been accidentally inserted into the body of the loop. As a result, only a(5) is initialized, and it gets the values that should have gone into a(1), a(2), etc.

```
for ii = 1:10
    ...
    ii = 5;    % Error!
    ...
    a(ii) = <calculation>
end
```

✳ **Good Programming Practice**

Never modify the value of a loop index within the body of the loop.

---

3. **Preallocating Arrays**. We learned in Chapter 2 that it is possible to extend an existing array simply by assigning a value to a higher array element. For example, the statement

```
arr = 1:4;
```

defines a 4-element array containing the values [1 2 3 4]. If the statement

```
arr(8) = 6;
```

is executed, the array will be automatically extended to eight elements and will contain the values [1 2 3 4 0 0 0 6]. Unfortunately, each time that an array is extended, MATLAB has to (1) create a new array, (2) copy the contents of the old array to the new longer array, (3) add the new value to the array, and then (4) delete the old array. This process is very time consuming for long arrays.

When a `for` loop stores values in a previously-undefined array, the loop forces MATLAB to go through this process each time the loop is executed. On the other hand, if the array is **preallocated** to its maximum size before the loop starts executing, no copying is required, and the code executes much faster. The code fragment shown below shows how to pre-allocate an array before the starting the loop.

```
square = zeros(1,100);
for ii = 1:100
    square(ii) = ii^2;
end
```

Good Programming Practice

Always preallocate all arrays used in a loop before executing the loop. This practice greatly increases the execution speed of the loop.

4. **Vectorizing Arrays.** It is often possible to perform calculations with either `for` loops or vectors. For example, the following code fragment calculates the squares, square roots, and cube roots of all integers between 1 and 100 using a `for` loop.

```
for ii = 1:100
    square(ii) = ii^2;
    square_root(ii) = ii^(1/2);
    cube_root(ii) = ii^(1/3);
end
```

The following code fragment performs the same calculation with vectors.

```
ii = 1:100;
square = ii.^2;
square_root = ii.^(1/2);
cube_root(ii) = ii.^(1/3);
```

Even though these two calculations produce the same answers, they are *not* equivalent. The version with the for loop can be *more than 15 times slower* than the vectorized version! This happens because the statements in the for loop must be interpreted[2] and executed a line at a time by MATLAB during each pass of the loop. In effect, MATLAB must interpret and execute 300 separate lines of code. In contrast, MATLAB only has to interpret and execute 4 lines in the vectorized case. Since MATLAB is designed to implement vectorized statements in a very efficient fashion, it is much faster in that mode.

In MATLAB, the process of replacing loops by vectorized statements is known as **vectorization**. Vectorization can yield dramatic improvements in performance for many MATLAB programs.

✳ Good Programming Practice

If it is possible to implement a calculation either with a for loop or by using vectors, implement the calculation with vectors. Your program will be much faster.

4.2.2 The MATLAB Just-in-Time (JIT) Compiler

A just-in-time (JIT) compiler was added to MATLAB 6.5 and later versions. The JIT compiler examines MATLAB code before it is executed, and where possible, compiles the code before executing it. Since the MATLAB code is compiled instead of being interpreted, it runs almost as fast as vectorized code. The JIT compiler can sometimes dramatically speed up the execution of for loops.

The JIT compiler is a very nice tool when it works, since it speeds up the loops without any action by the programmer. However, the JIT compiler has many limitations that prevent it from speeding up all loops. A full list of JIT compiler limitations appears in the MATLAB documentation, but some of the more important limitations are

1. The JIT compiler only accelerates loops containing double, logical, and char data types (plus integer data types that we haven't met yet). If other data types such as cell arrays or structures[3] appear in the loop, it will not be accelerated.
2. If an array in the loop has more than two dimensions, the loop will not be accelerated.
3. If the code in the loop calls external functions (other than built-in functions), it will not be accelerated.

---

[2]But see the next item about the MATLAB Just-in-Time compiler.
[3]We will learn about these data types in Chapter 7.

4. If the code in the loop changes the data type of a variable within a loop, the loop will not be accelerated.

Because of these limitations, a good programmer using vectorization can almost always create a faster program than one relying on the JIT compiler.

✳ Good Programming Practice

Do not rely on the JIT compiler to speed up your code. It has many limitations, and a programmer can typically do a better job with manual vectorization.

▶

Example 4.5—Comparing Loops and Vectors

To compare the execution speeds of loops and vectors, perform and time the following four sets of calculations.

1. Calculate the squares of every integer from 1 to 10,000 in a `for` loop without initializing the array of squares first.
2. Calculate the squares of every integer from 1 to 10,000 in a `for` loop, using the `zeros` function to preallocate the array of squares first, but calling an external function to perform the squaring. (This will disable the JIT compiler.)
3. Calculate the squares of every integer from 1 to 10,000 in a `for` loop, using the zeros function to preallocate the array of squares first, and calculating the square of the number in-line. (This will allow the JIT compiler to function.)
4. Calculate the squares of every integer from 1 to 10,000 with vectors.

SOLUTION This program must calculate the squares of the integers from 1 to 10,000 in each of the four ways described previously, timing the executions in each case. The timing can be accomplished using the MATLAB functions `tic` and `toc`. Function `tic` resets the built-in elapsed time counter, and function `toc` returns the elapsed time in seconds since the last call to function `tic`.

Since the real-time clocks in many computers have a fairly coarse granularity, it may be necessary to execute each set of instructions multiple times to get a valid average time.

A MATLAB program to compare the speeds of the four approaches is shown as follows:

```
% Script file: timings.m
%
% Purpose:
%    This program calculates the time required to
```

```
%    calculate the squares of all integers from 1 to
%    10,000 in four different ways:
%    1. Using a for loop with an uninitialized output
%       array.
%    2. Using a for loop with a pre-allocated output
%       array and NO JIT compiler.
%    3. Using a for loop with a pre-allocated output
%       array and the JIT compiler.
%    4. Using vectors.
%
% Record of revisions:
%    Date          Programmer        Description of change
%    ====          ==========        =====================
%    01/29/07     S. J. Chapman      Original code
%
% Define variables:
%    ii, jj       -- Loop index
%    average1     -- Average time for calculation 1
%    average2     -- Average time for calculation 2
%    average3     -- Average time for calculation 3
%    average4     -- Average time for calculation 4
%    maxcount     -- Number of times to loop calculation
%    square       -- Array of squares
```

% Perform calculation with an uninitialized array
% "square". This calculation is done only once
% because it is so slow.
```
maxcount = 1;               % Number of repetitions
tic;                        % Start timer
for jj = 1:maxcount
   clear square             % Clear output array
   for ii = 1:10000
      square(ii) = ii^2;    % Calculate square
   end
end
average1 = (toc)/maxcount;  % Calculate average time
```

% Perform calculation with a pre-allocated array
% "square", calling an external function to square
% the number. This calculation is averaged over 10
% loops.
```
maxcount = 10;              % Number of repetitions
tic;                        % Start timer
```

```
for jj = 1:maxcount
    clear square                  % Clear output array
    square = zeros(1,10000);     % Pre-initialize array
    for ii = 1:10000
        square(ii) = sqr(ii);    % Calculate square
    end
end
average2 = (toc)/maxcount;       % Calculate average time
```

```
% Perform calculation with a pre-allocated array
% "square". This calculation is averaged over 100
% loops.
maxcount = 100;                  % Number of repetitions
tic;                             % Start timer
for jj = 1:maxcount
    clear square                  % Clear output array
    square = zeros(1,10000);     % Pre-initialize array
    for ii = 1:10000
        square(ii) = ii^2;       % Calculate square
    end
end
average3 = (toc)/maxcount;       % Calculate average time
```

```
% Perform calculation with vectors. This calculation
% averaged over 1000 executions.
maxcount = 1000;                 % Number of repetitions
tic;                             % Start timer
for jj = 1:maxcount
    clear square                  % Clear output array
    ii = 1:10000;                % Set up vector
    square = ii.^2;              % Calculate square
end
average4 = (toc)/maxcount;       % Calculate average time
```

```
% Display results
fprintf('Loop / uninitialized array         = %8.4f\n', average1);
fprintf('Loop / initialized array / no JIT  = %8.4f\n', average2);
fprintf('Loop / initialized array / JIT     = %8.4f\n', average3);
fprintf('Vectorized                         = %8.4f\n', average4);
```

When this program is executed using MATLAB 7.4 on a 2.4 GHz Pentium IV computer, the results are:

```
» timings
Loop / uninitialized array        = 0.1111
Loop / initialized array / no JIT = 0.0922
```

```
Loop / initialized array / JIT    = 0.0002
Vectorized                        = 0.0001
```

The loop with the uninitialized array and the loop with the initialized array but no JIT were very slow compared with the loop executed with the JIT compiler or the vectorized loop. The vectorized loop was the fastest way to perform the calculation, but if the JIT compiler works for your loop, you get most of the acceleration without having to do anything! As you can see, designing loops to allow the JIT compiler to function[4] or replacing the loops with vectorized calculations can make an incredible difference in the speed of your MATLAB code! ◀

The M-Lint code checking tool can help you identify problems with uninitialized arrays that can slow the execution of a MATLAB program. For example, if we run M-Lint on program timings.m, the code checker will identify the uninitialized array and write out a warning message.

4.2.3 The break and continue Statements

There are two additional statements that can be used to control the operation of while loops and for loops: the break and continue statements. The break statement terminates the execution of a loop and passes control to the next statement after the end of the loop, while the continue statement terminates the current pass through the loop and returns control to the top of the loop.

If a break statement is executed in the body of a loop, the execution of the body will stop, and control will be transferred to the first executable statement after the loop. An example of the break statement in a for loop is shown here.

```
for ii = 1:5
  if ii == 3;
     break;
  end
  fprintf('ii = %d\n',ii);
end
disp(['End of loop!']);
```

When this program is executed, the output is

```
» test_break
ii = 1
ii = 2
End of loop!
```

---

[4]We will learn how to use the MATLAB Profiler later in this chapter. This tool can identify loops that do not get speeded up by the JIT compiler and tell you why they can't be accelerated.

(a)

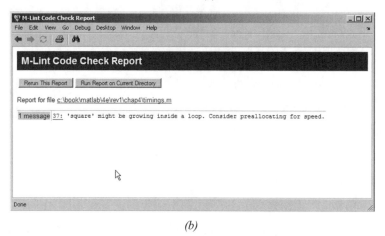

(b)

Figure 4.1 The M-Lint code checker can identify some problems that will slow down the execution of MATLAB loops.

Note that the break statement was executed on the iteration when ii was 3, and control transferred to the first executable statement after the loop without executing the fprintf statement.

If a continue statement is executed in the body of a loop, the execution of the current pass through the loop will stop, and control will return to the top of the loop. The controlling variable in the for loop will take on its next value, and the loop will be executed again. An example of the continue statement in a for loop is shown here.

```
for ii = 1:5
   if ii == 3;
       continue;
   end
   fprintf('ii = %d\n',ii);
end
disp(['End of loop!']);
```

When this program is executed, the output is

```
» test_continue
ii = 1
ii = 2
ii = 4
ii = 5
End of loop!
```

Note that the continue statement was executed on the iteration when ii was 3, and control will be transferred to the top of the loop without executing the fprintf statement.

The break and continue statements work with both while loops and for loops.

4.2.4 Nesting Loops

It is possible for one loop to be completely inside another loop. If one loop is completely inside another one, the two loops are called **nested loops**. The following example shows two nested for loops used to calculate and write out the product of two integers.

```
for ii = 1:3
   for jj = 1:3
       product = ii * jj;
       fprintf('%d * %d = %d\n',ii,jj,product);
   end
end
```

In this example, the outer for loop will assign a value of 1 to index variable ii, and then the inner for loop will be executed. The inner for loop will be executed three times with index variable jj having values 1, 2, and 3. When the entire inner for loop has been completed, the outer for loop will assign a value of 2 to index variable ii, and the inner for loop will be executed again. This process repeats until the outer for loop has executed three times, and the resulting output is

```
1 * 1 = 1
1 * 2 = 2
1 * 3 = 3
2 * 1 = 2
2 * 2 = 4
2 * 3 = 6
3 * 1 = 3
3 * 2 = 6
3 * 3 = 9
```

Note that the inner for loop executes completely before the index variable of the outer for loop is incremented.

When MATLAB encounters an end *statement, it associates that statement with the innermost currently open construct.* Therefore, the first end statement in the preceding closes the "for jj = 1:3" loop, and the second end statement closes the "for ii = 1:3" loop. This fact can produce hard-to-find errors if an end statement is accidentally deleted somewhere within a nested loop construct.

If for *loops are nested, they should have independent loop index variables.* If they have the same index variable, then the inner loop will change the value of the loop index that the outer loop just set.

If a break or continue statement appears inside a set of nested loops, then that statement refers to the *innermost* of the loops containing it. For example, consider the following program:

```
for ii = 1:3
   for jj = 1:3
      if jj == 3;
         break;
      end
      product = ii * jj;
      fprintf('%d * %d = %d\n',ii,jj,product);
   end
   fprintf('End of inner loop\n');
end
fprintf('End of outer loop\n');
```

If the inner loop counter jj is equal to 3, then the break statement will be executed. This will cause the program to exit the innermost loop. The program will print out "End of inner loop," the index of the outer loop will be increased by 1, and execution of the innermost loop will start over. The resulting output values are

```
1 * 1 = 1
1 * 2 = 2
End of inner loop
2 * 1 = 2
2 * 2 = 4
End of inner loop
3 * 1 = 3
3 * 2 = 6
End of inner loop
End of outer loop
```

4.3   Logical Arrays and Vectorization

We learned about the `logical` data type in Chapter 3. Logical data can have one of two possible values: `true` (1) or `false` (0). Scalars and arrays of `logical` data are created as the output of relational and logic operators.

For example, consider the following statements:

```
a = [1 2 3; 4 5 6; 7 8 9];
b = a > 5;
```

These statements produced two arrays a and b. Array a is a `double` array containing the values $\begin{bmatrix} 1 & 2 & 3 \\ 4 & 5 & 6 \\ 7 & 8 & 9 \end{bmatrix}$, while array b is a `logical` array containing the values $\begin{bmatrix} 0 & 0 & 0 \\ 0 & 0 & 1 \\ 1 & 1 & 1 \end{bmatrix}$. When the `whos` command is executed, the results are as shown as follows:

```
» whos
  Name       Size        Bytes        Class
    a         3x3           72         double array
    b         3x3            9         logical array
Grand total is 18 elements using 81 bytes
```

Logical arrays have a very important special property—*they can serve as a **mask** for arithmetic operations*. A mask is an array that selects the elements of another array for use in an operation. The specified operation will be applied to the selected elements and *not* to the remaining elements.

For example, suppose that arrays a and b are as defined previously. Then the statement a(b) = sqrt(a(b)) will take the square root of all elements for which the logical array b is true and leave all the other elements in the array unchanged.

```
» a(b) = sqrt(a(b))
a =
    1.0000    2.0000    3.0000
    4.0000    5.0000    2.4495
    2.6458    2.8284    3.0000
```

This is a very fast and very clever way of performing an operation on a subset of an array without needing loops and branches.

The following two code fragments both take the square root of all elements in array a whose value is greater than 5, but the vectorized approach is more compact and elegant than the loop approach.

```
for ii = 1:size(a,1)
    for jj = 1:size(a,2)
        if a(ii,jj) > 5
            a(ii,jj) = sqrt(a(ii,jj));
        end
    end
end


b = a > 5;
a(b) = sqrt(a(b));
```

4.3.1 Creating the Equivalent of if/else Constructs with Logical Arrays

Logical arrays can also be used to implement the equivalent of an if/else construct inside a set of for loops. As we saw in the preceding section, it is possible to apply an operation to selected elements of an array using a logical array as a mask. It is also possible to apply a different set of operations to the *unselected* elements of the array by simply adding the not operator (~) to the logical mask. For example, suppose that we wanted to take the square root of any elements in a two-dimensional array whose value is greater than 5, and to square the remaining elements in the array. The code for this operation using loops and branches is

```
for ii = 1:size(a,1)
    for jj = 1:size(a,2)
        if a(ii,jj) > 5
            a(ii,jj) = sqrt(a(ii,jj));
        else
            a(ii,jj) = a(ii,jj)^2;
        end
    end
end
```

The vectorized code for this operation is

```
b = a > 5;
a(b) = sqrt(a(b));
a(~b) = a(~b).^2;
```

The vectorized code is significantly faster than the loops-and-branches version.

Quiz 4.1

This quiz provides a quick check to see if you have understood the concepts introduced in Sections 4.1 through 4.3. If you have trouble with the quiz, reread the section, ask your instructor, or discuss the material with a fellow student. The answers to this quiz are found in the back of the book.

Examine the following `for` loops and determine how many times each loop will be executed.

1. `for index = 7:10`
2. `for jj = 7:-1:10`
3. `for index = 1:10:10`
4. `for ii = -10:3:-7`
5. `for kk = [0 5 ; 3 3]`

Examine the following loops and determine the value in `ires` at the end of each of the loops.

6.
```
        ires = 0;
        for index = 1:10
            ires = ires + 1;
        end
```

7.
```
        ires = 0;
        for index = 1:10
            ires =  ires + index;
        end
```

8.
```
        ires = 0;
        for index1 = 1:10
            for index2 = index1:10
                if index2 == 6
                    break;
                end
                ires = ires + 1;
            end
        end
```

9.
```
        ires = 0;
        for index1 = 1:10
            for index2 = index1:10
                if index2 == 6
                    continue;
                end
                ires = ires + 1;
            end
        end
```

10. Write the MATLAB statements to calculate the values of the function

$$f(t) = \begin{cases} \sin t & \text{for all } t \text{ where } \sin t > 0 \\ 0 & \text{elsewhere} \end{cases}$$

for $-6\pi \le t \le 6\pi$ at intervals of $\pi/10$. Do this twice, once using loops and branches and once using vectorized code.

4.4   The MATLAB Profiler

MATLAB includes a profiler, which can be used to identify the parts of a program that consume the most execution time. The profiler can identify "hot spots." where optimizing the code will result in major increases in speed.

The MATLAB profiler is started by selecting the "Tools/Open Profiler" option on the Edit/Debug Window. A Profiler Window opens, with a field containing the name of the program to profile and a pushbutton to start the profile process running.

After the profiler runs, a Profile Summary is displayed, showing how much time is spent in each function being profiled (see Figure 4.3a). Clicking on any profiled function brings up a more detailed display, showing exactly how much time was spent on each line when that function was executed (see Figure 4.3b). With this information, the programmer can identify the slow portions of the code and work to speed them up with vectorization and similar techniques. For example, the profiler will highlight loops that run slowly because they can't be handled by the JIT compiler.

Normally, the profiler should be run *after a program is working properly*. It is a waste of time to profile a program before it is working.

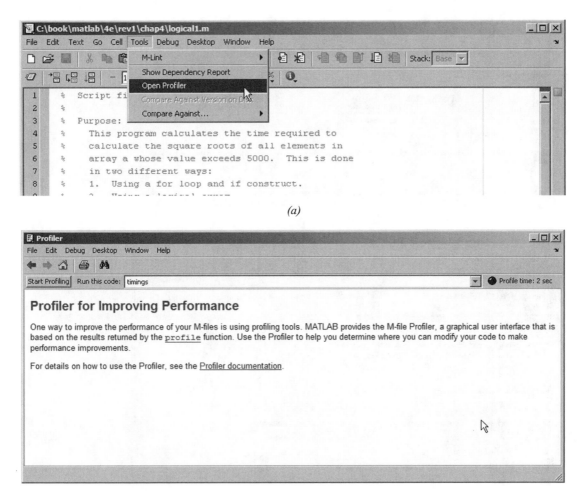

(a)

(b)

Figure 4.2 *(a)* The MATLAB Profiler is opened using the "Tools/Open Profile" menu option on the Edit/Debug Window. *(b)* The profiler has a box in which to type the name of the program to execute, and a pushbutton to start profiling.

✳ Good Programming Practice

Use the MATLAB Profiler to identify the parts of programs that consume the most CPU time. Optimizing those parts of the program will speed up the overall execution of the program.

(a)

(b)

Figure 4.3 *(a)* The Profile Summary, indicating the time spent in each profiled function.
(b) A detailed profile of function `timings`.

4.5   Additional Examples

▶

Example 4.6—Fitting a Line to a Set of Noisy Measurements

The velocity of a falling object in the presence of a constant gravitational field is given by the equation

$$v(t) = at + v_0 \qquad (4\text{-}3)$$

where $v(t)$ is the velocity at any time t, a is the acceleration due to gravity, and v_0 is the velocity at time 0. This equation is derived from elementary physics—it is known to every freshman physics student. If we plot velocity versus time for the falling object, our (v, t) measurement points should fall along a straight line. However, the same freshman physics student also knows that if we go into the laboratory and attempt to *measure* the velocity versus time of an object, our measurements will *not* fall along a straight line. They may come close, but they will never line up perfectly. Why not? Because we can never make perfect measurements. There is always some *noise* included in the measurements, which distorts them.

There are many cases in science and engineering where there are noisy sets of data such as this, and we wish to estimate the straight line that "best fits" the data. This problem is called the *linear regression* problem. Given a noisy set of measurements (x, y) that appear to fall along a straight line, how can we find the equation of the line

$$y = mx + b \qquad (4\text{-}4)$$

that "best fits" the measurements? If we can determine the regression coefficients m and b, then we can use this equation to predict the value of y at any given x by evaluating Equation (4-4) for that value of x.

A standard method for finding the regression coefficients m and b is the *method of least squares*. This method is named "least squares" because it produces the line $y = mx + b$ for which the sum of the squares of the differences between the observed y values and the predicted y values is as small as possible. The slope of the least squares line is given by

$$m = \frac{(\Sigma xy) - (\Sigma x)\bar{y}}{(\Sigma x^2) - (\Sigma x)\bar{x}} \qquad (4\text{-}5)$$

and the intercept of the least squares line is given by

$$b = \bar{y} - m\bar{x} \qquad (4\text{-}6)$$

where

Σx is the sum of the x values
Σx^2 is the sum of the squares of the x values
Σxy is the sum of the products of the corresponding x and y values
$\bar{x}$ is the mean (average) of the x values
$\bar{y}$ is the mean (average) of the y values

Write a program that will calculate the least-squares slope m and y-axis intercept b for a given set of noisy measured data points (x, y). The data points should be read from the keyboard, and both the individual data points and the resulting least-squares fitted line should be plotted.

SOLUTION

1. **State the problem.**

   Calculate the slope m and intercept b of a least-squares line that best fits an input data set consisting of an arbitrary number of (x, y) pairs. The input (x, y) data is read from the keyboard. Plot both the input data points and the fitted line on a single plot.

2. **Define the inputs and outputs.**

   The inputs required by this program are the number of points to read, plus the pairs of points (x, y).

   The outputs from this program are the slope and intercept of the least-squares fitted line, the number of points going into the fit, and a plot of the input data and the fitted line.

3. **Describe the algorithm.**

   This program can be broken down into six major steps:

   ```
   Get the number of input data points
   Read the input statistics
   Calculate the required statistics
   Calculate the slope and intercept
   Write out the slope and intercept
   Plot the input points and the fitted line
   ```

   The first major step of the program is to get the number of points to read in. To do this, we will prompt the user and read his or her answer with an input function. Next we will read the input (x, y) pairs one pair at a time using an input function in a for loop. Each pair of input value will be placed in an array ([x y]), and the function will return that array to the calling program. Note that a for loop is appropriate because we know in advance how many times the loop will be executed.

   The pseudocode for these steps is shown as follows:

   ```
   Print message describing purpose of the program
   n_points <- input('Enter number of [x y] pairs: ');
   for ii = 1:n_points
       temp <- input('Enter [x y] pair: ');
       x(ii) <- temp(1)
       y(ii) <- temp(2)
   end
   ```

   Next, we must accumulate the statistics required for the calculation. These statistics are the sums Σx, Σy, Σx^2, and Σxy. The pseudocode for these steps is

```
Clear the variables sum_x, sum_y, xum_x2, and sum_y2
for ii = 1:n_points
    sum_x <- sum_x + x(ii)
    sum_y <- sum_y + y(ii)
    sum_x2 <- sum_x2 + x(ii)^2
    sum_xy <- sum_xy + x(ii)*y(ii)
end
```

Next, we must calculate the slope and intercept of the least-squares line. The pseudocode for this step is just the MATLAB versions of Equations 4-4 and 4-5.

```
x_bar <- sum_x / n_points
y_bar <- sum_y / n_points
slope <- (sum_xy-sum_x * y_bar)/( sum_x2 - sum_x * x_bar)
y_int <- y_bar - slope * x_bar
```

Finally, we must write out and plot the results. The input data points should be plotted with circular markers and without a connecting line, and the fitted line should be plotted as a solid 2-pixel wide line. To do this, we will need to plot the points first, set hold on, plot the fitted line, and set hold off. We will add titles and a legend to the plot for completeness.

4. **Turn the algorithm into MATLAB statements.**
   The final MATLAB program follows:

```
%
% Purpose:
%   To perform a least-squares fit of an input data set
%   to a straight line, and print out the resulting slope
%   and intercept values. The input data for this fit
%   comes from a user-specified input data file.
%
% Record of revisions:
%     Date        Programmer            Description of change
%     ====        ==========            =====================
%   01/30/07    S. J. Chapman          Original code
%
% Define variables:
%   ii            -- Loop index
%   n_points      -- Number in input [x y] points
%   slope         -- Slope of the line
%   sum_x         -- Sum of all input x values
%   sum_x2        -- Sum of all input x values squared
%   sum_xy        -- Sum of all input x*y yalues
%   sum_y         -- Sum of all input y values
%   temp          -- Variable to read user input
%   x             -- Array of x values
```

```matlab
%   x_bar           -- Average x value
%   y               -- Array of y values
%   y_bar           -- Average y value
%   y_int           -- y-axis intercept of the line

disp('This program performs a least-squares fit of an ');
disp('input data set to a straight line.');
n_points = input('Enter the number of input [x y] points: ');

% Read the input data
for ii = 1:n_points
   temp = input('Enter [x y] pair: ');
   x(ii) = temp(1);
   y(ii) = temp(2);
end

% Accumulate statistics
sum_x = 0;
sum_y = 0;
sum_x2 = 0;
sum_xy = 0;
for ii = 1:n_points
   sum_x  = sum_x + x(ii);
   sum_y  = sum_y + y(ii);
   sum_x2 = sum_x2 + x(ii)^2;
   sum_xy = sum_xy + x(ii) * y(ii);
end

% Now calculate the slope and intercept.
x_bar = sum_x / n_points;
y_bar = sum_y / n_points;
slope = (sum_xy - sum_x * y_bar) / ( sum_x2 - sum_x * x_bar);
y_int = y_bar - slope * x_bar;

% Tell user.
disp('Regression coefficients for the least-squares line:');
fprintf('  Slope (m)     = %8.3f\n', slope);
fprintf('  Intercept (b)    = %8.3f\n', y_int);
fprintf('  No. of points    = %8d\n', n_points);

% Plot the data points as blue circles with no
% connecting lines.
plot(x,y,'bo');
hold on;
```

```
% Create the fitted line
xmin = min(x);
xmax = max(x);
ymin = slope * xmin + y_int;
ymax = slope * xmax + y_int;

% Plot a solid red line with no markers
plot([xmin xmax],[ymin ymax],'r-','LineWidth',2);
hold off;

% Add a title and legend
title ('\bfLeast-Squares Fit');
xlabel('\bf\itx');
ylabel('\bf\ity');
legend('Input data','Fitted line');
grid on
```

5. **Test the program.**
   To test this program, we will try a simple data set. For example, if every point in the input data set falls exactly along a line, then the resulting slope and intercept should be exactly the slope and intercept of that line. Thus the data set

   ```
   [1.1 1.1]
   [2.2 2.2]
   [3.3 3.3]
   [4.4 4.4]
   [5.5 5.5]
   [6.6 6.6]
   [7.7 7.7]
   ```

   should produce a slope of 1.0 and an intercept of 0.0. If we run the program with these values, the results are

```
» lsqfit
This program performs a least-squares fit of an
input data set to a straight line.
Enter the number of input [x y] points: 7
Enter [x y] pair: [1.1 1.1]
Enter [x y] pair: [2.2 2.2]
Enter [x y] pair: [3.3 3.3]
Enter [x y] pair: [4.4 4.4]
Enter [x y] pair: [5.5 5.5]
Enter [x y] pair: [6.6 6.6]
Enter [x y] pair: [7.7 7.7]
```

```
Regression coefficients for the least-squares line:
  Slope (m)       =  1.000
  Intercept (b)   =  0.000
  No. of points   =  7
```

Now let's add some noise to the measurements. The data set becomes

```
[1.1 1.01]
[2.2 2.30]
[3.3 3.05]
[4.4 4.28]
[5.5 5.75]
[6.6 6.48]
[7.7 7.84]
```

If we run the program with these values, the results are:

```
» lsqfit
This program performs a least-squares fit of an
input data set to a straight line.
```

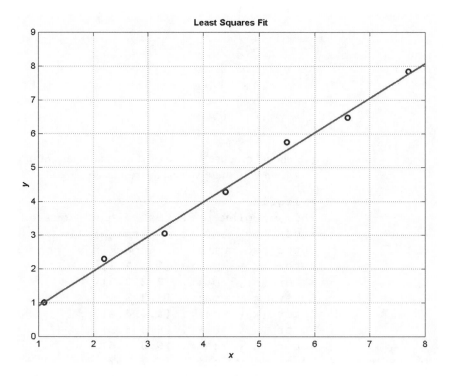

Figure 4.4 A noisy data set with a least-squares fitted line.

```
Enter the number of input [x y] points: 7
Enter [x y] pair: [1.1 1.01]
Enter [x y] pair: [2.2 2.30]
Enter [x y] pair: [3.3 3.05]
Enter [x y] pair: [4.4 4.28]
Enter [x y] pair: [5.5 5.75]
Enter [x y] pair: [6.6 6.48]
Enter [x y] pair: [7.7 7.84]
Regression coefficients for the least-squares line:
   Slope (m)       =  1.024
   Intercept (b)   = -0.120
   No of points    =  7
```

If we calculate the answer by hand, it is easy to show that the program gives the correct answers for our two test data sets. The noisy input data set and the resulting least-squares fitted line are shown in Figure 4.4.

◄

This example uses several of the plotting capabilities that we introduced in Chapter 3. It uses the `hold` command to allow multiple plots to be placed on the same axes, the `LineWidth` property to set the width of the least-squares fitted line, and escape sequences to make the title boldface and the axis labels bold italic.

►

Example 4.7—Physics—The Flight of a Ball

If we assume negligible air friction and ignore the curvature of the Earth, a ball that is thrown into the air from any point on the Earth's surface will follow a parabolic flight path (see Figure 4.5a). The height of the ball at any time t after it is thrown is given by Equation 4-7

$$y(t) = y_0 + v_{y0}t + \frac{1}{2}gt^2 \tag{4-7}$$

where y_0 is the initial height of the object above the ground, v_{y0} is the initial vertical velocity of the object, and g is the acceleration due to the Earth's gravity. The horizontal distance (range) traveled by the ball as a function of time after it is thrown is given by Equation (4-8)

$$x(t) = x_0 + v_{x0}t \tag{4-8}$$

where x_0 is the initial horizontal position of the ball on the ground, and v_{x0} is the initial horizontal velocity of the ball.

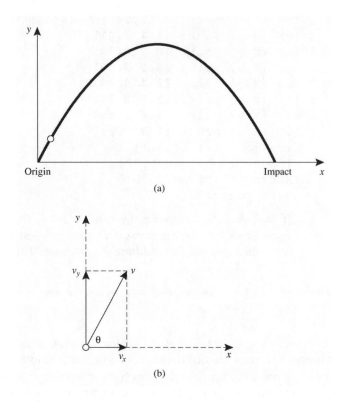

(a)

(b)

Figure 4.5 *(a)* When a ball is thrown upwards, it follows a parabolic trajectory. *(b)* The horizontal and vertical components of a velocity vector v at an angle θ with respect to the horizontal.

If the ball is thrown with some initial velocity v_0 at an angle of θ degrees with respect to the Earth's surface, then the initial horizontal and vertical components of velocity will be

$$v_{x0} = v_0 \cos \theta \qquad (4\text{-}9)$$

$$v_{y0} = v_0 \sin \theta \qquad (4\text{-}10)$$

Assume that the ball is initially thrown from position $(x_0, y_0) = (0, 0)$ with an initial velocity v_0 of 20 meters per second at an initial angle of θ degrees. Write a program that will plot the trajectory of the ball and also determine the horizontal distance traveled before it touches the ground again. The program should plot the trajectories of the ball for all angles θ from 5 to 85° in 10° steps, and it should determine the horizontal distance traveled for all angles θ from 0 to 90° in 1° steps. Finally, it should determine the angle θ that maximizes the range of the ball and plot that particular trajectory in a different color with a thicker line.

SOLUTION   To solve this problem, we must determine an equation for the time that the ball returns to the ground. Then, we can calculate the (x, y) position of the ball using Equations 4-7 through 4-10. If we do this for many times between 0 and the time that the ball returns to the ground, we can use those points to plot the ball's trajectory.

The time that the ball will remain in the air after it is thrown may be calculated from Equation 4-7. The ball will touch the ground at the time t for which $y(t) = 0$. Remembering that the ball will start from ground level ($y(0) = 0$), and solving for t, we get:

$$y(t) = y_0 + v_{y0}t + \frac{1}{2}gt^2 \tag{4-7}$$

$$0 = 0 + v_{y0}t + \frac{1}{2}gt^2$$

$$0 = \left(v_{y0} + \frac{1}{2}gt\right)t$$

so the ball will be at ground level at time $t_1 = 0$ (when we threw it), and at time

$$t_2 = -\frac{2v_{y0}}{g} \tag{4-11}$$

From the problem statement, we know that the initial velocity v_0 is 20 meters per second and that the ball will be thrown at all angles from 0° to 90° in 1° steps. Finally, any elementary physics textbook will tell us that the acceleration due to the earth's gravity is −9.81 meters per second squared.

Now let's apply our design technique to this problem.

1. **State the problem.**

   A proper statement of this problem would be: *Calculate the range that a ball would travel when it is thrown with an initial velocity of v_0 of 20 m/s at an initial angle θ. Calculate this range for all angles between 0° and 90° in 1° steps. Determine the angle θ that will result in the maximum range for the ball. Plot the trajectory of the ball for angles between 5° and 85° in 10° increments. Plot the maximum-range trajectory in a different color and with a thicker line. Assume that there is no air friction.*

2. **Define the inputs and outputs.**

   As the problem has been defined, no inputs are required. We know from the problem statement what v_0 and θ will be, so there is no need to input them. The outputs from this program will be a table showing the range of the ball for each angle θ, the angle θ for which the range is maximum, and a plot of the specified trajectories.

3. **Design the algorithm.**

   This program can be broken down into the following major steps:

```
Calculate the range of the ball for θ between 0 and 90°
Write a table of ranges
Determine the maximum range and write it out
```

```
Plot the trajectories for θ between 5 and 85°
Plot the maximum-range trajectory
```

Since we know the exact number of times that the loops will be repeated, `for` loops are appropriate for this algorithm. We will now refine the pseudocode for each of the preceding major steps.

To calculate the maximum range of the ball for each angle, we will first calculate the initial horizontal and vertical velocity from Equations (4-9) and (4-10). Then we will determine the time when the ball returns to Earth from Equation (4-11). Finally, we will calculate the range at that time from Equation (4-7). The detailed pseudocode for these steps is shown here. Note that we must convert all angles to radians before using the trig functions!

```
Create and initialize an array to hold ranges
for ii = 1:91
    theta <- ii - 1
    vxo <- vo * cos(theta*conv)
    vyo <- vo * sin(theta*conv)
    max_time <- -2 * vyo / g
    range(ii) <- vxo * max_time
end
```

Next, we must write a table of ranges. The pseudocode for this step is

```
Write heading
for ii = 1:91
    theta <- ii - 1
    print theta and range
end
```

The maximum range can be found with the `max` function. Recall that this function returns both the maximum value and its location. The pseudocode for this step is

```
[maxrange index] <- max(range)
Print out maximum range and angle (=index-1)
```

We will use nested `for` loops to calculate and plot the trajectories. To get all of the plots to appear on the screen, we must plot the first trajectory and then set `hold on` before plotting any other trajectories. After plotting the last trajectory, we must set `hold off`. To perform this calculation, we will divide each trajectory into 21 time steps and find the x and y positions of the ball for each time step. Then, we will plot those (x, y) positions. The pseudocode for this step is

```
for ii = 5:10:85

    % Get velocities and max time for this angle
    theta <- ii - 1
    vxo <- vo * cos(theta*conv)
```

```
vyo <- vo * sin(theta*conv)
max_time <- -2 * vyo / g

Initialize x and y arrays
for jj = 1:21
    time <- (jj-1) * max_time/20
    x(time) <- vxo * time
    y(time) <- vyo * time + 0.5 * g * time^2
end
plot(x,y) with thin green lines
Set "hold on" after first plot
end
Add titles and axis labels
```

Finally, we must plot the maximum range trajectory in a different color and with a thicker line.

```
vxo <- vo * cos(max_angle*conv)
vyo <- vo * sin(max_angle*conv)
max_time <- -2 * vyo / g

Initialize x and y arrays
for jj = 1:21
    time <- (jj-1) * max_time/20
    x(jj) <- vxo * time
    y(jj) <- vyo * time + 0.5 * g * time^2
end
plot(x,y) with a thick red line
hold off
```

4. **Turn the algorithm into MATLAB statements.**
   The final MATLAB program follows.

```
% Script file: ball.m
%
% Purpose:
%   This program calculates the distance traveled by a
%   ball thrown at a specified angle "theta" and a
%   specified velocity "vo" from a point on the surface of
%   the Earth, ignoring air friction and the Earth's
%   curvature. It calculates the angle yielding maximum
%   range, and also plots selected trajectories.
%
% Record of revisions:
%     Date         Programmer           Description of change
%     ====         ==========           =====================
%   01/30/07    S. J. Chapman           Original code
%
```

```
% Define variables:
%   conv          -- Degrees to radians conv factor
%   gravity       -- Accel. due to gravity (m/s^2)
%   ii, jj        -- Loop index
%   index         -- Location of maximum range in array
%   maxangle      -- Angle that gives maximum range (deg)
%   maxrange      -- Maximum range (m)
%   range         -- Range for a particular angle (m)
%   time          -- Time (s)
%   theta         -- Initial angle (deg)
%   traj_time     -- Total trajectory time (s)
%   vo            -- Initial velocity (m/s)
%   vxo           -- X-component of initial velocity (m/s)
%   vyo           -- Y-component of initial velocity (m/s)
%   x             -- X-position of ball (m)
%   y             -- Y-position of ball (m)

% Constants
conv = pi / 180;   % Degrees-to-radians conversion factor
g = -9.81;         % Accel. due to gravity
vo = 20;           % Initial velocity

%Create an array to hold ranges
range = zeros(1,91);

% Calculate maximum ranges
for ii = 1:91
   theta = ii - 1;
   vxo = vo * cos(theta*conv);
   vyo = vo * sin(theta*conv);
   max_time = -2 * vyo / g;
   range(ii) = vxo * max_time;
end

% Write out table of ranges
fprintf ('Range versus angle theta:\n');
for ii = 1:91
   theta = ii - 1;
   fprintf('  %2d    %8.4f\n',theta, range(ii));
end

% Calculate the maximum range and angle
[maxrange index] = max(range);
maxangle = index - 1;
fprintf ('\nMax range is %8.4f at %2d degrees.\n',...
        maxrange, maxangle);
```

```
% Now plot the trajectories
for ii = 5:10:85

    % Get velocities and max time for this angle
    theta = ii;
    vxo = vo * cos(theta*conv);
    vyo = vo * sin(theta*conv);
    max_time = -2 * vyo / g;
    % Calculate the (x,y) positions
    x = zeros(1,21);
    y = zeros(1,21);
    for jj = 1:21
        time = (jj-1) * max_time/20;
        x(jj) = vxo * time;
        y(jj) = vyo * time + 0.5 * g * time^2;
    end
    plot(x,y,'b');
    if ii == 5
        hold on;
    end
end

% Add titles and axis lables
title ('\bfTrajectory of Ball vs Initial Angle \theta');
xlabel ('\bf\itx \rm\bf(meters)');
ylabel ('\bf\ity \rm\bf(meters)');
axis ([0 45 0 25]);
grid on;
% Now plot the max range trajectory
vxo = vo * cos(maxangle*conv);
vyo = vo * sin(maxangle*conv);
max_time = -2 * vyo / g;

% Calculate the (x,y) positions
x = zeros(1,21);
y = zeros(1,21);
for jj = 1:21
    time = (jj-1) * max_time/20;
    x(jj) = vxo * time;
    y(jj) = vyo * time + 0.5 * g * time^2;
end
plot(x,y,'r','LineWidth',3.0);
hold off
```

The acceleration due to gravity at sea level can be found in any physics text. It is it is about 9.81 m/s^2, directed downward.

5. **Test the program.**

To test this program, we will calculate the answers by hand for a few of the angles and compare the results with the output of the program.

θ	$v_{x0} = v_0 \cos\theta$	$v_{y0} = v_0 \sin\theta$	$t_2 = -\dfrac{2v_{y0}}{g}$	$x = v_{x0}t_2$
0°	20 m/s	0 m/s	0 s	0 m
5°	19.92 m/s	1.74 m/s	0.355 s	7.08 m
40°	15.32 m/s	12.86 m/s	2.621 s	40.15 m
45°	14.14 m/s	14.14 m/s	2.883 s	40.77 m

When program `ball` is executed, a 91-line table of angles and ranges is produced. To save space, only a portion of the table is reproduced here.

```
» ball
Range versus angle theta:
    0    0.0000
    1    1.4230
    2    2.8443
    3    4.2621
    4    5.6747
    5    7.0805
  . . .
   40   40.1553
   41   40.3779
   42   40.5514
   43   40.6754
   44   40.7499
   45   40.7747
   46   40.7499
   47   40.6754
   48   40.5514
   49   40.3779
   50   40.1553
  . . .
   85    7.0805
   86    5.6747
   87    4.2621
   88    2.8443
   89    1.4230
   90    0.0000

Max range is  40.7747 at 45 degrees.
```

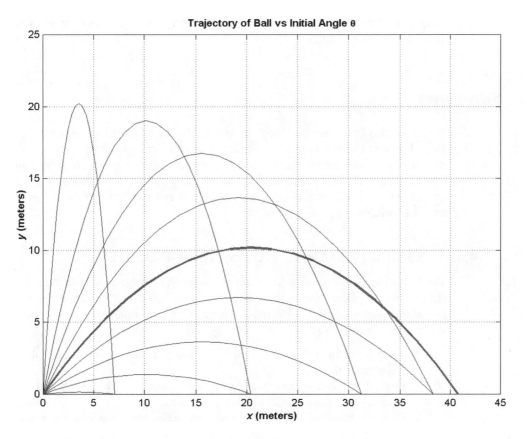

Figure 4.6 Possible trajectories for the ball.

The resulting plot is shown in Figure 4.6. The program output matches our hand calculation for the angles calculated above to the 4-digit accuracy of the hand calculation. Note that the maximum range occurred at an angle of 45°.

◄

This example uses several of the plotting capabilities that we introduced in Chapter 3. It uses the `axis` command to set the range of data to display, the `hold` command to allow multiple plots to be placed on the same axes, the `LineWidth` property to set the width of the line corresponding to the maximum-range trajectory, and escape sequences to create the desired title and *x*- and *y*-axis labels.

However, this program is not written in the most efficient manner, since there are a number of loops that could have been better replaced by vectorized statements. You will be asked to rewrite and improve ball.m in Exercise 4.11 at the end of this chapter.

4.6   Summary

There are two basic types of loops in MATLAB, the while loop and the for loop. The while loop is used to repeat a section of code in cases where we do not know in advance how many times the loop must be repeated. The for loop is used to repeat a section of code in cases where we know in advance how many times the loop should be repeated. It is possible to exit from either type of loop at any time using the break statement.

4.6.1   Summary of Good Programming Practice

The following guidelines should be adhered to when programming with loop constructs. If you follow them consistently, your code will contain fewer bugs, will be easier to debug, and will be more understandable to others who may need to work with it in the future.

1. Always indent code blocks in while and for constructs to make them more readable.
2. Use a while loop to repeat sections of code when you don't know in advance how often the loop will be executed.
3. Use a for loop to repeat sections of code when you know in advance how often the loop will be executed.
4. Never modify the values of a for loop index while inside the loop.
5. Always preallocate all arrays used in a loop before executing the loop. This practice greatly increases the execution speed of the loop.
6. If it is possible to implement a calculation either with a for loop or using vectors, implement the calculation with vectors. Your program will be much faster.
7. Do not rely on the JIT compiler to speed up your code. It has many limitations, and a programmer can typically do a better job with manual vectorization.
8. Use the MATLAB Profiler to identify the parts of programs that consume the most CPU time. Optimizing those parts of the program will speed up the overall execution of the program.

4.6.2   MATLAB Summary

The following summary lists all of the MATLAB commands and functions described in this chapter, along with a brief description of each one.

Commands and Functions

break	Stop the execution of a loop, and transfer control to the first statement after the end of the loop.
continue	Stop the execution of a loop, and transfer control to the top of the loop for the next iteration.
for loop	Loops over a block of statements a specified number of times.
tic	Resets elapsed time counter.
toc	Returns elapsed time since last call to tic.
while loop	Loops over a block of statements until a test condition becomes 0 (false).

4.7   Exercises

4.1   Write the MATLAB statements required to calculate $y(t)$ from the equation

$$y(t) = \begin{cases} -3t^2 + 5 & t \geq 0 \\ 3t^2 + 5 & t < 0 \end{cases}$$

for values of t between -9 and 9 in steps of 0.5. Use loops and branches to perform this calculation.

4.2   Rewrite the statements required to solve Exercise 4.1 using vectorization.

4.3   Write the MATLAB statements required to calculate and print out the squares of all the even integers between 0 and 50. Create a table consisting of each integer and its square, with appropriate labels over each column.

4.4   Write an M-file to evaluate the equation $y(x) = x^2 - 3x + 2$ for all values of x between -1 and 3, in steps of 0.1. Do this twice, once with a for loop and once with vectors. Plot the resulting function using a 3-point-thick dashed red line.

4.5   Write an M-file to calculate the factorial function N!, as defined in Example 4.2. Be sure to handle the special case of 0! Also, be sure to report an error if N is negative or not an integer.

4.6   Examine the following for statements and determine how many times each loop will be executed.

```
(a)  for ii = -32768:32767
(b)  for ii = 32768:32767
(c)  for kk = 2:4:3
(d)  for jj = ones(5,5)
```

4.7 Examine the following for loops and determine the value of `ires` at the end of each of the loops, and also the number of times each loop executes.

(*a*)
```
ires = 0;
for index = -10:10
    ires = ires + 1;
end
```

(*b*)
```
ires = 0;
for index = 10:-2:4
    if index == 6
        continue;
    end
    ires = ires + index;
end
```

(*c*)
```
ires = 0;
for index = 10:-2:4
    if index == 6
        break;
    end
    ires = ires + index;
end
```

(*d*)
```
ires = 0;
for index1 = 10:-2:4
    for index2 = 2:2:index1
        if index2 == 6
            break
        end
        ires = ires + index2;
    end
end
```

4.8 Examine the following `while` loops and determine the value of `ires` at the end of each of the loops and the number of times each loop executes.

(*a*)
```
ires = 1;
while mod(ires,10) ~= 0
    ires = ires + 1;
end
```

(*b*)
```
ires = 2;
while ires <= 200
    ires = ires^2;
end
```

(c)
```
ires = 2;
while ires > 200
    ires = ires^2;
end
```

4.9 What is contained in array arr1 after each of the following sets of statements have been executed?

(a)
```
arr1 = [1 2 3 4; 5 6 7 8; 9 10 11 12];
mask = mod(arr1,2) == 0;
arr1(mask) = -arr1(mask);
```

(b)
```
arr1 = [1 2 3 4; 5 6 7 8; 9 10 11 12];
arr2 = arr1 <= 5;
arr1(arr2) = 0;
arr1(~arr2) = arr1(~arr2).^2;
```

4.10 How can a logical array be made to behave as a logical mask for vector operations?

4.11 Modify program ball from Example 4.7 by replacing the inner for loops with vectorized calculations.

4.12 Modify program ball from Example 4.7 to read in the acceleration due to gravity at a particular location, and to calculate the maximum range of the ball for that acceleration. After modifying the program, run it with accelerations of -9.8 m/s^2, -9.7 m/s^2, and -9.6 m/s^2. What effect does the reduction in gravitational attraction have on the range of the ball? What effect does the reduction in gravitational attraction have on the best angle θ at which to throw the ball?

4.13 Modify program ball from Example 4.7 to read in the initial velocity with which the ball is thrown. After modifying the program, run it with initial velocities of 10 m/s, 20 m/s, and 30 m/s. What effect does changing the initial velocity v_0 have on the range of the ball? What effect does it have on the best angle θ at which to throw the ball?

4.14 Program lsqfit from Example 4.6 required the user to specify the number of input data points before entering the values. Modify the program so that it reads an arbitrary number of data values using a while loop, and stops reading input values when the user pressed the Enter key without typing any values. Test your program using the same two data sets that were used in Example 4.6. (*Hint:* The input function returns an empty array ([]) if a user presses Enter without supplying any data. You can use function isempty to test for an empty array, and stop reading data when one is detected.)

4.15 Modify program lsqfit from Example 4.6 to read its input values from an ASCII file named input1.dat. The data in the file will be organized in rows, with one pair of (x, y) values on each row, as shown here:

```
1.1    2.2
2.2    3.3
...
```

Test your program using the same two data sets that were used in Example 4.6. (*Hint:* Use the `load` command to read the data into an array named `input1`, and then store the first column of `input1` into array x and the second column of `input1` into array y.)

4.16 MATLAB Least-Squares Fit Function   MATLAB includes a standard function that performs a least-squares fit to a polynomial. Function `poly-fit` calculates the least-squares fit of a data set to a polynomial of order N:

$$p(x) = a_n x^n + a_{n-1} x^{n-1} + \cdots + a_1 x + a_0 \qquad (4\text{-}12)$$

where N can be any value greater than or equal to 1. Note that for $N = 1$, this polynomial is a linear equation, with the slope being the coefficient a_1 and the y-intercept being the coefficient a_0. The form of this function is

```
p = polyfit(x,y,n)
```

where x and y are vectors of x and y components, and n is the order of the fit.

Write a program that calculates the least-squares fit of a data set to a straight line using `polyfit`. Plot the input data points and the resulting fitted line. Compare the result produced by the program using polyfit with the result produced by `lsqfit` for the input data set in Example 4.6.

4.17 Program `doy` in Example 4.3 calculates the day of year associated with any given month, day, and year. As written, this program does not check to see if the data entered by the user is valid. It will accept nonsense values for months and days and do calculations with them to produce meaningless results. Modify the program so that it checks the input values for validity before using them. If the inputs are invalid, the program should tell the user what is wrong, and quit. The year should be a number greater than zero, the month should be a number between 1 and 12, and the day should be a number between 1 and a maximum that depends on the month. Use a `switch` construct to implement the bounds checking performed on the day.

4.18 Write a MATLAB program to evaluate the function

$$y(x) = \ln \frac{1}{1 - x}$$

for any user-specified value of x, where ln is the natural logarithm (logarithm to the base e). Write the program with a `while` loop, so that the program repeats the calculation for each legal value of x entered into the program. When an illegal value of x is entered, terminate the program. (Any $x \geq 1$ is considered an illegal value.)

4.19 Fibonacci Numbers   The nth Fibonacci number is defined by the following recursive equations:

$$f(1) = 1$$
$$f(2) = 2$$
$$f(n) = f(n - 1) + f(n - 2)$$

Figure 4.7 A semiconductor diode.

Therefore, $f(3) = f(2) + f(1) = 2 + 1 = 3$, and so forth for higher numbers. Write an M-file to calculate and write out the nth Fibonacci number for $n > 2$, where n is input by the user. Use a `while` loop to perform the calculation.

4.20 Current Through a Diode The current flowing through the semiconductor diode shown in Figure 4.7 is given by the equation

$$i_D = I_0 \left(e^{\frac{q v_D}{kT}} - 1 \right) \tag{4-13}$$

where i_D = the voltage across the diode, in volts
  v_D = the current flow through the diode, in amps
  I_0 = the leakage current of the diode, in amps
  q = the charge on an electron, 1.602×10^{-19} coulombs
  k = Boltzmann's constant, 1.38×10^{-23} joule/K
  T = temperature, in kelvins (K)

The leakage current I_0 of the diode is 2.0 μ A. Write a program to calculate the current flowing through this diode for all voltages from -1.0 V to $+0.6$ V, in 0.1 V steps. Repeat this process for the following temperatures: 75°F and 100°F, and 125°F. Create a plot of the current as a function of applied voltage, with the curves for the three different temperatures appearing as different colors.

4.21 Tension on a Cable A 200-pound object is to be hung from the end of a rigid 8-foot horizontal pole of negligible weight, as shown in Figure 4.8. The pole is attached to a wall by a pivot and is supported by an 8-foot cable that is attached to the wall at a higher point. The tension on this cable is given by the equation

$$T = \frac{W \cdot lc \cdot lp}{d\sqrt{lp^2 - d^2}} \tag{4-14}$$

where T is the tension on the cable, W is the weight of the object, lc is the length of the cable, lp is the length of the pole, and d is the distance along the pole at which the cable is attached. Write a program to determine the distance d at which to attach the cable to the pole in order to minimize the

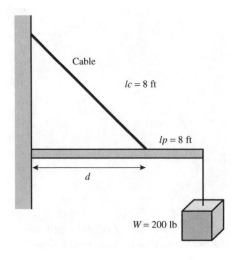

Figure 4.8 A 200 pound weight suspended from a rigid bar supported by a cable.

tension on the cable. To do this, the program should calculate the tension on the cable at regular one-foot intervals from $d = 1$ foot to $d = 7$ feet, and should locate the position d that produces the minimum tension. Also, the program should plot the tension on the cable as a function of d, with appropriate titles and axis labels.

4.22 Bacterial Growth Suppose that a biologist performs an experiment in which he or she measures the rate at which a specific type of bacterium reproduces asexually in different culture media. The experiment shows that in Medium A the bacteria reproduce once every 60 minutes, and in Medium B the bacteria reproduce once every 90 minutes. Assume that a single bacterium is placed on each culture medium at the beginning of the experiment. Write a program that calculates and plots the number of bacteria present in each culture at intervals of three hours from the beginning of the experiment until 24 hours have elapsed. Make two plots, one a linear xy plot and the other a linear-log (`semilogy`) plot. How do the numbers of bacteria compare on the two media after 24 hours?

4.23 Decibels Engineers often measure the ratio of two power measurements in *decibels*, or dB. The equation for the ratio of two power measurements in decibels is

$$dB = 10 \log_{10} \frac{P_2}{P_1} \qquad (4\text{-}15)$$

where P_2 is the power level being measured and P_1 is some reference power level. Assume that the reference power level P_1 is 1 watt, and write a program that calculates the decibel level corresponding to power levels between 1 and 20 watts, in 0.5 W steps. Plot the dB-versus-power curve on a log-linear scale.

4.24 Geometric Mean The *geometric mean* of a set of numbers x_1 through x_n is defined as the nth root of the product of the numbers:

$$\text{geometric mean} = \sqrt[n]{x_1 x_2 x_3 \ldots x_n} \qquad (4\text{-}16)$$

Write a MATLAB program that will accept an arbitrary number of positive input values and calculate both the arithmetic mean (i.e., the average) and the geometric mean of the numbers. Use a `while` loop to get the input values, and terminate the inputs when a user enters a negative number. Test your program by calculating the average and geometric mean of the four numbers 10, 5, 2, and 5.

4.25 RMS Average The *root-mean-square* (rms) *average* is another way of calculating a mean for a set of numbers. The rms average of a series of numbers is the square root of the arithmetic mean of the squares of the numbers:

$$\text{rms average} = \sqrt{\frac{1}{N}\sum_{i=1}^{N} x_i^2} \qquad (4\text{-}17)$$

Write a MATLAB program that will accept an arbitrary number of positive input values and calculate the rms average of the numbers. Prompt the user for the number of values to be entered, and use a `for` loop to read in the numbers. Test your program by calculating the rms average of the four numbers 10, 5, 2, and 5.

4.26 Harmonic Mean The *harmonic mean* is yet another way of calculating a mean for a set of numbers. The harmonic mean of a set of numbers is given by the equation:

$$\text{harmonic mean} = \frac{N}{\dfrac{1}{x_1} + \dfrac{1}{x_2} + \cdots + \dfrac{1}{x_N}} \qquad (4\text{-}18)$$

Write a MATLAB program that will read in an arbitrary number of positive input values and calculate the harmonic mean of the numbers. Use any method that you desire to read in the input values. Test your program by calculating the harmonic mean of the four numbers 10, 5, 2, and 5.

4.27 Write a single program that calculates the arithmetic mean (average), rms average, geometric mean, and harmonic mean for a set of positive numbers. Use any method that you desire to read in the input values. Compare these values for each of the following sets of numbers:

(*a*) 4, 4, 4, 4, 4, 4, 4
(*b*) 4, 3, 4, 5, 4, 3, 5
(*c*) 4, 1, 4, 7, 4, 1, 7
(*d*) 1, 2, 3, 4, 5, 6, 7

4.28 Mean Time Between Failure Calculations The reliability of a piece of electronic equipment is usually measured in terms of mean time between failures (MTBF), where MTBF is the average time that the piece of equipment can operate before a failure occurs in it. For large systems containing

Overall System

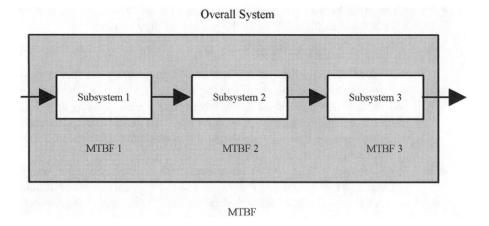

Figure 4.9 An electronic system containing three subsystems with known MTBFs.

many pieces of electronic equipment, it is customary to determine the MTBFs of each component and to calculate the overall MTBF of the system from the failure rates of the individual components. If the system is structured like the one shown in Figure 4.9, every component must work in order for the whole system to work, and the overall system MTBF can be calculated as

$$\text{MTBF}_{\text{sys}} = \frac{1}{\dfrac{1}{\text{MTBF}_1} + \dfrac{1}{\text{MTBF}_2} + \cdots + \dfrac{1}{\text{MTBF}_n}} \quad (4\text{-}19)$$

Write a program that reads in the number of series components in a system and the MTBFs for each component and then calculates the overall MTBF for the system. To test your program, determine the MTBF for a radar system consisting of an antenna subsystem with an MTBF of 2000 hours, a transmitter with an MTBF of 800 hours, a receiver with an MTBF of 3000 hours, and a computer with an MTBF of 5000 hours.

CHAPTER 5

User-Defined Functions

In Chapter 3, we learned the importance of good program design. The basic technique that we employed is **top-down design**. In top-down design, the programmer starts with a statement of the problem to be solved and the required inputs and outputs. Next, he or she describes the algorithm to be implemented by the program in broad outline and applies *decomposition* to break the algorithm down into logical subdivisions called sub-tasks. Then the programmer breaks down each sub-task until he or she winds up with many small pieces, each of which does a simple, clearly understandable job. Finally, the individual pieces are turned into MATLAB code.

Although we have followed this design process in our examples, the results have been somewhat restricted because we have had to combine the final MATLAB code generated for each sub-task into a single large program. There has been no way to code, verify, and test each sub-task independently before combining them into the final program.

Fortunately, MATLAB has a special mechanism designed to make sub-tasks easy to develop and debug independently before building the final program. It is possible to code each sub-task as a separate **function**, and each function can be tested and debugged independently of all of the other sub-tasks in the program.

Well-designed functions enormously reduce the effort required on a large programming project. Their benefits include

1. **Independent testing of sub-tasks**. Each sub-task can be written as an independent unit. The sub-task can be tested separately to ensure that it performs properly by itself before combining it into the larger program. This step is known as **unit testing**. It eliminates a major source of problems before the final program is even built.

2. **Reusable code**. In many cases, the same basic sub-task is needed in many parts of a program. For example, it may be necessary to sort a list of values into ascending order many times within a program, or even within other programs. It is possible to design, code, test, and debug a *single* function to do the sorting and then to reuse that function whenever sorting is required. This reusable code has two major advantages: it reduces the total programming effort required, and it simplifies debugging, since the sorting function needs to be debugged only once.

3. **Isolation from unintended side effects**. Functions receive input data from the program that invokes them through a list of variables called an **input argument list**, and return results to the program through an **output argument list**. Each function has its own workspace with its own variables, independent of all other functions and of the calling program. *The only variables in the calling program that can be seen by the function are those in the input argument list, and the only variables in the function that can be seen by the calling program are those in the output argument list.* This is very important, since accidental programming mistakes within a function can affect only the variables within function in which the mistake occurred.

Once a large program is written and released, it has to be *maintained*. Program maintenance involves fixing bugs and modifying the program to handle new and unforeseen circumstances. The programmer who modifies a program during maintenance is often not the person who originally wrote it. In poorly written programs, it is common for the programmer modifying the program to make a change in one region of the code and to have that change cause unintended side effects in a totally different part of the program. This happens because variable names are reused in different portions of the program. When the programmer changes the values left behind in some of the variables, those values are accidentally picked up and used in other portions of the code.

The use of well-designed functions minimizes this problem by **data hiding**. The variables in the main program are not visible to the function (except for those in the input argument list), and the variables in the main program cannot be accidentally modified by anything occurring in the function. Therefore, mistakes or changes in the function's variables cannot accidentally cause unintended side effects in the other parts of the program.

✳ Good Programming Practice

Break large program tasks into functions whenever practical to achieve the important benefits of independent component testing, reusability, and isolation from undesired side effects.

5.1   Introduction to MATLAB Functions

All of the M-files that we have seen so far have been **script files**. Script files are just collections of MATLAB statements that are stored in a file. When a script file is executed, the result is the same as it would be if all of the commands had been typed directly into the Command Window. Script files share the Command Window's workspace, so any variables that were defined before the script file starts are visible to the script file, and any variables created by the script file remain in the workspace after the script file finishes executing. A script file has no input arguments and returns no results, but script files can communicate with other script files through the data left behind in the workspace.

In contrast, a **MATLAB function** is a special type of M-file that runs in its own independent workspace. It receives input data through an **input argument list**, and returns results to the caller through an **output argument list**. The general form of a MATLAB function is

```
function [outarg1, outarg2, ...] = fname(inarg1, inarg2, ...)
% H1 comment line
% Other comment lines
...
(Executable code)
...
(return)
(end)
```

The `function` statement marks the beginning of the function. It specifies the name of the function and the input and output argument lists. The input argument list appears in parentheses after the function name, and the output argument list appears in brackets to the left of the equal sign. (If there is only one output argument, the brackets can be dropped.)

Each ordinary MATLAB function should be placed in a file with the same name (including capitalization) as the function, and the file extent ".m". For example, if a function is named `My_fun`, then that function should be placed in a file named `My_fun.m`.

The input argument list is a list of names representing values that will be passed from the caller to the function. These names are called **dummy arguments**. They are just placeholders for actual values that are passed from the caller when the function is invoked. Similarly, the output argument list contains a list of dummy arguments that are placeholders for the values returned to the caller when the function finishes executing.

A function is invoked by naming it in an expression together with a list of **actual arguments**. A function may be invoked by typing its name directly in the Command Window, or by including it in a script file or another function. The name in the calling program must *exactly match* the function name (including capitalization)[1]. When the function is invoked, the value of the first actual argument is used in place of the first dummy argument, and so forth for each other actual argument/dummy argument pair.

Execution begins at the top of the function and ends when either a `return` statement, an `end` statement, or the end of the function is reached. Because execution stops at the end of a function anyway, the `return` statement is not actually required in most functions and is rarely used. Each item in the output argument list must appear on the left side of a least one assignment statement in the function. When the function returns, the values stored in the output argument list are returned to the caller and may be used in further calculations.

The use of an `end` statement to terminate a function is a new feature as of MATLAB 7.0. In earlier versions of MATLAB, the `end` statement was used only to terminate structures such as `if`, `for`, `while`, and the like. It is optional in MATLAB 7 and later unless a file includes nested functions, which will be covered further on in this chapter. In this book, we will not use an `end` statement unless it is actually needed (in a set of nested functions).

The initial comment lines in a function serve a special purpose. The first comment line after the function statement is called the **H1 comment line**. It should always contain a one-line summary of the purpose of the function. The special significance of this line is that it is searched and displayed by the `lookfor` command. The remaining comment lines from the H1 line until the first blank line or the first executable statement are displayed by the `help` command. They should contain a brief summary of how to use the function.

A simple example of a user-defined function is shown here. Function `dist2` calculates the distance between points (x_1, y_1) and (x_2, y_2) in a Cartesian coordinate system.

```
function distance = dist2 (x1, y1, x2, y2)
%DIST2 Calculate the distance between two points
% Function DIST2 calculates the distance between
% two points (x1,y1) and (x2,y2) in a Cartesian
% coordinate system.
%
% Calling sequence:
%    distance = dist2(x1, y1, x2, y2)

% Define variables:
%    x1         -- x-position of point 1
```

---

[1] For example, suppose that a function has been declared with the name `My_Fun`, and placed in file `My_Fun.m`. Then this function should be called with the name `My_Fun`, not `my_fun` or `MY_FUN`. If the capitalization fails to match, this will produce an error on Linux, Unix, and Macintosh computers, and a warning on Windows-based computers.

```
%    y1          -- y-position of point 1
%    x2          -- x-position of point 2
%    y2          -- y-position of point 2
%    distance -- Distance between points

%    Record of revisions:
%      Date          Programmer            Description of change
%      ====          ==========            =====================
%    02/01/07      S. J. Chapman         Original code

% Calculate distance.
distance = sqrt((x2-x1).^2 + (y2-y1).^2);

end % function distance
```

This function has four input arguments and one output argument. A simple script file using this function is

```
% Script file: test_dist2.m
%
% Purpose:
%   This program tests function dist2.
%
% Record of revisions:
%      Date          Programmer            Description of change
%      ====          ==========            =====================
%    02/01/07      S. J. Chapman         Original code
%
% Define variables:
%    ax          -- x-position of point a
%    ay          -- y-position of point a
%    bx          -- x-position of point b
%    by          -- y-position of point b
%    result   -- Distance between the points

% Get input data.
disp('Calculate the distance between two points:');
ax = input('Enter x value of point a: ');
ay = input('Enter y value of point a: ');
bx = input('Enter x value of point b: ');
by = input('Enter y value of point b: ');

% Evaluate function
result = dist2 (ax, ay, bx, by);

% Write out result.
fprintf('The distance between points a and b is %f\n',result);
```

When this script file is executed, the results are

```
» test_dist2
Calculate the distance between two points:
Enter x value of point a: 1
Enter y value of point a: 1
Enter x value of point b: 4
Enter y value of point b: 5
The distance between points a and b is 5.000000
```

These results are correct, as we can verify from simple hand calculations.

Function dist2 also supports the MATLAB help subsystem. If we type "help dist2", the results are

```
» help dist2
DIST2 Calculate the distance between two points
  Function DIST2 calculates the distance between
  two points (x1,y1) and (x2,y2) in a Cartesian
  coordinate system.

  Calling sequence:
     res = dist2(x1, y1, x2, y2)
```

Similarly, "lookfor distance" produces the result

```
» lookfor distance
DIST2 Calculate the distance between two points
MAHAL Mahalanobis distance.
DIST Distances between vectors.
NBDIST Neighborhood matrix using vector distance.
NBGRID Neighborhood matrix using grid distance.
NBMAN Neighborhood matrix using Manhattan-distance.
```

To observe the behavior of the MATLAB workspace before, during, and after the function is executed, we will load function dist2 and the script file test_dist2 into the MATLAB debugger and set breakpoints before, during, and after the function call (see Figure 5.1). When the program stops at the breakpoint *before* the function call, the workspace is as shown in Figure 5.2(*a*). Note that variables ax, ay, bx, and by are defined in the workspace with the values that we have entered. When the program stops at the breakpoint *within* the function call, the function's workspace is active. It is as shown in Figure 5.2(*b*). Note that variables x1, x2, y1, y2, and distance are defined in the function's workspace, and the variables defined in the calling M-file are not present. When the program stops in the calling program at the breakpoint *after* the function call, the workspace is as shown in Figure 5.2(*c*). Now the original variables are back, with the variable result added to contain the value returned by the function.

```
Editor - C:\book\matlab\4e\rev1\chap5\test_dist2.m                          _|□|×|
File  Edit  Text  Go  Cell  Tools  Debug  Desktop  Window  Help                  ×|  ┌ ×
□ ☞ ▓   ✗  🖿 🖿 ⌐ ⌐ | 🖨 |  🗛 ← → ƒ. | 🗐 🗐 | 🗐 🗐 🗐 🗐 🗐 | Stack: test_dist2 ▼ |  🗔 🗔 🗖 🗗 🗖
⌗ | 🗐 🗐 🗐 |  ─ | 1.0   | +  |  ÷ | 1.1   | ×  | 🗙 🗙 | ⓞ

1      %  Script file: test_dist2.m
2      %
3      %  Purpose:
4      %    This program tests function dist2.
5      %
6      %  Record of revisions:
7      %    Date         Programmer            Description of change
8      %    ====         ==========            =====================
9      %    02/01/07     S. J. Chapman         Original code
10     %
11     %  Define variables:
12     %    ax      -- x-position of point a
13     %    ay      -- y-position of point a
14     %    bx      -- x-position of point b
15     %    by      -- y-position of point b
16     %    result -- Distance between the points
17
18     %  Get input data.
19 ─   disp('Calculate the distance between two points:');
20 ─   ax = input('Enter x value of point a:   ');
21 ─   ay = input('Enter y value of point a:   ');
22 ─   bx = input('Enter x value of point b:   ');
23 ─   by = input('Enter y value of point b:   ');
24
25     %  Evaluate function
26 ⬤➡  result = dist2 (ax, ay, bx, by);
27
28     %  Write out result.
29 ⬤   fprintf('The distance between points a and b is %f\n',result);
30

test_dist2.m  ×  dist2.m  ×
                                          script            Ln 25   Col 20   OVR
```

Figure 5.1 M-file `test_dist2` and function `dist2` are loaded into the debugger, with breakpoints set before, during, and after the function call.

These figures show that the workspace of the function is different from the workspace of the calling M-file.

5.2 Variable Passing in MATLAB: The Pass-by-Value Scheme

MATLAB programs communicate with their functions using a **pass-by-value** scheme. When a function call occurs, MATLAB makes a *copy* of the actual arguments and passes them to the function. This copying is very significant, because it means that even if the function modifies the input arguments, it won't affect the original data in the caller. This feature helps to prevent unintended side effects in which an error in the function might unintentionally modify variables in the calling program.

This behavior is illustrated in the function shown here. This function has two input arguments: a and b. During its calculations, it modifies both input arguments.

Figure 5.2 *(a)* The workspace before the function call. *(b)* The workspace during the function call. *(c)* The workspace after the function call.

```
function out = sample(a, b, c)
fprintf('In      sample: a = %f, b = %f %f\n',a,b);
a = b(1) + 2*a;
b = a .* b;
out = a + b(1);
fprintf('In      sample: a = %f, b = %f %f\n',a,b);
```

A simple test program to call this function is

```
a = 2; b = [6 4];
fprintf('Before sample: a = %f, b = %f %f\n',a,b);
out = sample(a,b);
fprintf('After  sample: a = %f, b = %f %f\n',a,b);
fprintf('After  sample: out = %f\n',out);
```

When this program is executed, the results are

```
» test_sample
Before sample: a = 2.000000, b = 6.000000 4.000000
In      sample: a = 2.000000, b = 6.000000 4.000000
In      sample: a = 10.000000, b = 60.000000 40.000000
After  sample: a = 2.000000, b = 6.000000 4.000000
After  sample: out = 70.000000
```

Note that a and b were both changed inside function sample, but those changes had *no effect on the values in the calling program.*

Users of the C language will be familiar with the pass-by-value scheme, since C uses it for scalar values passed to functions. However C does *not* use the pass-by-value scheme when passing arrays, so an unintended modification to a dummy array in a C function can cause side effects in the calling program. MATLAB improves on this by using the pass-by-value scheme for both scalars and arrays[2].

▶
▬▬▬▬▬▬▬▬▬▬▬▬▬▬▬▬▬▬▬▬▬▬▬▬▬▬▬▬▬▬▬▬▬▬▬▬▬

Example 5.1—Rectangular-to-Polar Conversion

The location of a point in a Cartesian plane can be expressed in either the rectangular coordinates (x, y) or the polar coordinates (r, θ), as shown in Figure 5.3. The relationships among these two sets of coordinates are given by the following equations:

$$x = r \cos \theta \tag{5-1}$$

$$y = r \sin \theta \tag{5-2}$$

---

[2]The implementation of argument passing in MATLAB is actually more sophisticated than this discussion indicates. As pointed out above, the copying associated with pass-by-value takes up a lot of time, but it provides protection against unintended side effects. MATLAB actually uses the best of both approaches: it analyzes each argument of each function and determines whether or not the function modifies that argument. If the function modifies the argument, then MATLAB makes a copy of it. If it does not modify the argument, then MATLAB simply points to the existing value in the calling program. This practice increases speed while still providing protection against side effects!

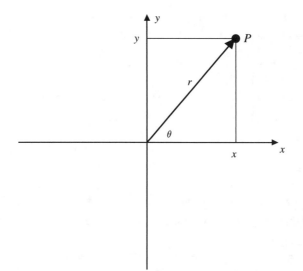

Figure 5.3   A point P in a Cartesian plane can be located by either the rectangular coordinates (x, y) or the polar coordinates (r, θ).

$$r = \sqrt{x^2 + y^2} \qquad (5\text{-}3)$$

$$\theta = \tan^{-1}\frac{y}{x} \qquad (5\text{-}4)$$

Write two functions `rect2polar` and `polar2rect` that convert coordinates from rectangular to polar form and vice versa, where the angle θ is expressed in degrees.

SOLUTION   We will apply our standard problem-solving approach to creating these functions. Note that MATLAB's trigonometric functions work in radians, so we must convert from degrees to radians and vice versa when solving this problem. The basic relationship between degrees and radians is

$$180° = \pi \text{ radians} \qquad (5\text{-}5)$$

1. **State the problem.**
   A succinct statement of the problem is

   > Write a function that converts a location on a Cartesian plane expressed in rectangular coordinates into the corresponding polar coordinates, with the angle θ is expressed in degrees. Also, write a function that converts a location on a Cartesian plane expressed in polar coordinates with the angle θ expressed in degrees into the corresponding rectangular coordinates.

2. **Define the inputs and outputs.**

   The inputs to function `rect2polar` are the rectangular (x, y) location of a point. The outputs of the function are the polar (r, θ) location of the point. The inputs to function `polar2rect` are the polar (r, θ) location of a point. The outputs of the function are the rectangular (x, y) location of the point.

3. **Describe the algorithm.**

   These functions are very simple, so we can directly write the final pseudocode for them. The pseudocode for function `polar2rect` is

   ```
   x <- r * cos(theta * pi/180)
   y <- r * sin(theta * pi/180)
   ```

   The pseudocode for function `rect2polar` will use the function `atan2`, because that function works over all four quadrants of the Cartesian plane. (Look up that function in the MATLAB Help Browser!)

   ```
   r <- sqrt( x.^2 + y .^2 )
   theta <- 180/pi * atan2(y,x)
   ```

4. **Turn the algorithm into MATLAB statements.**

   The MATLAB code for the selection `polar2rect` function is

```
function [x, y] = polar2rect(r,theta)
%POLAR2RECT Convert rectangular to polar coordinates
% Function POLAR2RECT accepts the polar coordinates
% (r,theta), where theta is expressed in degrees,
% and converts them into the rectangular coordinates
% (x,y).
%
% Calling sequence:
%   [x, y] = polar2rect(r,theta)

% Define variables:
%   r               -- Length of polar vector
%   theta           -- Angle of vector in degrees
%   x               -- x-position of point
%   y               -- y-position of point

% Record of revisions:
%     Date          Programmer          Description of change
%     ====          ==========          =====================
%   02/01/07      S. J. Chapman         Original code

x = r * cos(theta * pi/180);
y = r * sin(theta * pi/180);
```

The MATLAB code for the selection `rect2polar` function is

```
function [r, theta] = rect2polar(x,y)
%RECT2POLAR Convert rectangular to polar coordinates
% Function RECT2POLAR accepts the rectangular coordinates
% (x,y) and converts them into the polar coordinates
% (r,theta), where theta is expressed in degrees.
%
% Calling sequence:
%   [r, theta] = rect2polar(x,y)

% Define variables:
%   r          -- Length of polar vector
%   theta      -- Angle of vector in degrees
%   x          -- x-position of point
%   y          -- y-position of point

% Record of revisions:
%    Date           Programmer          Description of change
%    ====           ==========          =====================
%   02/01/07       S. J. Chapman        Original code

r = sqrt( x.^2 + y .^2 );
theta = 180/pi * atan2(y,x);
```

Note that these functions both include help information, so they will work properly with MATLAB's help subsystem and with the `lookfor` command.

5. **Test the program.**

To test these functions, we will execute them directly in the MATLAB Command Window. We will test the functions using the 3-4-5 triangle, which is familiar to most people from high school. The smaller angle within a 3-4-5 triangle is approximately 36.87°. We will also test the function in all four quadrants of the Cartesian plane to ensure that the conversions are correct everywhere.

```
» [r, theta] = rect2polar(4,3)
r =
     5
theta =
   36.8699
» [r, theta] = rect2polar(-4,3)
r =
     5
theta =
   143.1301
```

```
» [r, theta] = rect2polar(-4,-3)
r =
       5
theta =
    -143.1301
» [r, theta] = rect2polar(4,-3)
r =
       5
theta =
    -36.8699
» [x, y] = polar2rect(5,36.8699)
x =
    4.0000
y =
    3.0000
» [x, y] = polar2rect(5,143.1301)
x =
   -4.0000
y =
    3.0000
» [x, y] = polar2rect(5,-143.1301)
x =
   -4.0000
y =
   -3.0000
» [x, y] = polar2rect(5,-36.8699)
x =
    4.0000
y =
   -3.0000
»
```

These functions appear to be working correctly in all quadrants of the Cartesian plane. ◄

►

Example 5.2—Sorting Data

In many scientific and engineering applications, it is necessary to take a random input data set and to sort it so that the numbers in the data set are either all in *ascending order* (lowest to highest) or all in *descending order* (highest to lowest). For example, suppose that you were a zoologist studying a large population of animals and that you wanted to identify the largest 5 percent of the animals in the population. The most straightforward way to approach this problem would be to sort the sizes of all of the animals in the population into ascending order and take the top 5 percent of the values.

Sorting data into ascending or descending order seems to be an easy job. After all, we do it all the time. It is simple matter for us to sort the data (10, 3, 6, 4, 9) into the order (3, 4, 6, 9, 10). How do we do it? We first scan the input data list (10, 3, 6, 4, 9) to find the smallest value in the list (3), and then scan the remaining input data (10, 6, 4, 9) to find the next smallest value (4), and so on until the complete list is sorted.

In fact, sorting can be a very difficult job. As the number of values to be sorted increases, the time required to perform the simple sort just described increases rapidly, since we must scan the input data set once for each value sorted. For very large data sets, this technique just takes too long to be practical. Even worse, how would we sort the data if there were too many numbers to fit into the main memory of the computer? The development of efficient sorting techniques for large data sets is an active area of research and is the subject of whole courses all by itself.

In this example, we will confine ourselves to the simplest possible algorithm to illustrate the concept of sorting. This simplest algorithm is called the **selection sort**. It is just a computer implementation of the mental math described previously. The basic algorithm for the selection sort is

1. Scan the list of numbers to be sorted to locate the smallest value in the list. Place that value at the front of the list by swapping it with the value currently at the front of the list. If the value at the front of the list is already the smallest value, then do nothing.
2. Scan the list of numbers from position 2 to the end to locate the next smallest value in the list. Place that value in position 2 of the list by swapping it with the value currently at that position. If the value in position 2 is already the next smallest value, then do nothing.
3. Scan the list of numbers from position 3 to the end to locate the third smallest value in the list. Place that value in position 3 of the list by swapping it with the value currently at that position. If the value in position 3 is already the third smallest value, then do nothing.
4. Repeat this process until the next-to-last position in the list has been reached. After the next-to-last position in the list has been processed, the sort is complete.

Note that if we are sorting N values, this sorting algorithm requires $N-1$ scans through the data to accomplish the sort.

This process is illustrated in Figure 5.4. Since there are five values in the data set to be sorted, we will make four scans through the data. During the first pass through the entire data set, the minimum value is 3, so the 3 is swapped with the 10 which was in position 1. Pass 2 searches for the minimum value in positions 2 through 5. That minimum is 4, so the 4 is swapped with the 10 in

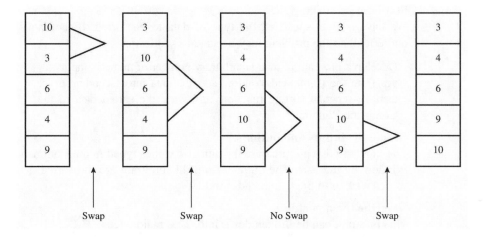

Figure 5.4   An example problem demonstrating the selection sort algorithm.

position 2. Pass 3 searches for the minimum value in positions 3 through 5. That minimum is 6, which is already in position 3, so no swapping is required. Finally, pass 4 searches for the minimum value in positions 4 through 5. That minimum is 9, so the 9 is swapped with the 10 in position 4, and the sort is completed.   ◄

💣 Programming Pitfalls

The selection sort algorithm is the easiest sorting algorithm to understand, but it is computationally inefficient. *It should never be applied to sort large data sets* (say, sets with more than 1000 elements). Over the years, computer scientists have developed much more efficient sorting algorithms. The `sort` and `sortrows` functions built into MATLAB are extremely efficient and should be used for all real work.

We will now develop a program to read in a data set from the Command Window, sort it into ascending order, and display the sorted data set. The sorting will be done by a separate user-defined function.

SOLUTION   This program must be able to ask the user for the input data, sort the data, and write out the sorted data. The design process for this problem is given as follows.

1. **State the problem.**

   We have not yet specified the type of data to be sorted. If the data is numeric, then the problem may be stated as follows:

   > Develop a program to read an arbitrary number of numeric input values from the Command Window, sort the data into ascending order using a separate sorting function, and write the sorted data to the Command Window.

2. **Define the inputs and outputs.**

   The inputs to this program are the numeric values typed in the Command Window by the user. The outputs from this program are the sorted data values written to the Command Window.

3. **Describe the algorithm.**

   This program can be broken down into three major steps:

   ```
   Read the input data into an array
   Sort the data in ascending order
   Write the sorted data
   ```

   The first major step is to read in the data. We must prompt the user for the number of input data values and then read in the data. Since we will know how many input values there are to read, a `for` loop is appropriate for reading in the data. The detailed pseudocode is

   ```
   Prompt user for the number of data values
   Read the number of data values
   Preallocate an input array
   for ii = 1:number of values
      Prompt for next value
      Read value
   end
   ```

   Next we have to sort the data in a separate function. We will need to make `nvals-1` passes through the data, finding the smallest remaining value each time. We will use a pointer to locate the smallest value in each pass. Once the smallest value has been found, it will be swapped to the top of the list of it is not already there. The detailed pseudocode is

   ```
   for ii = 1:nvals-1
      % Find the minimum value in a(ii) through a(nvals)
      iptr <- ii
      for jj == ii+1 to nvals
         if a(jj) < a(iptr)
            iptr <- jj
         end
      end
   ```

```
% iptr now points to the min value, so swap a(iptr)
% with a(ii) if iptr ~= ii.
if i ~= iptr
   temp <- a(i)
   a(i) <- a(iptr)
   a(iptr) <- temp
end
end
```

The final step is writing out the sorted values. No refinement of the pseudocode is required for that step. The final pseudocode is the combination of the reading, sorting, and writing steps.

4. **Turn the algorithm into MATLAB statements.**
   The MATLAB code for the selection sort function is

```
function out = ssort(a)
%SSORT Selection sort data in ascending order
% Function SSORT sorts a numeric data set into
% ascending order. Note that the selection sort
% is relatively inefficient. DO NOT USE THIS
% FUNCTION FOR LARGE DATA SETS. Use MATLAB's
% "sort" function instead.

% Define variables:
%   a           -- Input array to sort
%   ii          -- Index variable
%   iptr        -- Pointer to min value
%   jj          -- Index variable
%   nvals       -- Number of values in "a"
%   out         -- Sorted output array
%   temp        -- Temp variable for swapping

% Record of revisions:
%     Date          Programmer          Description of change
%     ====          ==========          =====================
%   02/02/07      S. J. Chapman         Original code

% Get the length of the array to sort
nvals = size(a,2);

% Sort the input array
for ii = 1:nvals-1
   % Find the minimum value in a(ii) through a(n)
   iptr = ii;
   for jj = ii+1:nvals
      if a(jj) < a(iptr)
         iptr = jj;
      end
   end
```

```
% iptr now points to the minimum value, so swap a(iptr)
% with a(ii) if ii ~= iptr.
if ii ~= iptr
    temp     = a(ii);
    a(ii)    = a(iptr);
    a(iptr)  = temp;

    end
end

% Pass data back to caller
out = a;
```

The program to invoke the selection sort function is

```
% Script file: test_ssort.m
%
% Purpose:
%   To read in an input data set, sort it into ascending
%   order using the selection sort algorithm, and to
%   write the sorted data to the Command Window. This
%   program calls function "ssort" to do the actual
%   sorting.
%
% Record of revisions:
%      Date         Programmer        Description of change
%      ====         ==========        =====================
%    02/02/07     S. J. Chapman        Original code
%
% Define variables:
%   array      -- Input data array
%   ii         -- Index variable
%   nvals      -- Number of input values
%   sorted     -- Sorted data array

% Prompt for the number of values in the data set
nvals = input('Enter number of values to sort: ');
% Preallocate array
array = zeros(1,nvals);

% Get input values
for ii = 1:nvals

    % Prompt for next value
    string = ['Enter value ' int2str(ii) ': '];
    array(ii) = input(string);

end
```

```
% Now sort the data
sorted = ssort(array);

% Display the sorted result.
fprintf('\nSorted data:\n');
for ii = 1:nvals
   fprintf(' %8.4f\n',sorted(ii));
end
```

5. **Test the program.**

To test this program, we will create an input data set and run the program with it. The data set should contain a mixture of positive and negative numbers as well as at least one duplicated value to see if the program works properly under those conditions.

```
» test_ssort
Enter number of values to sort: 6
Enter value 1: -5
Enter value 2: 4
Enter value 3: -2
Enter value 4: 3
Enter value 5: -2
Enter value 6: 0

Sorted data:
  -5.0000
  -2.0000
  -2.0000
   0.0000
   3.0000
   4.0000
```

The program gives the correct answers for our test data set. Note that it works for both positive and negative numbers as well as for repeated numbers. ◀

5.3 Optional Arguments

Many MATLAB functions support optional input arguments and output arguments. For example, we have seen calls to the `plot` function with as few as two or as many as seven input arguments. On the other hand, the function `max` supports either one or two output arguments. If there is only one output argument, `max` returns the maximum value of an array. If there are two output arguments, `max` returns both the maximum value and the location of the maximum value in an array. How do MATLAB functions know how many input and output arguments are present, and how do they adjust their behavior accordingly?

There are eight special functions that can be used by MATLAB functions to get information about their optional arguments and to report errors in those arguments. Six of these functions are introduced here, and the remaining two will be introduced in Chapter 7 after we learn about the cell array data type. The functions introduced now are

- `nargin`—This function returns the number of actual input arguments that were used to call the function.
- `nargout`—This function returns the number of actual output arguments that were used to call the function.
- `nargchk`—This function returns a standard error message if a function is called with too few or too many arguments.
- `error`—Display error message and abort the function producing the error. This function is used if the argument errors are fatal.
- `warning`—Display warning message and continue function execution. This function is used if the argument errors are not fatal and execution can continue.
- `inputname`—This function returns the actual name of the variable that corresponds to a particular argument number.

When functions `nargin` and `nargout` are called within a user-defined function, these functions return the number of actual input arguments and the number of actual output arguments that were used when the user-defined function was called.

Function `nargchk` generates a string containing a standard error message if a function is called with too few or too many arguments. The syntax of this function is

```
message = nargchk(min_args,max_args,num_args);
```

where `min_args` is the minimum number of arguments, `max_args` is the maximum number of arguments, and `num_args` is the actual number of arguments. If the number of arguments is outside the acceptable limits, a standard error message is produced. If the number of arguments is within acceptable limits, then an empty string is returned.

Function `error` is a standard way to display an error message and abort the user-defined function causing the error. The syntax of this function is `error('msg')`, where `msg` is a character string containing an error message. When `error` is executed, it halts the current function and returns to the keyboard, displaying the error message in the Command Window. If the message string is empty, `error` does nothing and execution continues. This function works well with `nargchk`, which produces a message string when an error occurs and an empty string when there is no error.

Function `warning` is a standard way to display a warning message that includes the function and line number where the problem occurred but lets execution continue. The syntax of this function is `warning('msg')`, where `msg` is a character string containing a warning message. When `warning` is executed, it displays the warning message in the Command Window and lists the function name and line number where the warning came from. If the message string is empty, `warning` does nothing. In either case, execution of the function continues.

Function `inputname` returns the name of the actual argument used when a function is called. The syntax of this function is

```
name = inputname(argno);
```

where `argno` is the number of the argument. If the argument is a variable, then its name is returned. If the argument is an expression, then this function will return an empty string. For example, consider the function

```
function myfun(x,y,z)
name = inputname(2);
disp(['The second argument is named ' name]);
```

When this function is called, the results are

```
» myfun(dog,cat)
The second argument is named cat
» myfun(1,2+cat)
The second argument is named
```

Function `inputname` is useful for displaying argument names in warning and error messages.

▶

Example 5.3—Using Optional Arguments

We will illustrate the use of optional arguments by creating a function that accepts an (x, y) value in rectangular coordinates and produces the equivalent polar representation consisting of a magnitude and an angle in degrees. The function will be designed to support two input arguments, x and y. However, if only one argument is supplied, the function will assume that the y value is zero and proceed with the calculation. The function will normally return both the magnitude and the angle in degrees, but if only one output argument is present, it will return only the magnitude. This function is

```
function [mag, angle] = polar_value(x,y)
%POLAR_VALUE Converts (x,y) to (r,theta)
% Function POLAR_VALUE converts an input (x,y)
% value into (r,theta), with theta in degrees.
% It illustrates the use of optional arguments.
```

```
% Define variables:
%   angle    -- Angle in degrees
%   msg      -- Error message
%   mag      -- Magnitude
%   x        -- Input x value
%   y        -- Input y value (optional)

% Record of revisions:
%    Date          Programmer            Description of change
%    ====          ==========            =====================
%   02/03/07    S. J. Chapman            Original code

% Check for a legal number of input arguments.
msg = nargchk(1,2,nargin);
error(msg);

% If the y argument is missing, set it to 0.
if nargin < 2
   y = 0;
end

% Check for (0,0) input arguments, and print out
% a warning message.
if x == 0 & y == 0
   msg = 'Both x any y are zero: angle is meaningless!';
   warning(msg);
end

% Now calculate the magnitude.
mag = sqrt(x.^2 + y.^2);

% If the second output argument is present, calculate
% angle in degrees.
if nargout == 2
   angle = atan2(y,x) * 180/pi;
end
```

We will test this function by calling it repeatedly from the Command Window. First, we will try to call the function with too few or too many arguments.

```
» [mag angle] = polar_value
??? Error using ==> polar_value
Not enough input arguments.
» [mag angle] = polar_value(1,-1,1)
??? Error using ==> polar_value
Too many input arguments.
```

The function provides proper error messages in both cases. Next, we will try to call the function with one or two input arguments.

```
» [mag angle] = polar_value(1)
mag =
     1
angle =
     0
» [mag angle] = polar_value(1,-1)
mag =
     1.4142
angle =
   -45
```

The function provides the correct answer in both cases. Next, we will try to call the function with one or two output arguments.

```
» mag = polar_value(1,-1)
mag =
     1.4142
» [mag angle] = polar_value(1,-1)
mag =
     1.4142
angle =
     -45
```

The function provides the correct answer in both cases. Finally, we will try to call the function with both *x* and *y* equal to zero.

```
» [mag angle] = polar_value(0,0)

Warning: Both x any y are zero: angle is meaningless!
> In d:\book\matlab\chap5\polar_value.m at line 32
mag =
     0
angle =
     0
```

In this case, the function displays the warning message, but execution continues. ◀

Note that a MATLAB function may be declared to have more output arguments than are actually used, and this is not an error. The function does not actually have to check nargout to determine if an output argument is present. For example, consider the following function:

```
function [z1, z2] = junk(x,y)
z1 = x + y;
z2 = x - y;
```

This function can be called successfully with one or two output arguments.

```
» a = junk(2,1)
a =
    3
» [a b] = junk(2,1)
a =
    3
b =
    1
```

The reason for checking `nargout` in a function is to prevent useless work. If a result is going to be thrown away anyway, why bother to calculate it in the first place? A programmer can speed up the operation of a program by not bothering with useless calculations.

Quiz 5.1

This quiz provides a quick check to see if you have understood the concepts introduced in Sections 5.1 through 5.3. If you have trouble with the quiz, reread the section, ask your instructor, or discuss the material with a fellow student. The answers to this quiz are found in the back of the book.

1. What are the differences between a script file and a function?
2. How does the `help` command work with user-defined functions?
3. What is the significance of the H1 comment line in a function?
4. What is the pass-by-value scheme? How does it contribute to good program design?
5. How can a MATLAB function be designed to have optional arguments?

For questions 6 and 7, determine whether the function calls are correct or not. If they are in error, specify what is wrong with them.

6. `out = test1(6);`

```
function res = test1(x,y)
res = sqrt(x.^2 + y.^2);
```

7. `out = test2(12);`

```
function res = test2(x,y)
error(nargchk(1,2,nargin));
if nargin == 2
   res = sqrt(x.^2 + y.^2);
else
   res = x;
end
```

5.4 Sharing Data Using Global Memory

We have seen that programs exchange data with the functions they call through argument lists. When a function is called, each actual argument is copied and the copy is used by the function.

In addition to the argument list, MATLAB functions can exchange data with each other and with the base workspace through global memory. **Global memory** is a special type of memory that can be accessed from any workspace. If a variable is declared to be global in a function, then it will be placed in the global memory instead of the local workspace. If the same variable is declared to be global in another function, then that variable will refer to the *same memory location* as the variable in the first function. Each script file or function that declares the global variable will have access to the same data values, *so global memory provides a way to share data between functions.*

A global variable is declared with the **global statement**. The form of a `global` statement is

```
global var1 var2 var3 . . .
```

where `var1`, `var2`, `var3`, and so on are the variables to be placed in global memory. By convention, global variables are declared in all capital letters, but this is not actually a requirement.

✳ **Good Programming Practice**

Declare global variables in all capital letters to make them easy to distinguish from local variables.

Each global variable must be declared to be global before it is used for the first time in a function—it is an error to declare a variable to be global after it has already been created in the local workspace[3]. To avoid this error, it is customary to declare global variables immediately after the initial comments and before the first executable statement in a function.

✳ **Good Programming Practice**

Declare global variables immediately after the initial comments and before the first executable statement of each function that uses them.

---

[3]If a variable is declared `global` after it has already been defined in a function, MATLAB will issue a warning message and then change the local value to match the global value. You should never rely on this capability, though, because future versions of MATLAB will not allow it.

Global variables are especially useful for sharing very large volumes of data among many functions, because the entire data set does not have to be copied each time that a function is called. The downside of using global memory to exchange data among functions is that the functions will work only for that specific data set. A function that exchanges data through input arguments can be reused by simply calling it with different arguments, but a function that exchanges data through global memory must actually be modified to allow it to work with a different data set.

Global variables are also useful for sharing hidden data among a group of related functions while keeping it invisible from the invoking program unit.

✳ Good Programming Practice

You may use global memory to pass large amounts of data among functions within a program.

►

Example 5.4—Random Number Generator

It is impossible to make perfect measurements in the real world. There will always be some *measurement noise* associated with each measurement. This fact is an important consideration in the design of systems to control the operation of such real-world devices as airplanes, refineries, and nuclear reactors. A good engineering design must take these measurement errors into account so that the noise in the measurements will not lead to unstable behavior (no plane crashes, refinery explosions, or meltdowns!).

Most engineering designs are tested by running *simulations* of the operation of the system before it is ever built. These simulations involve creating mathematical models of the behavior of the system, and feeding the models a realistic string of input data. If the models respond correctly to the simulated input data, then we can have reasonable confidence that the real-world system will respond correctly to the real-world input data.

The simulated input data supplied to the models must be corrupted by a simulated measurement noise, which is just a string of random numbers added to the ideal input data. The simulated noise is usually produced by a *random number generator*.

A random number generator is a function that will return a different and apparently random number each time it is called. Since the numbers are in fact generated by a deterministic algorithm, they only appear to be random[4]. However, if the algorithm used to generate them is complex enough, the numbers will be random enough to use in the simulation.

---

[4]For this reason, some people refer to these functions as *pseudorandom number generators*.

One simple random number generator algorithm is described in the following text[5]. It relies on the unpredictability of the modulus function when applied to large numbers. Recall from Chapter 4 that the modulus function mod returns the remainder after the division of two numbers. If the first number comes from an unknown series and the second number is also unknown, then the resulting series of remainders will appear to be random.

For example, consider Equation (5-6). This equation defines the next value in a series by multiplying the previous value by a linear equation and then taking the modulus of the resulting value.

$$n_{i+1} = \mathrm{mod}\,(8121\,n_i + 28411, 134456) \tag{5-6}$$

Assume that n_i is a nonnegative integer. Then because of the modulus function, n_{i+1} will be a number between 0 and 134455 inclusive. Next, n_{i+1} can be fed into the equation to produce a number n_{i+2} that is also between 0 and 134455. This process can be repeated forever to produce a series of numbers in the range [0,134455]. If we didn't know the numbers 8121, 28411, and 134456 in advance, it would be impossible to guess the order in which the values of n would be produced. Furthermore, it turns out that there is an equal (or uniform) probability that any given number will appear in the sequence. Because of these properties, Equation 5-6 can serve as the basis for a simple random number generator with a uniform distribution.

We will now use Equation 5-6 to design a random number generator whose output is a real number in the range [0.0, 1.0)[6].

SOLUTION   We will write a function that generates one random number in the range $0 \leq ran < 1.0$ each time it is called. The random number will be based on the equation

$$ran_i = \frac{n_i}{134456} \tag{5-7}$$

where n_i is a number in the range 0 to 134455 produced by Equation 5-7.

The particular sequence produced by Equations 5-6 and 5-7 will depend on the initial value of n_0 (called the *seed*) of the sequence. We must provide a way for the user to specify n_0 so that the sequence may be varied from run to run.

1. **State the problem.**

    Write a function random0 that will generate and return an array ran containing one or more numbers with a uniform probability distribution

---

[5]This algorithm is adapted from the discussion found in Chapter 7 of *Numerical Recipes: The Art of Scientific Programming*, by Press, Flannery, Teukolsky, and Vetterling, Cambridge University Press, 1986.
[6]The notation [0.0, 1.0) implies that the range of the random numbers is between 0.0 and 1.0, including the number 0.0, but excluding the number 1.0.

in the range $0 \le \text{ran} < 1.0$, based on the sequence specified by Equations 5-6 and 5-7. The function should have one or two input arguments (m and n) specifying the size of the array to return. If there is one argument, the function should generate a square array of size m × m. If there are two arguments, the function should generate an array of size m × n. The initial value of the seed n_0 will be specified by a call to a function called seed.

2. **Define the inputs and outputs.**
   There are two functions in this problem: seed and random0. The input to function seed is an integer to serve as the starting point of the sequence. There is no output from this function. The input to function random0 is one or two integers specifying the size of the array of random numbers to be generated. If only argument m is supplied, the function should generate a square array of size m × m. If both arguments m and n are supplied, the function should generate an array of size n × m. The output from the function is the array of random values in the range [0.0, 1.0).

3. **Describe the algorithm.**
   The pseudocode for function random0 is

```
function ran = random0 (m, n)
Check for valid arguments
Set n <- m if not supplied
Create output array with "zeros" function
for ii = 1:number of rows
   for jj = 1:number of columns
      iseed <- mod (8121 * iseed + 28411, 134456)
      ran(ii,jj) <- iseed / 134456
   end
end
```

where the value of iseed is placed in global memory so that it is saved between calls to the function. The pseudocode for function seed is trivial.

```
function seed (new_seed)
new_seed <- round(new_seed)
iseed <- abs(new_seed)
```

The round function is used in case the user fails to supply an integer, and the absolute value function is used in case the user supplies a negative seed. The user will not have to know in advance that only positive integers are legal seeds.

The variable iseed will be placed in global memory so that it may be accessed by both functions.

4. **Turn the algorithm into MATLAB statements.**
Function random0 is

```
function ran = random0(m,n)
%RANDOM0 Generate uniform random numbers in [0,1)
% Function RANDOM0 generates an array of uniform
% random numbers in the range [0,1). The usage
% is:
%
% random0(m)    -- Generate an m x m array
% random0(m,n)  -- Generate an m x n array

% Define variables:
%   ii       -- Index variable
%   iseed    -- Random number seed (global)
%   jj       -- Index variable
%   m        -- Number of columns
%   msg      -- Error message
%   n        -- Number of rows
%   ran      -- Output array
%
% Record of revisions:
%     Date          Programmer          Description of change
%     ====          ==========          =====================
%   02/04/07      S. J. Chapman        Original code

% Declare global values
global iseed       % Seed for random number generator

% Check for a legal number of input arguments.
msg = nargchk(1,2,nargin);
error(msg);

% If the n argument is missing, set it to m.
if nargin < 2
   n = m;
end

% Initialize the output array
ran = zeros(m,n);

% Now calculate random values
for ii = 1:m
   for jj = 1:n
      iseed = mod(8121*iseed + 28411, 134456 );
      ran(ii,jj) = iseed / 134456;
   end
end
```

5. **Test the resulting MATLAB programs.**
   If the numbers generated by these functions are truly uniformly distributed random numbers in the range $0 \le \text{ran} < 1.0$, then the average of many numbers should be close to 0.5 and the standard deviation of the numbers

   should be close to $\dfrac{1}{\sqrt{12}}$.

   Furthermore, if the range between 0 and 1 is divided into a number of equal-size bins, the number of random values falling in each bin should be about the same. A **histogram** is a plot of the number of values falling in each bin. MATLAB function `hist` will create and plot a histogram from an input data set, so we will use it to verify the distribution of random number generated by `random0`.
   To test the results of these functions, we will perform the following tests:

   1. Call `seed` with `new_seed` set to 1024.
   2. Call `random0(4)` to see that the results appear random.
   3. Call `random0(4)` to verify that the results differ from call to call.
   4. Call `seed` again with `new_seed` set to 1024.
   5. Call `random0(4)` to see that the results are the same as in (2). This verifies that the seed is properly being reset.
   6. Call `random0(2,3)` to verify that both input arguments are being used correctly.
   7. Call `random0(1,100000)` and calculate the average and standard deviation of the resulting data set using MATLAB functions `mean` and `std`. Compare the results to 0.5 and $\dfrac{1}{\sqrt{12}}$.
   8. Create a histogram of the data from (7) to see if approximately equal numbers of values fall in each bin.

   We will perform these tests interactively, checking the results as we go.

```
» seed(1024)
» random0(4)
ans =
    0.0598    1.0000    0.0905    0.2060
    0.2620    0.6432    0.6325    0.8392
    0.6278    0.5463    0.7551    0.4554
    0.3177    0.9105    0.1289    0.6230
» random0(4)
ans =
    0.2266    0.3858    0.5876    0.7880
    0.8415    0.9287    0.9855    0.1314
    0.0982    0.6585    0.0543    0.4256
    0.2387    0.7153    0.2606    0.8922
```

```
» seed(1024)
» random0(4)
ans =
      0.0598        1.0000        0.0905        0.2060
      0.2620        0.6432        0.6325        0.8392
      0.6278        0.5463        0.7551        0.4554
      0.3177        0.9105        0.1289        0.6230
» random0(2,3)
ans =
      0.2266        0.3858        0.5876
      0.7880        0.8415        0.9287
» arr = random0(1,100000);
» mean(arr)
ans =
      0.5001
» std(arr)
ans =
      0.2887
```

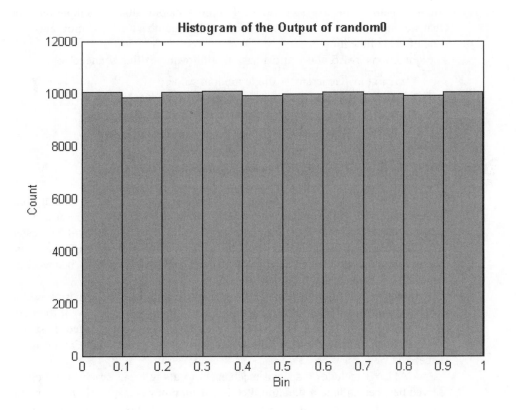

Figure 5.5 Histogram of the output of function random0.

```
» hist(arr,10)
» title('\bfHistogram of the Output of random0');
» xlabel('Bin');
» ylabel('Count');
```

The results of these tests look reasonable, so the function appears to be working. The average of the data set was 0.5001, which is quite close to the theoretical value of 0.5000, and the standard deviation of the data set was 0.2887 to the accuracy displayed. The histogram is shown in Figure 5.5, and the distribution of the random values is roughly even across all of the bins.

◄

MATLAB includes two standard functions that generate random values from different distributions. They are

- rand—Generates random values from a uniform distribution on the range [0, 1).
- randn—Generates random values from a normal distribution.

Both of them are much faster and much more "random" than the simple function that we have created. If you really need random numbers in your programs, use one of these functions.

Functions rand and randn have the following calling sequences:

- rand()—Generates a single random value.
- rand(n)—Generates an $n \times n$ array of random values.
- rand(m,n)—Generates an $m \times n$ array of random values.

5.5 Preserving Data between Calls to a Function

When a function finishes executing, the special workspace created for that function is destroyed, so the contents of all local variables within the function will disappear. The next time that the function is called, a new workspace will be created and all of the local variables will be returned to their default values. This behavior is usually desirable, since it ensures that MATLAB functions behave in a repeatable fashion every time they are called.

However, it is sometimes useful to preserve some local information within a function between calls to the function. For example, we might want to create a counter to count the number of times the function has been called. If such a counter were destroyed every time the function exited, the count would never exceed 1!

MATLAB includes a special mechanism to allow local variables to be preserved between calls to a function. **Persistent memory** is a special type of memory that can be accessed only from within the function but is preserved unchanged between calls to the function.

A persistent variable is declared with the **persistent statement**. The form of a global statement is

```
persistent var1 var2 var3 . . .
```

where *var1*, *var2*, *var3*, and so on are the variables to be placed in persistent memory.

✳ Good Programming Practice

Use persistent memory to preserve the values of local variables within a function between calls to the function.

▶

Example 5.5—Running Averages

It is sometimes desirable to calculate running statistics on a data set on the fly as the values are being entered. The built-in MATLAB functions mean and std could perform this function, but we would have to pass the entire data set to them for recalculation after each new data value is entered. A better result can be achieved by writing a special function that keeps track of the appropriate running sums between calls and needs only the latest value to calculate the current average and standard deviation.

The average or arithmetic mean of a set of numbers is defined as

$$\bar{x} = \frac{1}{N}\sum_{i=1}^{N} x_i \tag{5-8}$$

where x_i is sample i out of N samples. The standard deviation of a set of numbers is defined as

$$s = \sqrt{\frac{N\sum_{i=1}^{N} x_i^2 - \left(\sum_{i=1}^{N} x_i\right)^2}{N(N-1)}} \tag{5-9}$$

Standard deviation is a measure of the amount of scatter on the measurements; the greater the standard deviation, the more scattered the points in the data set are. If we can keep track of the number of values N, the sum of the values $\sum x$, and the sum of the squares of the values $\sum x^2$, then we can calculate the average and standard deviation at any time from Equations 5-8 and 5-9.

Write a function to calculate the running average and standard deviation of a data set as it is being entered.

SOLUTION   This function must be able to accept input values one at a time and keep running sums of N, $\sum x$, and $\sum x^2$, which will be used to calculate the current average and standard deviation. It must store the running sums in global memory so that they are preserved between calls. Finally, there must be a mechanism to reset the running sums.

1. **State the problem.**

   Create a function to calculate the running average and standard deviation of a data set as new values are entered. The function must also include a feature to reset the running sums when desired.

2. **Define the inputs and outputs.**

   There are two types of inputs required by this function:

   1. The character string `'reset'` to reset running sums to zero
   2. The numeric values from the input data set, present one value per function call

   The outputs from this function are the mean and standard deviation of the data supplied to the function so far.

3. **Design the algorithm.**

   This function can be broken down into four major steps:

   ```
   Check for a legal number of arguments
   Check for a 'reset', and reset sums if present
   Otherwise, add current value to running sums
   Calculate and return running average and std dev
       if enough data is available. Return zeros if
       not enough data is available.
   ```

   The detailed pseudocode for these steps is

   ```
   Check for a legal number of arguments
   if x == 'reset'
      n <- 0
      sum_x <- 0
      sum_x2 <- 0
   else
      n <- n + 1
      sum_x <- sum_x + x
      sum_x2 <- sum_x2 + x^2
   end

   % Calculate ave and sd
   if n == 0
      ave <- 0
      std <- 0
   elseif n == 1
      ave <- sum_x
      std <- 0
   else
   ```

```
       ave <- sum_x / n
       std <- sqrt((n*sum_x2 - sum_x^2) / (n*(n-1)))
    end
```

4. **Turn the algorithm into MATLAB statements.**
   The final MATLAB function is

```
function [ave, std] = runstats(x)
%RUNSTATS Generate running ave / std deviation
% Function RUNSTATS generates a running average
% and standard deviation of a data set. The
% values x must be passed to this function one
% at a time. A call to RUNSTATS with the argument
% 'reset' will reset the running sums.

% Define variables:
%   ave        -- Running average
%   msg        -- Error message
%   n          -- Number of data values
%   std        -- Running standard deviation
%   sum_x      -- Running sum of data values
%   sum_x2     -- Running sum of data values squared
%   x          -- Input value
%
% Record of revisions:
%     Date          Programmer        Description of change
%     ====          ==========        =====================
%   02/05/07      S. J. Chapman       Original code

% Declare persistent values
persistent n                % Number of input values
persistent sum_x            % Running sum of values
persistent sum_x2           % Running sum of values squared

% Check for a legal number of input arguments.
msg = nargchk(1,1,nargin);
error(msg);

% If the argument is 'reset', reset the running sums.
if x == 'reset'
   n = 0;
   sum_x = 0;
   sum_x2 = 0;
else
   n = n + 1;
   sum_x = sum_x + x;
   sum_x2 = sum_x2 + x^2;
end
```

```
% Calculate ave and sd
if n == 0
   ave = 0;
   std = 0;
elseif n == 1
   ave = sum_x;
   std = 0;
else
   ave = sum_x / n;
   std = sqrt((n*sum_x2 - sum_x^2) / (n*(n-1)));
end
```

5. **Test the program.**
To test this function, we must create a script file that resets `runstats`, reads input values, calls `runstats`, and displays the running statistics. An appropriate script file is

```
% Script file: test_runstats.m
%
% Purpose:
%   To read in an input data set and calculate the
%   running statistics on the data set as the values
%   are read in. The running stats will be written
%   to the Command Window.
%
% Record of revisions:
%     Date          Programmer            Description of change
%     ====          ==========            =====================
%   02/05/07      S. J. Chapman          Original code
%
% Define variables:
%   array  -- Input data array
%   ave    -- Running average
%   std    -- Running standard deviation
%   ii     -- Index variable
%   nvals  -- Number of input values
%   std    -- Running standard deviation

% First reset running sums
[ave std] = runstats('reset');

% Prompt for the number of values in the data set nvals =
input('Enter number of values in data set: ');

% Get input values
for ii = 1:nvals
```

```
% Prompt for next value
string = ['Enter value ' int2str(ii) ': '];
x = input(string);

% Get running statistics
[ave std] = runstats(x);

% Display running statistics
fprintf('Average = %8.4f; Std dev = %8.4f\n',ave, std);
end
```

To test this function, we will calculate running statistics by hand for a set of five numbers, and compare the hand calculations to the results from the program. If a data set is created with the following five input values

$$3., \quad 2., \quad 3., \quad 4., \quad 2.8 \qquad\qquad (2\text{-}8)$$

then the running statistics calculated by hand would be those shown in the following table.

Value	n	$\sum x$	$\sum x^2$	Average	Std_dev
3.0	1	3.0	9.0	3.00	0.000
2.0	2	5.0	13.0	2.50	0.707
3.0	3	8.0	22.0	2.67	0.577
4.0	4	12.0	38.0	3.00	0.816
2.8	5	14.8	45.84	2.96	0.713

The output of the test program for the same data set is

```
» test_runstats
Enter number of values in data set: 5
Enter value 1:   3
Average =   3.0000; Std dev =    0.0000
Enter value 2:   2
Average =   2.5000; Std dev =    0.7071
Enter value 3:   3
Average =   2.6667; Std dev =    0.5774
Enter value 4:   4
Average =   3.0000; Std dev =    0.8165
Enter value 5:   2.8
Average =   2.9600; Std dev =    0.7127
```

so the results check to the accuracy shown in the hand calculations.   ◄

5.6 Function Functions

"Function functions" is the rather awkward name that MATLAB gives to a function whose input arguments include the names of other functions. The functions that are passed to the "function function" are normally used during that function's execution.

For example, MATLAB contains a function function called `fzero`. This function locates a zero of the function that is passed to it. The statement `fzero('cos',[0 pi])` locates a zero of the function `cos` between 0 and π, and `fzero('exp(x)-2',[0 1])` locates a zero of the function "`exp(x)-2`" between 0 and 1. When these statements are executed, the result is

```
» fzero('cos',[0 pi])
ans =
    1.5708
» fzero('exp(x)-2',[0 1])
ans =
    0.6931
```

The keys to the operation of function functions are two special MATLAB functions—`eval` and `feval`. Function `eval` *evaluates a character string* as though it had been typed in the Command Window, and function `feval` *evaluates a named function* at a specific input value.

Function `eval` evaluates a character string as though it had been typed in the Command Window. This function gives MATLAB functions a chance to construct executable statements during execution. The form of the `eval` function is

```
eval(string)
```

For example, the statement x = eval('sin(pi/4)') produces the result

```
» x = eval('sin(pi/4)')
x =
    0.7071
```

An example in which a character string is constructed and evaluated using the `eval` function is

```
x = 1;
str = ['exp(' num2str(x) ') -1'];
res = eval(str);
```

In this case, `str` contains the character string `'exp(1) -1'`, which `eval` evaluates to get the result 1.7183.

Function `feval` evaluates a *named function* defined by an M-file at a specified input value. The general form of the `feval` function is

```
feval(fun,value)
```

For example, the statement x = feval('sin',pi/4) produces the result

```
» x = feval('sin',pi/4)
```

Table 5.1   Common MATLAB Function Functions

Function Name	Description
`fminbnd`	Minimize a function of one variable.
`fzero`	Find a zero of a function of one variable.
`quad`	Numerically integrate a function.
`ezplot`	Easy to use function plotter.
`fplot`	Plot a function by name.

```
x =
   0.7071
```

Some of the more common MATLAB function functions are listed in Table 5.1. Type `help fun_name` to learn how to use each of these functions.

Example 5.6—Creating a Function Function

Create a function function that will plot any MATLAB function of a single variable between specified starting and ending values.

SOLUTION   This function has two input arguments, the first one containing the name of the function to plot and the second one containing a two-element vector with the range of values to plot.

1. **State the problem.**
   Create a function to plot any MATLAB function of a single variable between two user-specified limits.

2. **Define the inputs and outputs.**
   There are two inputs required by this function:

   1. A character string containing the name of a function
   2. A two-element vector containing the first and last values to plot

   The output from this function is a plot of the function specified in the first input argument.

3. **Design the algorithm.**
   This function can be broken down into four major steps:

   ```
   Check for a legal number of arguments
   Check that the second argument has two elements
   Calculate the value of the function between the
       start and stop points
   Plot and label the function
   ```

   The detailed pseudocode for the evaluation and plotting steps is

   ```
   n_steps <- 100
   step_size <- (xlim(2) - xlim(1)) / n_steps
   ```

```
x <- xlim(1):step_size:xlim(2)
y <- feval(fun,x)
plot(x,y)
title(['\bfPlot of function ' fun '(x)'])
xlabel('\bfx')
ylabel(['\bf' fun '(x)'])
```

4. **Turn the algorithm into MATLAB statements.**
   The final MATLAB function is

```
function quickplot(fun,xlim)
%QUICKPLOT Generate quick plot of a function
% Function QUICKPLOT generates a quick plot
% of a function contained in a external M-file,
% between user-specified x limits.

% Define variables:
%   fun        -- Name of function to plot in a char string
%   msg        -- Error message
%   n_steps    -- Number of steps to plot
%   step_size  -- Step size
%   x          -- X-values to plot
%   y          -- Y-values to plot
%   xlim       -- Plot x limits
%
% Record of revisions:
%    Date          Programmer       Description of change
%    ====          ==========       =====================
%   02/07/07      S. J. Chapman     Original code

% Check for a legal number of input arguments.
msg = nargchk(2,2,nargin);
error(msg);

% Check the second argument to see if it has two
% elements. Note that this double test allows the
% argument to be either a row or a column vector.
if ( size(xlim,1) == 1 & size(xlim,2) == 2 ) | ...
   ( size(xlim,1) == 2 & size(xlim,2) == 1 )

   % Ok--continue processing.
   n_steps = 100;
   step_size = (xlim(2) - xlim(1)) / n_steps;
   x = xlim(1):step_size:xlim(2);
   y = feval(fun,x);
   plot(x,y);
   title(['\bfPlot of function ' fun '(x)']);
   xlabel('\bfx');
   ylabel(['\bf' fun '(x)']);
```

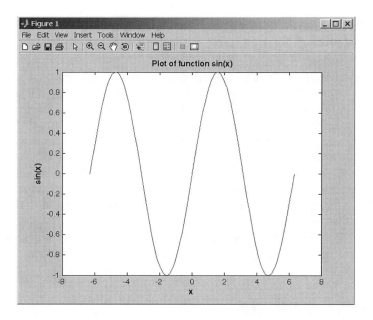

Figure 5.6 Plot of sin x versus x generated by function `quickplot`.

```
else
    % Else wrong number of elements in xlim.
    error('Incorrect number of elements in xlim.');
end
```

5. **Test the program.**

   To test this function, we must call it with correct and incorrect input arguments, verifying that it handles both correct inputs and errors properly. The results are

   ```
   » quickplot('sin')
   ??? Error using ==> quickplot
   Not enough input arguments.
   ```

   ```
   » quickplot('sin',[-2*pi 2*pi],3)
   ??? Error using ==> quickplot
   Too many input arguments.
   ```

   ```
   » quickplot('sin',-2*pi)
   ??? Error using ==> quickplot
   Incorrect number of elements in xlim.
   ```

   ```
   » quickplot('sin',[-2*pi 2*pi])
   ```

   The last call was correct, and it produced the plot shown in Figure 5.6.

5.7 Subfunctions, Private Functions, and Nested Functions

MATLAB includes several special types of functions that behave differently than the ordinary functions we have used so far. Ordinary functions can be called by any other function, as long as they are in the same directory or in any directory on the MATLAB path.

The **scope** of a function is defined as the locations within MATLAB from which the function can be accessed. The scope of an ordinary MATLAB function is the current working directory. If the function lies in a directory on the MATLAB path, then the scope extends to all MATLAB functions in a program, because they all check the path when trying to find a function with a given name.

In contrast, the scope of the other function types that we will discuss in the rest of this chapter is more limited in one way or another.

5.7.1 Subfunctions

It is possible to place more than one function in a single file. If more than one function is present in a file, the top function is a normal or **primary function**, while the ones below it are **subfunctions**. The primary function should have the same name as the file it appears in. Subfunctions look just like ordinary functions, but they are accessible only to the other functions within the same file. In other words, the scope of a subfunction is the other functions within the same file (see Figure 5.7).

Subfunctions are often used to implement "utility" calculations for a main function. For example, the file mystats.m shown subsequently contains a primary function—mystats—and two subfunctions—mean and median. Function mystats is a normal MATLAB function, so it can be called by any other MATLAB function in the same directory. If this file is in a directory included in the MATLAB search path, it can be called by any other MATLAB function, even if the other function is not in the same directory. By contrast, the scope of functions mean and median is restricted to other functions within the same file. Function mystats can call them and they can call each other, but a function outside of the file cannot. They are "utility" functions that perform a part the job of the main function mystats.

```
function [avg, med] = mystats(u)
% MYSTATS Find mean and median with internal functions.
% Function MYSTATS calculates the average and median
% of a data set using subfunctions.

n = length(u);
avg = mean(u,n);
```

```
med = median(u,n);

function a = mean(v,n)
% Subfunction to calculate average.
a = sum(v)/n;

function m = median(v,n)
% Subfunction to calculate median.
w = sort(v);
if rem(n,2) == 1
   m = w((n+1)/2);
else
   m = (w(n/2)+ w(n/2+1))/2;
end
```

5.7.2   Private Functions

Private functions are functions that reside in subdirectories with the special name `private`. They are visible only to other functions in the `private` directory, or to functions in the parent directory. In other words, the scope of these functions is restricted to the private directory and to the parent directory that contains it.

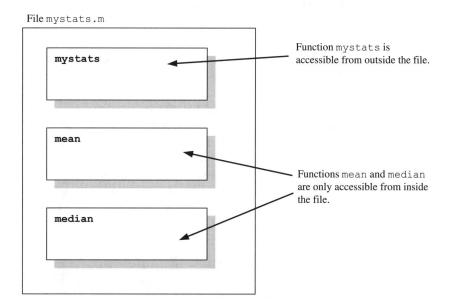

Figure 5.7   The first function in a file is called the primary function. It should have the same name as the file it appears in, and it is accessible from outside the file. The remaining functions in the file are subfunctions; they are only accessible from within the file.

For example, assume the directory `testing` is on the MATLAB search path. A subdirectory of `testing` called `private` can contain functions that only the functions in `testing` can call. Because private functions are invisible outside of the parent directory, they can use the same names as functions in other directories. This is useful if you want to create your own version of a particular function while retaining the original in another directory. Because MATLAB looks for private functions before standard M-file functions, it will find a private function named `test.m` before a nonprivate function named `test.m`.

You can create your own private directories simply by creating a subdirectory called `private` under the directory containing your functions. Do not place these private directories on your search path.

When a function is called from within an M-file, MATLAB first checks the file to see if the function is a subfunction defined in the same file. If not, it checks for a private function with that name. If it is not a private function, MATLAB checks the current directory for the function name. If it is not in the current directory, MATLAB checks the standard search path for the function.

If you have special-purpose MATLAB functions that should be used only by other functions and never be called directly by the user, consider hiding them as subfunctions or private functions. Hiding the functions will prevent their accidental use, and will also prevent conflicts with other public functions of the same name.

5.7.3   Nested Functions

Nested functions are functions that are defined *entirely within the body of another function*, called the **host function**. They are visible only to the host function in which they are embedded and to other nested functions embedded at the same level within the same host function.

A nested function has access to any variables defined with it, *plus any variables defined within the host function* (see Figure 5.8). The only exception occurs if a variable in the nested function has the same name as a variable within the host function. In that case, the variable within the host function is not accessible.

Note that if a file contains one or more nested functions, then *every function in the file* must be terminated with an `end` statement. This is the only time the `end` statement is required at the end of a function—at all other times it is optional.

Programming Pitfalls

If a file contains one or more nested functions, then *every function in the file* must be terminated with an `end` statement. It is an error to omit `end` statements in this case.

Figure 5.8   Nested functions are defined within a host function, and they inherit variables defined within the host function.

The following program illustrates the use of variables in nested functions. It contains a host function `test_nested_1` and a nested function `fun1`. When the program starts, variables a, b, x, and y are initialized as shown in the host function, and their values are displayed. Then the program calls `fun1`. Since `fun1` is nested, it inherits a, b, and x from the host function. Note that it does *not* inherit y, because `fun1` defines a local variable with that name. When the values of the variables are displayed at the end of `fun1`, we see that a has been increased by 1 (due to the assignment statement), and that y is set to 5. When execution returns to the host function, a is still increased by 1, showing that the variable a in the host function and the variable a in the nested function are really the same. On the other hand, y is again 9, because the variable y in the host function is not the same as the variable y in the nested function.

```
function res = test_nested_1

% This is the top level function.
% Define some variables.
a = 1; b = 2; x = 0; y = 9;

% Display variables before call to fun1
fprintf('Before call to fun1:\n');
fprintf('a, b, x, y = %2d %2d %2d %2d\n', a, b, x, y);

% Call nested function fun1
x = fun1(x);

% Display variables after call to fun1
fprintf('\nAfter call to fun1:\n');
fprintf('a, b, x, y = %2d %2d %2d %2d\n', a, b, x, y);
```

```
% Declare a nested function
function res = fun1(y)

% Display variables at start of call to fun1
fprintf('\nAt start of call to fun1:\n');
fprintf('a, b, x, y = %2d %2d %2d %2d\n', a, b, x, y);

  y = y + 5;
  a = a + 1;
  res = y;

% Display variables at end of call to fun1
fprintf('\nAt end of call to fun1:\n');
fprintf('a, b, x, y = %2d %2d %2d %2d\n', a, b, x, y);

end % function fun1

end % function test_nested_1
```

When this program is executed, the results are

```
» test_nested_1
Before call to fun1:
a, b, x, y = 1 2 0 9

At start of call to fun1:
a, b, x, y = 1 2 0 0

At end of call to fun1:
a, b, x, y = 2 2 0 5

After call to fun1:
a, b, x, y = 2 2 5 9
```

Like subfunctions, nested functions can be used to perform special-purpose calculations within a host function.

✳ Good Programming Practice

Use subfunctions, private functions, or nested functions to hide special-purpose calculations that should not be generally accessible to other functions. Hiding the functions will prevent their accidental use and will also prevent conflicts with other public functions of the same name.

5.7.4 Order of Function Evaluation

In a large program, there could possibly be multiple functions (subfunctions, private functions, nested functions, and public functions) with the same name. When a function with a given name is called, how do we know which copy of the function will be executed?

The answer this question is that MATLAB locates functions in a specific order as follows:

1. First, MATLAB checks to see if there is a nested function with the specified name. If so, it is executed.
2. MATLAB checks to see if there is a subfunction with the specified name. If so, it is executed.
3. MATLAB checks for a private function with the specified name. If so, it is executed.
4. MATLAB checks for a function with the specified name in the current directory. If so, it is executed.
5. MATLAB checks for a function with the specified name on the MATLAB path. MATLAB will stop searching and execute the first function with the right name found on the path.

5.8 Summary

In Chapter 5, we presented an introduction to user-defined functions. Functions are special types of M-files that receive data through input arguments and return results through output arguments. Each function has its own independent workspace. Each normal function (one that is not a subfunction) should appear in a separate file with the same name as the function, *including capitalization.*

Functions are called by naming them in the Command Window or another M-file. The names used should match the function name exactly, including capitalization. Arguments are passed to functions using a pass-by-value scheme, meaning that MATLAB copies each argument and passes the copy to the function. This copying is important, because the function can freely modify its input arguments without affecting the actual arguments in the calling program.

MATLAB functions can support varying numbers of input and output arguments. Function `nargin` reports the number of actual input arguments used in a function call, and function `nargout` reports the number of actual output arguments used in a function call.

Data can also be shared between MATLAB functions by placing the data in global memory. Global variables are declared using the `global` statement.

Global variables may be shared by all functions that declare them. By convention, global variable names are written in all capital letters.

Internal data within a function can be preserved between calls to that function by placing the data in persistent memory. Persistent variables are declared using the `persistent` statement.

Function functions are MATLAB functions whose input arguments include the names of other functions. The functions whose names are passed to the function function are normally used during that function's execution. Examples are some root-solving and plotting functions.

Subfunctions are additional functions placed within a single file. Subfunctions are accessible only from other functions within the same file. Private functions are functions placed in a special subdirectory called `private`. They are accessible only to functions in the parent directory. Nested functions are functions completely defined within the body of another function (called the host function). Nested functions have access to the variables in the host function as well as to their own local variables. Subfunctions, private functions, and nested functions can be used to restrict access to MATLAB functions.

5.8.1 Summary of Good Programming Practice

The following guidelines should be adhered to when working with MATLAB functions.

1. Break large program tasks into smaller, more understandable functions whenever possible.
2. Declare global variables in all capital letters to make them easy to distinguish from local variables.
3. Declare global variables immediately after the initial comments and before the first executable statement each function that uses them.
4. You may use global memory to pass large amounts of data among functions within a program.
5. Use persistent memory to preserve the values of local variables within a function between calls to the function.
6. Use subfunctions, private functions, or nested functions to hide special-purpose calculations that should not be generally accessible to other functions. Hiding the functions will prevent their accidental use, and will also prevent conflicts with other public functions of the same name.

5.8.2 MATLAB Summary

The following summary lists all of the MATLAB commands and functions described in this chapter, along with a brief description of each one.

Commands and Functions

error	Displays error message and aborts the function producing the error. This function is used if the argument errors are fatal.
eval	Evaluates a character string as though it had been typed in the Command Window.
ezplot	Easy-to-use function plotter.
feval	Calculates the value of a function $f(x)$ defined by an M-file at a specific x.
fmin	Minimize a function of one variable.
fplot	Plot a function by name.
fzero	Find a zero of a function of one variable.
global	Declares global variables.
hist	Calculate and plot a histogram of a data set.
inputname	Returns the actual name of the variable that corresponds to a particular argument number.
nargchk	Returns a standard error message if a function is called with too few or too many arguments.
nargin	Returns the number of actual input arguments that were used to call the function.
nargout	Returns the number of actual output arguments that were used to call the function.
persistent	Declares persistent variables.
quad	Numerically integrate a function.
rand	Generates random values from a uniform distribution.
randn	Generates random values from a normal distribution.
return	Stop executing a function and return to caller.
warning	Displays a warning message and continues function execution. This function is used if the argument errors are not fatal and execution can continue.

5.9   Exercises

5.1   What is the difference between a script file and a function?

5.2   When a function is called, how is data passed from the caller to the function, and how are the results of the function returned to the caller?

5.3   What are the advantages and disadvantages of the pass-by-value scheme used in MATLAB?

5.4   Modify the selection sort function developed in this chapter so that it accepts a second optional argument, which may be either 'up' or 'down'. If the argument is 'up', sort the data in ascending order. If the argument is 'down', sort the data in descending order. If the argument is missing, the default case is to sort the data in ascending order. (Be sure to handle the case of invalid arguments, and be sure to include the proper help information in your function.)

5.5 Modify function random0 so that it can accept 0, 1, or 2 calling arguments. If it has no calling arguments, it should return a single random value. If it has 1 or 2 calling arguments, it should behave as it currently does.

5.6 As function random0 is currently written, it will fail if function seed is not called first. Modify function random0 so that it will function properly with some default seed even if function seed is never called.

5.7 Write a function that uses function random0 to generate a random value in the range $[-1.0, 1.0]$. Make random0 a subfunction of your new function.

5.8 Write a function that uses function random0 to generate a random value in the range [low, high), where low and high are passed as calling arguments. Make random0 a private function called by your new function.

5.9 **Dice Simulation** It is often useful to be able to simulate the throw of a fair die. Write a MATLAB function dice that simulates the throw of a fair die by returning some random integer between 1 and 6 every time that it is called. (*Hint:* Call random0 to generate a random number. Divide the possible values out of random0 into six equal intervals and return the number of the interval that a given random value falls into.)

5.10 **Road Traffic Density** Function random0 produces a number with a *uniform* probability distribution in the range [0.0, 1.0). This function is suitable for simulating random events if each outcome has an equal probability of occurring. However, in many events the probability of occurrence is *not* equal for every event, and a uniform probability distribution is not suitable for simulating such events.

For example, when traffic engineers studied the number of cars passing a given location in a time interval of length t, they discovered that the probability of k cars passing during the interval is given by the equation

$$P(k, t) = e^{-\lambda t}\frac{(\lambda t)^k}{k!} \quad \text{for } t \geq 0, \ \lambda > 0, \quad \text{and } k = 0, 1, 2, \ldots \quad (5\text{-}10)$$

This probability distribution is known as the *Poisson distribution*; it occurs in many applications in science and engineering. For example, the number of calls k to a telephone switchboard in time interval t, the number of bacteria k in a specified volume t of liquid, and the number of failures k of a complicated system in time interval t all have Poisson distributions.

Write a function to evaluate the Poisson distribution for any k, t, and λ. Test your function by calculating the probability of 0, 1, 2, . . ., 5 cars passing a particular point on a highway in 1 minute, given that λ is 1.6 per minute for that highway. Plot the Poisson distribution for $t = 1$ and $\lambda = 1.6$.

5.11 **Recursion and the Factorial Function** A function is said to be *recursive* if the function calls itself. MATLAB functions are designed to allow recursive operation. To test this feature, write a MATLAB function to evaluate the factorial function, which is defined as follows:

$$N! = \begin{cases} N(N-1)! & N \geq 1 \\ 1 & N = 0 \end{cases} \quad (5\text{-}11)$$

where N is a positive integer. The function should check to make sure that there is a single argument N, and that N is a nonnegative integer. If it is not, generate an `error` using the error function. If the input argument is a nonnegative integer, the function should evaluate $N!$ using Equation (5-11).

5.12 **Binomial Coefficients** The probability that exactly k successes will occur out of a series of n independent yes/no trials, each of which has a probability of success p, is given by Equation (5-12):

$$P(k; n, p) = \binom{n}{k} p^k (1-p)^{n-k} \tag{5-12}$$

where the expression $\binom{n}{k}$ is called the *binomial coefficient*. The value of the binomial coefficient is given by the Equation (5-13):

$$\binom{n}{k} = \frac{n!}{k!(n-k)!} \tag{5-13}$$

For example, suppose that we performed an experiment in which we tossed a fair coin ($p_{\text{heads}} = 0.5$) ten times. In that case, the probability that we would get exactly four heads is

$$P(4;10,0.5) = \frac{10!}{4!(10-4)!} 0.5^4 (1-0.5)^{10-4} = 0.2051 \tag{5-14}$$

and the probability that we would get exactly five heads is

$$P(5;10,0.5) = \frac{10!}{5!(10-5)!} 0.5^5 (1-0.5)^{10-5} = 0.2461 \tag{5-15}$$

Write a function that will determine the probability that exactly k successes will occur out of n trials for a specified probability of success p. The function should include two private functions, one to calculate the values of the binomial coefficient and one to calculate the value of the factorial function $n!$. Test your function using the two example values calculated in Equations (5-14) and (5-15).

5.12 **The Binomial Distribution** *The binomial distribution* is the probability distribution of the number of successes in a sequence of n independent yes/no trials, each of which has a probability of success p. In other words, it is a list of every possible output from the n trials: the probability of no successes, the probability of exactly one success, the probability of exactly two successes, and so forth up to the probability of exactly n successes. This distribution can be calculated for any number of trials n with probability of success p by calling Equation (5-12) repeatedly with values $k = 0, 1, 2, \ldots, n$.

Write a program to calculate and plot the binomial distribution for any given n and p. Use the function created in Exercise 5.11 to create the values in distribution.

5.13 **Colored Dice** Suppose that a game contained a set of six-sided dice with two opposite faces colored red, two opposite faces colored green, and two opposite faces colored yellow. If one of these dice is tossed 20 times, what is the probability that there will be 5 or less green results? What is the probability that there will be 10 or less green results?

5.14 **Defending Ships against Missile Attacks** Antiship missiles (ASMs) are sometimes used to attack naval ships, and the ships being attacked use surface-to-air missiles (SAMs) to try to kill the attacking missiles before they hit the ship. Answer the following questions about this situation using the function developed in Exercise 5.11.

(a) If the ship uses three SAMs to attack an incoming ASM and the probability of success of each attack is 0.3, what is the probability that the ship will destroy the ASM before it is hit?

(b) If the ship uses three SAMs to attack an incoming ASM and the probability of success of each attack is 0.5, what is the probability that the ship will destroy the ASM before it is hit?

(c) If the ship uses three SAMs to attack an incoming ASM and the probability of success of each attack is 0.7, what is the probability that the ship will destroy the ASM before it is hit?

(d) If the ship uses three SAMs to attack an incoming ASM and the probability of success of each attack is 0.9, what is the probability that the ship will destroy the ASM before it is hit?

5.15 **Defending Ships against Missile Attacks** Suppose a designer wanted to ensure that there is a 90% probability that the ship will survive an ASM attack without being hit. How many SAMs should the ship fire to destroy the incoming ASM if the individual probability of success of for a SAM is (a) 30%? (b) 50%? (c) 70%?

5.16 Write three MATLAB functions to calculate the hyperbolic sine, cosine, and tangent functions:

$$\sinh(x) = \frac{e^x - e^{-x}}{2} \qquad \cosh(x) = \frac{e^x + e^{-x}}{2} \qquad \tanh(x) = \frac{e^x - e^{-x}}{e^x + e^{-x}}$$

Use your functions to plot the shapes of the hyperbolic sine, cosine, and tangent functions.

5.17 Write a single MATLAB function `hyperbolic` to calculate the hyperbolic sine, cosine, and tangent functions as defined in the previous problem. The function should have two arguments. The first argument will be a string containing the function names `'sinh'`, `'cosh'`, or `'tanh'`, and the second argument will be the value of x at which to evaluate the function. The file should also contain three subfunctions `sinh1`, `cosh1`, and `tanh1` to perform the actual calculations, and the primary function should call the proper subfunction depending on the value in the string. [*Note:* Be sure to handle the case of an incorrect number of arguments, and also the case of an invalid string. In either case, the function should generate an error.]

5.18 **Cross Product** Write a function to calculate the cross product of two vectors $\mathbf{V}_1$ and $\mathbf{V}_2$:

$$\mathbf{V}_1 \times \mathbf{V}_2 = (V_{y1}V_{z2} - V_{y2}V_{z1})\,\mathbf{i} + (V_{z1}V_{x2} - V_{z2}V_{x1})\,\mathbf{j} \\ + (V_{x1}V_{y2} - V_{x2}V_{y1})\,\mathbf{k}$$

where $\mathbf{V}_1 = V_{x1}\,\mathbf{i} + V_{y1}\,\mathbf{j} + V_{z1}\,\mathbf{k}$ and $\mathbf{V}_2 = V_{x2}\,\mathbf{i} + V_{y2}\,\mathbf{j} + V_{z2}\,\mathbf{k}$. Note that this function will return a real array as its result. Use the function to calculate the cross product of the two vectors $\mathbf{V}_1 = [-2, 4, 0.5]$ and $\mathbf{V}_2 = [0.5, 3, 2]$.

5.19 **Sort with Carry** It is often useful to sort an array `arr1` into ascending order, while simultaneously carrying along a second array `arr2`. In such a sort, each time an element of array `arr1` is exchanged with another element of `arr1`, the corresponding elements of array `arr2` are also swapped. When the sort is over, the elements of array `arr1` are in ascending order, while the elements of array `arr2` that were associated with particular elements of array `arr1` are still associated with them. For example, suppose we have the following two arrays:

Element	arr1	arr2
1.	6.	1.
2.	1.	0.
3.	2.	10.

After sorting array `arr1` while carrying along array `arr2`, the contents of the two arrays will be:

Element	arr1	arr2
1.	1.	0.
2.	2.	10.
3.	6.	1.

Write a function to sort one real array into ascending order while carrying along a second one. Test the function with the following two 9-element arrays:

```
a = [1, 11, -6, 17, -23, 0, 5, 1, -1];
b = [31, 101, 36, -17, 0, 10, -8, -1, -1];
```

5.20 Use the Help Browser to look up information about the standard MATLAB function `sortrows`, and compare the performance of `sortrows` with the sort-with-carry function created in the previous exercise. To do this, create two copies of a 1000 × 2 element array containing random values, and sort column 1 of each array while carrying along column 2 using both functions. Determine the execution times of each sort function using `tic` and `toc`. How does the speed of your function compare with the speed of the standard function `sortrows`?

5.21 Figure 5.9 shows two ships steaming on the ocean. Ship 1 is at position (x_1, y_1) and steaming on heading θ_1. Ship 2 is at position (x_2, y_2) and steaming on heading θ_2. Suppose that Ship 1 makes radar contact with an object at range r_1 and bearing ϕ_1. Write a MATLAB function that will calculate the range r_2 and bearing ϕ_2 at which Ship 2 should see the object.

5.22 **Minima and Maxima of a Function**   Write a function that attempts to locate the maximum and minimum values of an arbitrary function $f(x)$ over a certain range. The function being evaluated should be passed to the function as a calling argument. The function should have the following input arguments:

> `first_value`—The first value of x to search
> `last_value`—The last value of x to search
> `num_steps`—The number of steps to include in the search
> `func`—The name of the function to search

The function should have the following output arguments:

> `xmin`—The value of x at which the minimum was found
> `min_value`—The minimum value of $f(x)$ found
> `xmax`—The value of x at which the maximum was found
> `max_value`—The maximum value $f(x)$ found

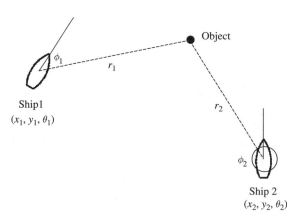

Figure 5.9   Two ships at positions (x_1, y_1) and (x_2, y_2) respectively. Ship 1 is traveling at heading θ_1, and Ship 2 is traveling at heading θ_2.

Be sure to check that there are a valid number of input arguments, and that the MATLAB `help` and `lookfor` commands are properly supported.

5.23 Write a test program for the function generated in the previous exercise. The test program should pass to the function function the user-defined function $f(x) = x^3 - 5x^2 + 5x + 2$, and search for the minimum and maximum in 200 steps over the range $-1 \leq x \leq 3$. It should print out the resulting minimum and maximum values.

5.24 **Derivative of a Function**   The *derivative* of a continuous function $f(x)$ is defined by the equation

$$\frac{d}{dx}f(x) = \lim_{\Delta x \to 0} \frac{f(x + \Delta x) - f(x)}{\Delta x} \tag{5-16}$$

In a sampled function, this definition becomes

$$f'(x_i) = \frac{f(x_{i+1}) - f(x_i)}{\Delta x} \tag{5-17}$$

where $\Delta x = x_{i+1} - x_i$. Assume that a vector `vect` contains `nsamp` samples of a function taken at a spacing of `dx` per sample. Write a function that will calculate the derivative of this vector from Equation (5-17). The function should check to make sure that `dx` is greater than zero to prevent divide-by-zero errors in the function.

To check your function, you should generate a data set whose derivative is known and compare the result of the function with the known correct answer. A good choice for a test function is $\sin x$. From elementary calculus, we know that $\dfrac{d}{dx} (\sin x) = \cos x$. Generate an input vector containing 100 values of the function $\sin x$ starting at $x = 0$ and using a step size Δx of 0.05. Take the derivative of the vector with your function, and then compare the resulting answers to the known correct answer. How close did your function come to calculating the correct value for the derivative?

5.25 **Derivative in the Presence of Noise**   We will now explore the effects of input noise on the quality of a numerical derivative. First, generate an input vector containing 100 values of the function $\sin x$ starting at $x = 0$ and using a step size Δx of 0.05, just as you did in the previous problem. Next, use function `random0` to generate a small amount of random noise with a maximum amplitude of ± 0.02, and add that random noise to the samples in your input vector. Note that the peak amplitude of the noise is only 2% of the peak amplitude of your signal, since the maximum value of $\sin x$ is 1. Now take the derivative of the function using the derivative function that you developed in the last problem. How close to the theoretical value of the derivative did you come?

5.26 **Linear Least-Squares Fit**   Develop a function that will calculate slope m and intercept b of the least-squares line that best fits an input data set. The input data points (x, y) will be passed to the function in

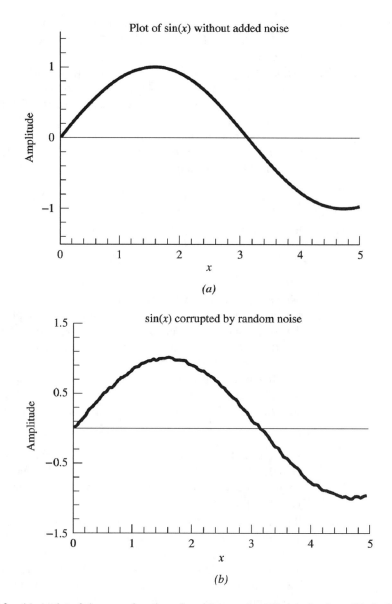

Figure 5.10 *(a)* A plot of sin *x* as a function of *x* with no noise added to the data. *(b)* A plot of sin *x* as a function of *x* with a 2% peak amplitude uniform random noise added to the data.

two input arrays, x and y. (The equations describing the slope and intercept of the least-squares line given in Example 4.6 in the previous chapter.) Test your function using a test program and the following 20-point input data set:

Sample Data to Test Least Squares Fit Routine

No.	x	y	No.	x	y
1	−4.91	−8.18	11	−0.94	0.21
2	−3.84	−7.49	12	0.59	1.73
3	−2.41	−7.11	13	0.69	3.96
4	−2.62	−6.15	14	3.04	4.26
5	−3.78	−5.62	15	1.01	5.75
6	−0.52	−3.30	16	3.60	6.67
7	−1.83	−2.05	17	4.53	7.70
8	−2.01	−2.83	18	5.13	7.31
9	0.28	−1.16	19	4.43	9.05
10	1.08	0.52	20	4.12	10.95

5.27 **Correlation Coefficient of Least Squares Fit** Develop a function that will calculate both the slope m and intercept b of the least-squares line that best fits an input data set, and also the correlation coefficient of the fit. The input data points (x, y) will be passed to the function in two input arrays, x and y. The equations describing the slope and intercept of the least-squares line are given in Example 4.6, and the equation for the correlation coefficient is

$$r = \frac{n\left(\sum xy\right) - \left(\sum x\right)\left(\sum y\right)}{\sqrt{\left[\left(n\sum x^2\right) - \left(\sum x\right)^2\right]\left[\left(n\sum y^2\right) - \left(\sum y\right)^2\right]}} \quad (5\text{-}18)$$

where

$\sum x$ is the sum of the x values
$\sum y$ is the sum of the y values
$\sum x^2$ is the sum of the squares of the x values
$\sum y^2$ is the sum of the squares of the y values
$\sum xy$ is the sum of the products of the corresponding x and y values
n is the number of points included in the fit

Test your function using a test driver program and the 20-point input data set given in the previous problem.

5.28 **The Birthday Problem** The Birthday Problem is stated as follows: if there is a group of n people in a room, what is the probability that two or more of them have the same birthday? It is possible to determine the answer to this question by simulation. Write a function that calculates the probability that two or more of n people will have the same birthday, where n is a calling argument. (*Hint*: To do this, the function should create an array of size n and generate n birthdays in the range 1 to 365 randomly. It should then check to see if any of the *n* birthdays are identical. The function should perform this experiment at least 5000 times and calculate the fraction of those times in which two or more people had the same birthday.) Write a test program that calculates and prints out the probability that two or more of *n* people will have the same birthday for *n* = 2, 3, . . . , 40.

5.29 Use function random0 to generate a set of three arrays of random numbers. The three arrays should be 100, 1000, and 2000 elements long. Then, use functions tic and toc to determine the time that it takes function ssort to sort each array. How does the elapsed time to sort increase as a function of the number of elements being sorted? (*Hint*: On a fast computer, you will need to sort each array many times and calculate the average sorting time in order to overcome the quantization error of the system clock.)

5.30 **Gaussian (Normal) Distribution** Function random0 returns a uniformly distributed random variable in the range [0, 1), which means that there is an equal probability of any given number in the range occurring on a given call to the function. Another type of random distribution is the Gaussian distribution, in which the random value takes on the classic bell-shaped curve shown in Figure 5.11. A Gaussian distribution with an average of 0.0 and a standard deviation of 1.0 is called a *standardized normal distribution*, and the probability of any given value occurring in the standardized normal distribution is given by the equation

$$p(x) = \frac{1}{\sqrt{2\pi}} e^{-x^2/2} \tag{5-19}$$

It is possible to generate a random variable with a standardized normal distribution starting from a random variable with a uniform distribution in the range [−1, 1) as follows:

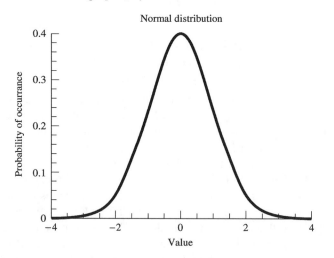

Figure 5.11 A Normal probability distribution.

1. Select two uniform random variables x_1 and x_2 from the range $[-1, 1)$ such that $x_1^2 + x_2^2 < 1$. To do this, generate two uniform random variables in the range $[-1, 1)$, and see if the sum of their squares happens to be less than 1. If so, use them. If not, try again.
2. Then each of the values y_1 and y_2 in the equations below will be a normally-distributed random variable.

$$y_1 = \sqrt{\frac{-2\ln r}{r}}\, x_1 \qquad\qquad (5\text{-}20)$$

$$y_2 = \sqrt{\frac{-2\ln r}{r}}\, x_2 \qquad\qquad (5\text{-}21)$$

where

$$r = x_1^2 + x_2^2 \qquad\qquad (5\text{-}22)$$

Write a function that returns a normally-distributed random value each time that it is called. Test your function by getting 1000 random values, calculating the standard deviation, and plotting a histogram of the distribution. How close to 1.0 was the standard deviation?

5.31 Gravitational Force The gravitational force F between two bodies of masses m_1 and m_2 is given by the equation

$$F = \frac{Gm_1m_2}{r^2} \qquad\qquad (5\text{-}23)$$

where G is the gravitation constant (6.672×10^{-11} N m^2 / kg^2), m_1 and m_2 are the masses of the bodies in kilograms, and r is the distance between the two bodies. Write a function to calculate the gravitational force between two bodies given their masses and the distance between them. Test you function by determining the force on an 800 kg satellite in orbit 38,000 km above the Earth. (The mass of the Earth is 5.98×10^{24} kg.)

5.32 Rayleigh Distribution The Rayleigh distribution is another random number distribution that appears in many practical problems. A Rayleigh-distributed random value can be created by taking the square root of the sum of the squares of two normally-distributed random values. In other words, to generate a Rayleigh-distributed random value r, get two normally distributed random values (n_1 and n_2), and perform the following calculation:

$$r = \sqrt{n_1^2 + n_2^2} \qquad\qquad (5\text{-}24)$$

(*a*) Create a function `rayleigh(n,m)` that returns an n $\times$ m array of Rayleigh-distributed random numbers. If only one argument is supplied [`rayleigh(n)`], the function should return an n $\times$ n array of Rayleigh-distributed random numbers. Be sure to design your function with input argument checking, and with proper documentation for the MATLAB help system.

(b) Test your function by creating an array of 20,000 Rayleigh-distributed random values and plotting a histogram of the distribution. What does the distribution look like?

(c) Determine the mean and standard deviation of the Rayleigh distribution.

5.33 **Constant False Alarm Rate (CFAR)** A simplified radar receiver chain is shown in Figure 5.12a. When a signal is received in this receiver, it contains both the desired information (returns from targets) and thermal noise. After the detection step in the receiver, we would like to be able to pick out received target returns from the thermal noise background. We can do this by setting a threshold level and then declaring that we see a target whenever the signal crosses that threshold. Unfortunately, it is occasionally possible for the receiver noise to cross the detection threshold even if no target is present. If that happens, we will declare the noise spike to be a target, creating a *false alarm*. The detection threshold needs to be set as low as possible so that we can detect weak targets, but it must not be set too low, or we get many false alarms.

After video detection, the thermal noise in the receiver has a Rayleigh distribution. Figure 5.12b shows 100 samples of a Rayleigh-distributed noise with a mean amplitude of 10 volts. Note that there would be one false alarm even if the detection threshold were as high as 26! The probability distribution of these noise samples is shown in Figure 5.12c.

Detection thresholds are usually calculated as a multiple of the mean noise level, so that if the noise level changes, the detection threshold will change with it to keep false alarms under control. This is known as *constant false alarm rate* (CFAR) detection. A detection threshold is typically quoted in decibels. The relationship between the threshold in dB and the threshold in volts is

$$\text{Threshold (volts)} = \text{Mean Noise Level (volts)} \times 10^{\frac{\text{dB}}{20}} \qquad (5\text{-}25)$$

or

$$\text{dB} = 20 \log_{10}\left(\frac{\text{Threshold (volts)}}{\text{Mean Noise Level (volts)}}\right) \qquad (5\text{-}26)$$

The false alarm rate for a given detection threshold is calculated as

$$P_{fa} = \frac{\text{Number of False Alarms}}{\text{Total Number of Samples}} \qquad (5\text{-}27)$$

Write a program that generates 1,000,000 random noise samples with a mean amplitude of 10 volts and a Rayleigh noise distribution. Determine the false alarm rates when the detection threshold is set to 5, 6, 7, 8, 9, 10, 11, 12, and 13 dB above the mean noise level. At what level should the threshold be set to achieve a false alarm rate of 10^{-4}?

(a)

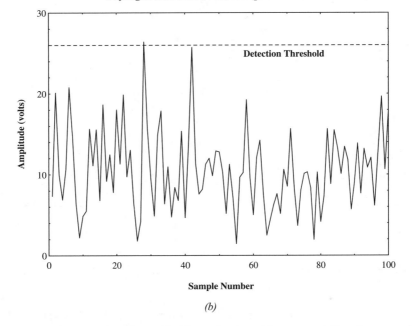

(b)

Figure 5.12 *(a)* A typical radar receiver. *(b)* Thermal noise with a mean of 10 volts output from the detector. The noise sometimes crosses the detection threshold. *(c)* Probability distribution of the noise out of the detector.

5.34 **Probability of Detection (P_d) versus Probability of False Alarm (P_{fa})** The signal strength returned by a radar target usually fluctuates over time. The target will be detected if its signal strength exceeds the detection threshold for any given look. The probability that the target will be detected can be calculated as:

$$P_d = \frac{\text{Number of Target Detections}}{\text{Total Number of Looks}} \qquad (5\text{-}28)$$

Suppose that a specific radar looks repeatedly in a given direction. On each look, the range between 10 km and 20 km is divided into 100 independent

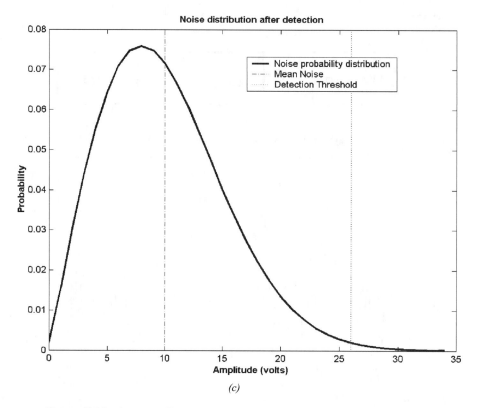

(c)

Figure 5.12 (*continued*)

range samples (called *range gates*). One of these range gates contains a target whose amplitude has a normal distribution with a mean amplitude of 7 volts and a standard deviation of 1 volt. All 100 of the range gates contain system noise with a mean amplitude of 2 volts and a Rayleigh distribution. Determine both the probability of target detection P_d and the probability of a false alarm P_{fa} on any given look for detection thresholds of 8.0, 8.5, 9.0, 9.5, 10.0, 10.5, 11.0, 11.5, and 12.0 dB. What threshold would you use for detection in this radar? (*Hint*: Perform the experiment many times for each threshold and average the results to determine valid probabilities.)

C H A P T E R 6

Additional Data Types and Plot Types

In earlier chapters, we were introduced to three fundamental MATLAB data types: `double`, `logical`, and `char`. In this chapter, we will learn additional details about these data types, and then we will study some additional MATLAB data types.

First, we will learn how to create, manipulate, and plot complex values in the `double` data type. Then, we will learn more about using the `char` data type and how to extend MATLAB arrays of any type to more than two dimensions.

Next, we will learn about some additional data types. The MATLAB data types are shown in Figure 6.1. We will learn about the `single` and integer data types in this chapter and discuss the remaining ones on the figure later in this book.

The chapter concludes with a discussion of additional types of plots available in MATLAB.

6.1 Complex Data

Complex numbers are numbers with both a real and an imaginary component. Complex numbers occur in many problems in science and engineering. For example, complex numbers are used in electrical engineering to represent alternating current voltages, currents, and impedances. The differential equations that describe the behavior of most electrical and mechanical systems also give rise to complex numbers. Because they are so ubiquitous, it is impossible to work as an engineer without a good understanding of the use and manipulation of complex numbers.

A complex number has the general form

$$c = a + bi \tag{6-1}$$

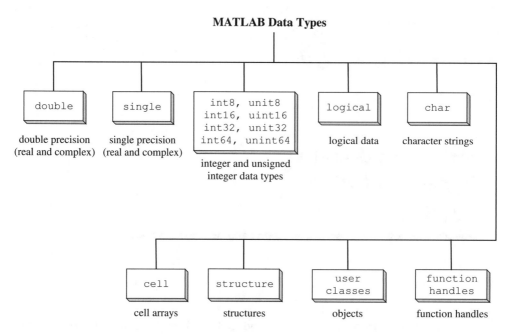

Figure 6.1   MATLAB data types.

where c is a complex number, a and b are both real numbers, and i is $\sqrt{-1}$. The number a is called the *real part* and b is called the *imaginary part* of the complex number c. Since a complex number has two components, it can be plotted as a point on a plane (see Figure 6.2). The horizontal axis of the plane is the real axis, and the vertical axis of the plane is the imaginary axis, so that any complex number $a + bi$ can be represented as a single point a units along the real axis and b units along the imaginary axis. A complex number represented this way is said to be in *rectangular coordinates*, since the real and imaginary axes define the sides of a rectangle.

A complex number can also be represented as a vector of length z and angle θ pointing from the origin of the plane to the point P (see Figure 6.3). A complex number represented this way is said to be in *polar coordinates*.

$$c = a + bi = z\angle\theta$$

The relationships among the rectangular and polar coordinate terms a, b, z, and θ are:

$$a = z\cos\theta \tag{6-2}$$

$$b = z\sin\theta \tag{6-3}$$

$$z = \sqrt{a^2 + b^2} \tag{6-4}$$

$$\theta = \tan^{-1}\frac{b}{a} \tag{6-5}$$

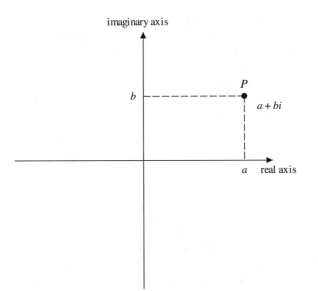

Figure 6.2   Representing a complex number in Rectangular Coordinates.

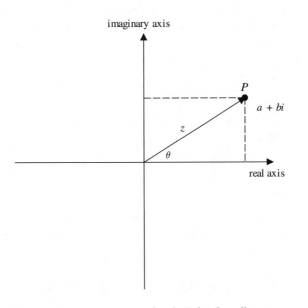

Figure 6.3   Representing a complex number in Polar Coordinates.

MATLAB uses rectangular coordinates to represent complex numbers. Each complex number consists of a pair of real numbers (a, b). The first number (a) is the real part of the complex number, and the second number (b) is the imaginary part of the complex number.

If complex numbers c_1 and c_2 are defined as $c_1 = a_1 + b_1 i$ and $c_2 = a_2 + b_2 i$, then the addition, subtraction, multiplication, and division of c_1 and c_2 are defined as follows.

$$c_1 + c_2 = (a_1 + a_2) + (b_1 + b_2)i \tag{6-6}$$

$$c_1 - c_2 = (a_1 - a_2) + (b_1 - b_2)i \tag{6-7}$$

$$c_1 \times c_2 = (a_1 a_2 - b_1 b_2) + (a_1 b_2 + b_1 a_2)i \tag{6-8}$$

$$\frac{c_1}{c_2} = \frac{a_1 a_2 + b_1 b_2}{a_2^2 + b_2^2} + \frac{b_1 a_2 - a_1 b_2}{a_2^2 + b_2^2} i \tag{6-9}$$

When two complex numbers appear in a binary operation, MATLAB performs the required additions, subtractions, multiplications, or divisions between the two complex numbers, using versions of the previously stated formulas.

6.1.1   Complex Variables

A complex variable is created automatically when a complex value is assigned to a variable name. This easiest way to create a complex value is to use the intrinsic values i or j, both of which are predefined to be $\sqrt{-1}$. For example, the following statement stores the complex value $4 + i3$ into variable c1.

```
» c1 = 4 + i*3
c1 =
    4.0000 + 3.0000i
```

Alternatively, the imaginary part can be specified by simply appending an i or j to the end of a number.

```
» c1 = 4 + 3i
c1 =
    4.0000 + 3.0000i
```

The function isreal can be used to determine whether a given array is real or complex. If any element of an array has an imaginary component, then the array is complex, and isreal(array) returns a 0.

6.1.2   Using Complex Numbers with Relational Operators

It is possible to compare two complex numbers with the == relational operator to see if they are equal to each other, and to compare them with the ~= operator to see if they are not equal to each other. Both of these operators produce the expected results. For example, if $c_1 = 4 + i3$ and $c_2 = 4 - i3$, then the relational operation $c_1 == c_2$ produces a 0 and the relational operation $c_1 \sim= c_2$ produces a 1.

Color Figures
These pages include color versions of selected figures from the text. These figures are included in color because the color illustrates some essential point about each figure that could not be displayed in greyscale.

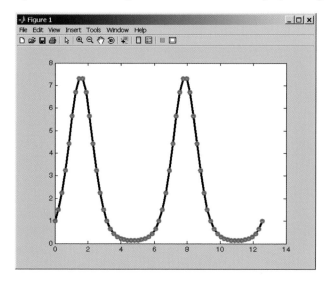

Figure 3.8   A plot illustrating the use of the `LineWidth` and `Marker` properties.

Figure 3.17   An Edit/Debug window with a MATLAB program loaded.

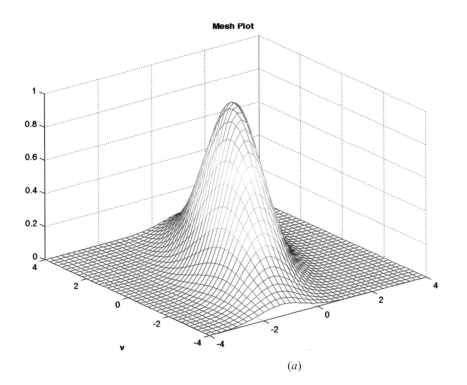

(a)

Figure 6.12 (a) A mesh plot of the function $z(x, y) = e^{-0.5[x^2 + 0.5(x-y)^2]}$. (b) A surface plot of the same function.

(b)

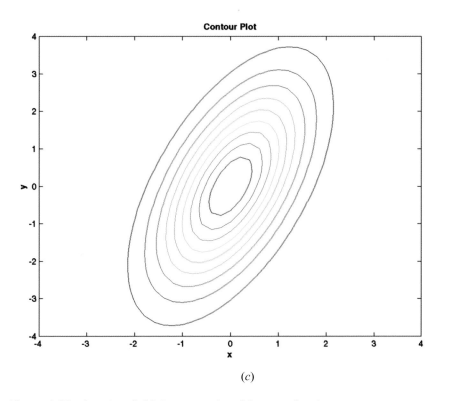

(*c*)

Figure 6.12 (*continued*) (*c*) A contour plot of the same function.

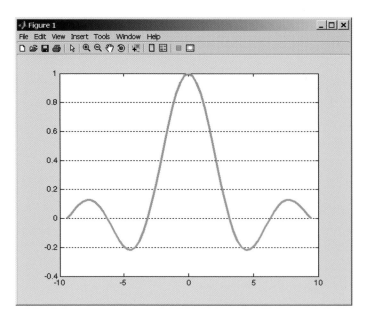

Figure 9.4   Plot of sinc x versus x.

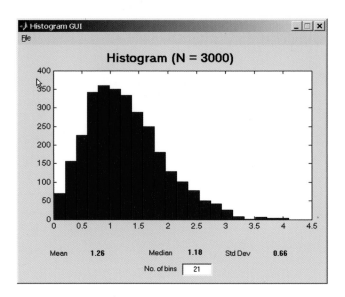

Figure 10.45   The GUI produced by program `histGUI`.

However, *comparisons with the* $>$, $<$, $>=$, *or* $<=$ *operators do not produce the expected results.* When complex numbers are compared with these relational operators, only the *real parts* of the numbers are compared. For example, if $c_1 = 4 + i3$ and $c_2 = 3 + i8$, then the relational operation $c_1 > c_2$ produces a true (1) even though the magnitude of c_1 is really smaller than the magnitude of c_2.

If you ever need to compare two complex numbers with these operators, you will probably be more interested in the total magnitude of the number than we are in the magnitude of only its real part. The magnitude of a complex number can be calculated with the `abs` intrinsic function (see below) or directly from Equation 6-4.

$$|c| = \sqrt{a^2 + b^2} \qquad (6\text{-}4)$$

If we compare the *magnitudes* of c_1 and c_2, the results are more reasonable: $\text{abs}(c_1) > \text{abs}(c_2)$ produces a 0, since the magnitude of c_2 is greater than the magnitude of c_1.

Programming Pitfalls

Be careful when using the relational operators with complex numbers. The relational operators $>$, $>=$, $<$, and $<=$ compare only the *real parts* of complex numbers, not their magnitudes. If you need these relational operators with a complex number, it will probably be more sensible to compare the total magnitudes rather than only the real components.

6.1.3   Complex Functions

MATLAB includes many functions that support complex calculations. These functions fall into three general categories:

1. **Type conversion functions.** These functions convert data from the complex data type to the real (`double`) data type. Function `real` converts the *real part* of a complex number into the double data type and throws away the imaginary part of the complex number. Function `imag` converts the *imaginary part* of a complex number into a real number.

2. **Absolute value and angle functions.** These functions convert a complex number to its polar representation. Function `abs(c)` calculates the absolute value of a complex number using the equation

$$abs(c) = \sqrt{a^2 + b^2}$$

where $c = a + bi$. Function `angle(c)` calculates the angle of a complex number using the equation

```
angle(c) = atan2(imag(c),real(c))
```

producing an answer in the range $-\pi \le \theta \le \pi$.

Table 6.1   Some Functions that Support Complex Numbers

Function	Description
conj(c)	Computes the complex conjugate of a number c. If $c = a + bi$, then conj(c) = $a - bi$
real(c)	Returns the real portion of the complex number c.
imag(c)	Returns the imaginary portion of the complex number c.
isreal(c)	Returns true (1) if no element of array c has an imaginary component. Therefore, ~isreal(c) returns true (1) if an array is complex.
abs(c)	Returns the magnitude of the complex number c.
angle(c)	Returns the angle of the complex number c, computed from the expression atan2(imag(c), real(c)).

3. **Mathematical functions.** Most elementary mathematical functions are defined for complex values. These functions include exponential functions, logarithms, trigonometric functions, and square roots. The functions sin, cos, log, sqrt, and so forth will work as well with complex data as they will with real data.

Some of the intrinsic functions that support complex numbers are listed in Table 6.1.

▶

Example 6.1—The Quadratic Equation (Revisited)

The availability of complex numbers often simplifies the calculations required to solve problems. For example, when we solved the quadratic equation in Example 3.2, it was necessary to take three separate branches through the program depending on the sign of the discriminant. With complex numbers available, the square root of a negative number presents no difficulties, so we can greatly simplify these calculations.

Write a general program to solve for the roots of a quadratic equation, regardless of type. Use complex variables so that no branches will be required based on the value of the discriminant.

SOLUTION

1. **State the problem.**
   Write a program that will solve for the roots of a quadratic equation, whether they are distinct real roots, repeated real roots, or complex roots, without requiring tests on the value of the discriminant.

2. **Define the inputs and outputs.**

   The inputs required by this program are the coefficients a, b, and c of the quadratic equation

   $$ax^2 + bx + c = 0$$

   The output from the program will be the roots of the quadratic equation, whether they are real, repeated, or complex.

3. **Describe the algorithm.**

   This task can be broken down into three major sections, whose functions are input, processing, and output.

   ```
   Read the input data
   Calculate the roots
   Write out the roots
   ```

   We will now break each of the foregoing major sections into smaller, more detailed pieces. In this algorithm, the value of the discriminant is unimportant in determining how to proceed. The resulting pseudocode is

   ```
   Prompt the user for the coefficients a, b, and c.
   Read a, b, and c
   discriminant <- b^2 - 4 * a * c
   x1 <- ( -b + sqrt(discriminant) ) / ( 2 * a )
   x2 <- ( -b - sqrt(discriminant) ) / ( 2 * a )
   Print 'The roots of this equation are: '
   Print 'x1 = ', real(x1), ' +i ', imag(x1)
   Print 'x2 = ', real(x2), ' +i ', imag(x2)
   ```

4. **Turn the algorithm into MATLAB statements.**

   The final MATLAB code is

```
% Script file: calc_roots2.m
%
% Purpose:
%   This program solves for the roots of a quadratic equation
%   of the form a*x**2 + b*x + c = 0. It calculates the answers
%   regardless of the type of roots that the equation possesses.
%
% Record of revisions:
%     Date          Programmer           Description of change
%     ====          ==========           =====================
%   02/24/07       S. J. Chapman         Original code
%
% Define variables:
%   a               -- Coefficient of x^2 term of equation
%   b               -- Coefficient of x term of equation
%   c               -- Constant term of equation
```

```
%    discriminant -- Discriminant of the equation
%    x1           -- First solution of equation
%    x2           -- Second solution of equation

% Prompt the user for the coefficients of the equation
disp ('This program solves for the roots of a quadratic ');
disp ('equation of the form A*X^2 + B*X + C = 0. ');
a = input ('Enter the coefficient A: ');
b = input ('Enter the coefficient B: ');
c = input ('Enter the coefficient C: ');

% Calculate discriminant
discriminant = b^2 - 4 * a * c;

% Solve for the roots
x1 = ( -b + sqrt(discriminant) ) / ( 2 * a );
x2 = ( -b - sqrt(discriminant) ) / ( 2 * a );

% Display results
disp ('The roots of this equation are:');
fprintf ('x1 = (%f) +i (%f)\n', real(x1), imag(x1));
fprintf ('x2 = (%f) +i (%f)\n', real(x2), imag(x2));
```

5. **Test the program.**

   Next, we must test the program using real input data. We will test cases in which the discriminant is greater than, less than, and equal to 0 to be certain that the program is working properly under all circumstances. From Equation (3-1), it is possible to verify the solutions to the following equations:

$$x^2 + 5x + 6 = 0 \qquad x = -2, \text{ and } x = -3$$
$$x^2 + 4x + 4 = 0 \qquad x = -2$$
$$x^2 + 2x + 5 = 0 \qquad x = -1 \pm 2i$$

   When the preceding coefficients are fed into the program, the results are

```
» calc_roots2
This program solves for the roots of a quadratic
equation of the form A*X^2 + B*X + C = 0.
Enter the coefficient A: 1
Enter the coefficient B: 5
Enter the coefficient C: 6
The roots of this equation are:
x1 = (-2.000000) +i (0.000000)
x2 = (-3.000000) +i (0.000000)
» calc_roots2
This program solves for the roots of a quadratic
equation of the form A*X^2 + B*X + C = 0.
Enter the coefficient A: 1
```

```
Enter the coefficient B: 4
Enter the coefficient C: 4
The roots of this equation are:
x1 = (-2.000000) +i (0.000000)
x2 = (-2.000000) +i (0.000000)
» calc_roots2
This program solves for the roots of a quadratic
equation of the form A*X^2 + B*X + C = 0.
Enter the coefficient A: 1
Enter the coefficient B: 2
Enter the coefficient C: 5
The roots of this equation are:
x1 = (-1.000000) +i (2.000000)
x2 = (-1.000000) +i (-2.000000)
```

The program gives the correct answers for our test data in all three possible cases. Note how much simpler this program is compared to the quadratic root solver found in Example 3.1. The complex data type has greatly simplified our program.
◄

6.1.4   Plotting Complex Data

Complex data has both real and imaginary components, and plotting complex data with MATLAB is a bit different from plotting real data. For example, consider the function

$$y(t) = e^{-0.2t}(\cos t + i \sin t) \tag{6-10}$$

If this function is plotted with the conventional `plot` command, only the real data will be plotted—the imaginary part will be ignored. The following statements produce the plot shown in Figure 6.4, together with a warning message that the imaginary part of the data is being ignored.

```
t = 0:pi/20:4*pi;
y = exp(-0.2*t).*(cos(t)+i*sin(t));
plot(t,y,'LineWidth',2);
title('\bfPlot of Complex Function vs Time');
xlabel('\bf\itt');
ylabel('\bf\ity(t)');
```

If both the real and imaginary parts of the function are of interest, then the user has several choices. Both parts can be plotted as a function of time on the same axes using the following statements (see Figure 6.5).

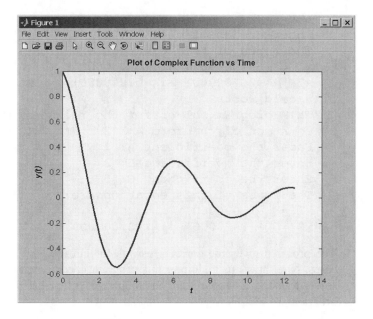

Figure 6.4 Plot of $y(t) = e^{-0.2t}(\cos t + i \sin t)$ using the command plot(t,y).

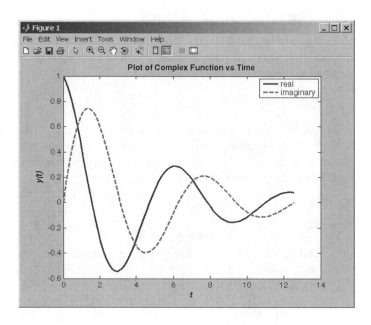

Figure 6.5 Plot of real and imaginary parts of $y(t)$ versus time.

```
t = 0:pi/20:4*pi;
y = exp(-0.2*t).*(cos(t)+i*sin(t));
plot(t,real(y),'b-','LineWidth',2);
hold on;
plot(t,imag(y),'r--','LineWidth',2);
title('\bfPlot of Complex Function vs Time');
xlabel('\bf\itt');
ylabel('\bf\ity(t)');
legend ('real','imaginary');
hold off;
```

Alternatively, the real part of the function can be plotted versus the imaginary part. If a single complex argument is supplied to the `plot` function, it automatically generates a plot of the real part versus the imaginary part. The statements to generate this plot are shown here, and the result is shown in Figure 6.6.

```
t = 0:pi/20:4*pi;
y = exp(-0.2*t).*(cos(t)+i*sin(t));
plot(y,'b-','LineWidth',2);
title('\bfPlot of Complex Function');
xlabel('\bfReal Part');
ylabel('\bfImaginary Part');
```

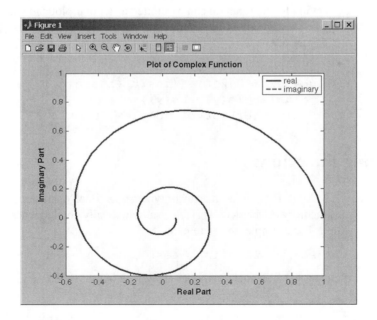

Figure 6.6 Plot of real versus imaginary parts of $y(t)$.

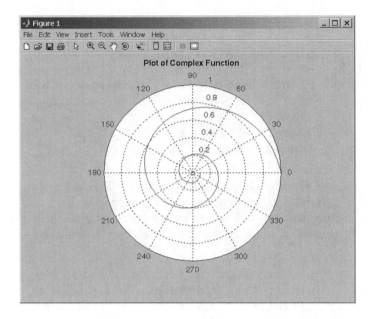

Figure 6.7 Polar plot of magnitude of $y(t)$ versus angle.

Finally, the function can be plotted as a polar plot showing magnitude versus angle. The statements to generate this plot are shown here, and the result is shown in Figure 6.7.

```
t = 0:pi/20:4*pi;
y = exp(-0.2*t).*(cos(t)+i*sin(t));
polar(angle(y),abs(y));
title('\bfPlot of Complex Function');
```

6.2 String Functions

A MATLAB string is an array of type `char`. Each character is stored in two bytes of memory. A character variable is automatically created when a string is assigned to it. For example, the statement

```
str = 'This is a test';
```

creates a 14-element character array. The output of **whos** for this array is

```
» whos str
  Name      Size      Bytes      Class      Attributes

  str       1x14      28         char
```

A special function `ischar` can be used to check for character arrays. If a given variable is of type character, then `ischar` returns a true (1) value. If it is not, `ischar` returns a false (0) value.

The following subsections describe MATLAB functions useful for manipulating character strings.

6.2.1 String Conversion Functions

Variables may be converted from the `char` data type to the `double` data type using the `double` function. Thus the statement `double(str)` yields the result

```
» x = double(str)
x =
 Columns 1 through 12
   84 104 105 115  32 105 115  32  97  32 116 101
 Columns 13 through 14
 115 116
```

Variables can also be converted from the `double` data type to the `char` data type using the `char` function. If x is the 14-element array created previously, then the statement `char(x)` yields the result

```
» z = char(x)
z =
This is a test
```

6.2.2 Creating Two-Dimensional Character Arrays

It is possible to create two-dimensional character arrays, but *each row of such an array must have exactly the same length*. If one of the rows is shorter than the other rows, the character array is invalid and will produce an error. For example, the following statements are illegal because the two rows have different lengths.

```
name = ['Stephen J. Chapman';'Senior Engineer'];
```

The easiest way to produce two-dimensional character arrays is with the `char` function. This function will automatically pad all strings to the length of the largest input string.

```
» name = char('Stephen J. Chapman','Senior Engineer')
name =
Stephen J. Chapman
Senior Engineer
```

Two-dimensional character arrays can also be created with the function `strvcat`, which is described later.

> ☀ **Good Programming Practice**
>
> Use the `char` function to create two-dimensional character arrays without worrying about padding each row to the same length.

It is possible to remove any extra trailing blanks from a string when it is extracted from an array using the `deblank` function. For example, the following statements remove the second line from array `name` and compare the results with and without blank trimming.

```
» line2 = name(2,:)
line2 =
Senior Engineer
» line2_trim = deblank(name(2,:))
line2_trim =
Senior Engineer
» size(line2)
ans =
     1    18
» size(line2_trim)
ans =
     1    15
```

6.2.3 Concatenating Strings

Function `strcat` concatenates two or more strings horizontally, ignoring any trailing blanks but preserving blanks within the strings. This function produces the following result:

```
» result = strcat('String 1 ','String 2')
result =
String 1String 2
```

The result is `'String 1String 2'`. Note that the trailing blanks in the first string were ignored.

Function `strvcat` concatenates two or more strings vertically, automatically padding the strings to make a valid two-dimensional array. This function produces the following result:

```
» result = strvcat('Long String 1 ','String 2')
result =
Long String 1
String 2
```

6.2.4   Comparing Strings

Strings and substrings can be compared in several ways:

- Two strings, or parts of two strings, can be compared for equality.
- Two individual characters can be compared for equality.
- Strings can be examined to determine whether each character is a letter or whitespace.

Comparing Strings for Equality

You can use four MATLAB functions to compare two strings as a whole for equality. They are

- `strcmp` determines whether two strings are identical.
- `strcmpi` determines whether two strings are identical ignoring case.
- `strncmp` determines whether the first n characters of two strings are identical.
- `strncmpi` determines whether the first n characters of two strings are identical ignoring case.

Function `strcmp` compares two strings, including any leading and trailing blanks, and returns a true (1) if the strings are identical[1]. Otherwise, it returns a false (0). Function `strcmpi` is the same as `strcmp`, except that it ignores the case of letters (that is, it treats `'a'` as equal to `'A'`.)

Function `strncmp` compares the first n characters of two strings, including any leading blanks, and returns a true (1) if the characters are identical. Otherwise, it returns a false (0). Function `strncmpi` is the same as `strncmp`, except that it ignores the case of letters.

To understand these functions, consider the two strings:

```
str1 = 'hello';
str2 = 'Hello';
str3 = 'help';
```

Strings `str1` and `str2` are not identical, but they differ only in the case of one letter. Therefore, `strcmp` returns false (0), while `strcmpi` returns true (1).

```
» c = strcmp(str1,str2)
c =
      0
» c = strcmpi(str1,str2)
c =
      1
```

---

[1]**Caution:** The behavior of this function is different from that of the `strcmp` in C. C programmers can be tripped up by this difference.

Strings str1 and str3 are also not identical, and both strcmp and strcmpi will return a false (0). However, the first three characters of str1 and str3 *are* identical, so invoking strncmp with any value up to 3 returns a true (1):

```
» c = strncmp(str1,str3,2)
c =
1
```

Comparing Individual Characters for Equality and Inequality

You can use MATLAB relational operators on character arrays to test for equality *one character at a time*, as long as the arrays you are comparing have equal dimensions or one is a scalar. For example, you can use the equality operator (==) to determine which characters in two strings match:

```
» a = 'fate';
» b = 'cake';
» result = a == b
result =
0 1 0 1
```

All of the relational operators (>, >=, <, <=, ==, ~=) compare the ASCII values of corresponding characters.

Unlike C, MATLAB does not have an intrinsic function to define a "greater than" or "less than" relationship between two strings taken as a whole. We will create such a function in an example at the end of this section.

Categorizing Characters Within a String

There are three functions for categorizing characters on a character-by-character basis inside a string:

- isletter determines whether a character is a letter.
- isspace determines if a character is whitespace (blank, tab, or new line).
- isstrprop ('str', 'category') is a more general function. It determines whether a character falls into a user-specified category, such as alphabetic, alphanumeric, uppercase, lowercase, numeric, control and so on.

To understand these functions, let's create a string named mystring:

```
mystring = 'Room 23a';
```

We will use this string to test the categorizing functions.

Function isletter examines each character in the string, producing a logical output vector of the same length as mystring that contains a true (1) in each location corresponding to a character and a false (0) in the other locations. For example,

```
» a = isletter(mystring)
a =
1 1 1 1 0 0 0 1
```

The first four and the last elements in a are true (1) because the corresponding characters of mystring are letters.

Function isspace also examines each character in the string, producing a logical output vector of the same length as mystring that contains a true (1) in each location corresponding to whitespace, and a false (0) in the other locations. "Whitespace" is any character that separates tokens in MATLAB: a space, a tab, a linefeed, carriage return, etc. For example,

```
» a = isspace(mystring)
a =
0 0 0 0 1 0 0 0
```

The fifth element in a is true (1) because the corresponding character of mystring is a space.

Function isstrprop was added in MATLAB 7.0. It is a more flexible replacement for isletter, isspace, and several other functions. This function has two arguments, 'str' and 'category'. The first argument is the string to characterize, and the second argument is the type of category to check for. Some possible categories are given in Table 6.2.

This function examines each character in the string, producing a logical output vector of the same length as the input string that contains a true (1) in

Table 6.2 Selected Categories for Function isstrprop

Category	Description
'alpha'	Return true (1) for each character of the string that is alphabetic, and false (0) otherwise.
'alphanum'	Return true (1) for each character of the string that is alphanumeric, and false (0) otherwise.
	[*Note:* This category replaces function isletter.]
'cntrl'	Return true (1) for each character of the string is that is a control character, and false (0) otherwise.
'digit'	Return true (1) for each character of the string that is a number, and false (0) otherwise.
'lower'	Return true (1) for each character of the string that is a lower case letter, and false (0) otherwise.
'wspace'	Return true (1) for each character of the string that is whitespace, and false (0) otherwise.
	[*Note:* This category replaces function isspace.]
'upper'	Return true (1) for each character of the string that is an upper case letter, and false (0) otherwise.
'xdigit'	Return true (1) for each character of the string that is a hexadecimal digit, and false (0) otherwise.

each location that matches the category and a false (0) in the other locations. For example, the following function checks to see which characters in `mystring` are numbers:

```
» a = isstrprop(mystring,'digit')
a =
 0  0  0  0  0  1  1  0
```

Also, the following function checks to see which characters in `mystring` are lowercase letters:

```
» a = isstrprop(mystring,'lower')
a =
 0  1  1  1  0  0  0  1
```

✳ Good Programming Practice

Use function `isstrprop` to determine the characteristics of each character in a string array. This function replaces the older functions `isletter` and `isspace`, which may be deleted in a future version of MATLAB.

6.2.5   Searching and Replacing Characters Within a String

MATLAB provides several functions for searching and replacing characters within a string. Consider a string named `test`:

```
test = 'This is a test!';
```

Function `findstr` returns the starting position of all occurrences of the shorter of two strings within a longer string. For example, to find all occurrences of the string `'is'` inside `test`,

```
» position = findstr(test,'is')
position =
     3     6
```

The string `'is'` occurs twice within `test`, starting at positions 3 and 6.

Function `strmatch` is another matching function. This one looks at the beginning characters of the *rows* of a two-dimensional character array and returns a list of those rows that start with the specified character sequence. The form of this function is

```
result = strmatch(str,array);
```

For example, suppose that we create a two-dimensional character array with the function `strvcat`:

```
array = strvcat('maxarray','min value','max value');
```

Then the following statement will return the row numbers of all rows beginning with the letters `'max'`:

```
» result = strmatch('max',array)
result =
              1
              3
```

Function `strrep` performs the standard search-and-replace operation. It finds all occurrences if one string within another one and replaces them by a third string. The form of this function is

```
result = strrep(str,srch,repl)
```

where `str` is the string being checked, `srch` is the character string to search for, and `repl` is the replacement character string. For example,

```
» test = 'This is a test!'
» result = strrep(test,'test','pest')
result =
This is a pest!
```

The `strtok` function returns the characters before the first occurrence of a delimiting character in an input string. The default delimiting characters are the set of whitespace characters. The form of `strtok` is

```
[token,remainder] = strtok(string,delim)
```

where `string` is the input character string, `delim` is the (optional) set of delimiting characters, `token` is the first set of characters delimited by a character in `delim`, and `remainder` is the rest of the line. For example,

```
» [token,remainder] = strtok('This is a test!')
token =
This
remainder =
is a test!
```

You can use the `strtok` function to parse a sentence into words; for example,

```
function all_words = words(input_string)
remainder = input_string;
all_words = '';
while (any(remainder))
    [chopped,remainder] = strtok(remainder);
    all_words = strvcat(all_words,chopped);
end
```

6.2.6 Uppercase and Lowercase Conversion

Functions upper and lower convert all of the alphabetic characters within a string to uppercase and lowercase, respectively. For example,

```
» result = upper('This is test 1!')
result =
THIS IS TEST 1!
» result = lower('This is test 2!')
result =
this is test 2!
```

Note that the alphabetic characters were converted to the proper case, while the numbers and punctuation were unaffected.

6.2.7 Trimming Whitespace from Strings

There are two functions that trim leading and/or trailing whitespace from a string. Whitespace characters consists of the spaces, newlines, carriage returns, tabs, vertical tabs, and formfeeds.

Function deblank removes any extra *trailing* whitespace from a string, and function strtrim removes any extra *leading and trailing* whitespace from a string.

For example, the following statements create a 21-character string with leading and trailing whitespace. Function deblank trims the trailing whitespace characters in the string only, while function strtrim trims both the leading and the trailing whitespace characters.

```
» test_string = '   This is a test.   '
test_string =
   This is a test.
» length(test_string)
ans =
    21
» test_string_trim1= deblank(test_string)
test_string_trim1 =
   This is a test.
» length(test_string_trim1)
ans =
    18
» test_string_trim2 = strtrim(test_string)
test_string_trim2 =
This is a test.
» length(test_string_trim2)
ans =
    15
```

6.2.8 Numeric-to-String Conversions

MATLAB contains several functions to convert numeric values into character strings. We have already seen two such functions, num2str and int2str. Consider a scalar x:

```
x = 5317;
```

By default, MATLAB stores the number x as a 1 × 1 double array containing the value 5317. The int2str (integer to string) function converts this scalar into a 1-by-4 char array containing the string '5317':

```
» x = 5317;
» y = int2str(x);
» whos
  Name      Size       Bytes      Class        Attributes

  x         1x1        8          double
  y         1x4        8          char

Grand total is 5 elements using 16 bytes
```

Function num2str converts a double value into a string, even if it does not contain an integer. It provides more control of the output string format than int2str. An optional second argument sets the number of digits in the output string or specifies an actual format to use. The format specifications in the second argument as similar to those used by fprintf. For example,

```
» p = num2str(pi)
p =
3.1416
» p = num2str(pi,7)
p =
3.141593
» p = num2str(pi,'%10.5e')
p =
3.14159e+000
```

Both int2str and num2str are handy for labeling plots. For example, the following lines use num2str to prepare automated labels for the x-axis of a plot:

```
function plotlabel(x,y)
plot(x,y)
str1 = num2str(min(x));
str2 = num2str(max(x));
out = ['Value of f from ' str1 ' to ' str2];
xlabel(out);
```

There are also conversion functions designed to change numeric values into strings representing a decimal value in another base, such as a binary or hexadecimal representation. For example, the dec2hex function converts a decimal value into the corresponding hexadecimal string:

```
dec_num = 4035;
hex_num = dec2hex(dec_num)
hex_num =
FC3
```

Other functions of this type include hex2num, hex2dec, bin2dec, dec2bin, base2dec, and dec2base. MATLAB includes on-line help for all of these functions.

MATLAB function mat2str converts an array to a string that MATLAB can evaluate. This string is useful input for a function such as eval, which evaluates input strings just as if they were typed at the MATLAB command line. For example, if we define array a as

```
» a = [1 2 3; 4 5 6]
a =
        1        2        3
        4        5        6
```

then the function mat2str will return a string containing the result

```
» b = mat2str(a)
b =
[1 2 3; 4 5 6]
```

Finally, MATLAB includes a special function sprintf that is identical to function fprintf, except that the output goes into a character string instead of the Command Window. This function provides complete control over the formatting of the character string. For example,

```
» str = sprintf('The value of pi = %8.6f.',pi)
str =
The value of pi = 3.141593.
```

This function is extremely useful in creating complex titles and labels for plots.

6.2.9 String-to-Numeric Conversions

MATLAB also contains several functions to change character strings into numeric values. The most important of these function are eval, str2double, and sscanf.

Function eval evaluates a string containing a MATLAB expression and returns the result. The expression can contain any combination of MATLAB functions, variables, constants, and operations. For example, the string a containing the characters '2 * 3.141592' can be converted to numeric form by the following statements:

```
» a = '2 * 3.141592';
» b = eval(a)
b =
        6.2832
» whos
  Name        Size       Bytes      Class        Attributes

  a           1x12       24         char
  b           1x1        8          double
```

Function `str2double` converts character strings into an equivalent `double` value[2]. For example, the string a containing the characters `'3.141592'` can be converted to numeric form by the following statements:

```
» a = '3.141592';
» b = str2double(a)
b =
        3.1416
```

Strings can also be converted to numeric form using the function `sscanf`. This function converts a string into a number according to a format conversion character. The simplest form of this function is

```
value = sscanf(string,format)
```

where `string` is the string to scan, and `format` specifies the type of conversion to occur. The two most common conversion specifiers for `sscanf` are `'%d'` for decimals and `'%g'` for floating-point numbers. This function will be covered in much greater detail in Chapter 8.

The following examples illustrate the use of `sscanf`.

```
» a = '3.141592';
» value1 = sscanf(a,'%g')
value1 =
        3.1416
» value2 = sscanf(a,'%d')
value2 =
        3
```

6.2.10 Summary

The common MATLAB string functions are summarized in Table 6.3.

---

[2]MATLAB also contains a function `str2num` that can convert a string into a number. For a variety of reasons mentioned in the MATLAB documentation, function `str2double` is better than function `str2num`. You should recognize function `str2num` when you see it, but always use function `str2double` in any new code that you write.

Table 6.3   Common MATLAB String Functions

Category	Function	Description
General	char	(1) Convert numbers to the corresponding character values. (2) Create a two dimensional character array from a series of strings.
	double	Convert characters to the corresponding numeric codes.
	blanks	Create a string of blanks.
	deblank	Remove trailing whitespace from a string.
	strtrim	Remove leading and trailing whitespace from a string.
String tests	ischar	Returns true (1) for a character array.
	isletter	Returns true (1) for letters of the alphabet.
	isspace	Returns true (1) for whitespace.
	isstrprop	Returns true (1) for characters matching the specified property.
String operations	strcat	Concatenate strings.
	strvcat	Concatenate strings vertically.
	strcmp	Returns true (1) if two strings are identical.
	strcmpi	Returns true (1) if two strings are identical, ignoring case.
	strncmp	Returns true (1) if first n characters of two strings are identical.
	strncmpi	Returns true (1) if first n characters of two strings are identical, ignoring case.
	findstr	Find one string within another one.
	strjust	Justify string.
	strmatch	Find matches for string.
	strrep	Replace one string with another.
	strtok	Find token in string.
	upper	Convert string to uppercase.
	lower	Convert string to lowercase.
Number-to-string conversion	int2str	Convert integer to string.
	num2str	Convert number to string.
	mat2str	Convert matrix to string.
	sprintf	Write formatted data to string.
String-to-number conversion	eval	Evaluate the result of a MATLAB expression.
	str2double	Convert string to a double value.
	str2num	Convert string to number.
	sscanf	Read formatted data from string.
Base number conversion	hex2num	Convert IEEE hexadecimal string to double.
	hex2dec	Convert hexadecimal string to decimal integer.
	dec2hex	Convert decimal to hexadecimal string.
	bin2dec	Convert binary string to decimal integer.
	dec2bin	Convert decimal integer to binary string.
	base2dec	Convert base B string to decimal integer.
	dec2base	Convert decimal integer to base B string.

Example 6.2—String Comparison Function

In C, function strmcp compares two strings according to the order of their characters in the ASCII table (called the **lexicographic order** of the characters), and returns a −1 if the first string is lexicographically less than the second string, a 0 if the strings are equal, and a +1 if the first string is lexicographically greater than the second string. This function is extremely useful for such purposes as sorting strings in alphabetic order.

Create a new MATLAB function c_strcmp that compares two strings in a similar fashion to the C function and returns similar results. The function should ignore trailing blanks in doing its comparisons. Note that the function must be able to handle the situation where the two strings are of different lengths.

SOLUTION

1. **State the problem.**
   Write a function that will compare two strings str1 and str2, and return the following results:

   - −1   if str1 is lexicographically less than str2.
   - 0   if str1 is lexicographically less than str2.
   - +1   if str1 is lexicographically greater than str2.

   The function must work properly if str1 and str2 do not have the same length, and the function should ignore trailing blanks.

2. **Define the inputs and outputs.**
   The inputs required by this function are two strings, str1 and str2. The output from the function will be a −1, 0, or 1, as appropriate.

3. **Describe the algorithm.**
   This task can be broken down into four major sections:

```
Verify input strings
Pad strings to be equal length
Compare characters from beginning to end, looking
     for the first difference
Return a value based on the first difference
```

We will now break each of the preceding major sections into smaller, more detailed pieces. First, we must verify that the data passed to the function is correct. The function must have exactly two arguments, and the arguments must both be characters. The pseudocode for this step is

```
% Check for a legal number of input arguments.
msg = nargchk(2,2,nargin)
error(msg)
```

```
% Check to see if the arguments are strings
if either argument is not a string
   error('str1 and str2 must both be strings')
else

   (add code here)

end
```

Next, we must pad the strings to equal lengths. The easiest way to do this is to combine both strings into a two-dimensional array using `strvcat`. Note that this step effectively results in the function's ignoring trailing blanks, because both strings are padded out to the same length. The pseudocode for this step is

```
% Pad strings
strings = strvcat(str1,str2)
```

Now we must compare each character until we find a difference and return a value based on that difference. One way to do this is to use relational operators to compare the two strings, creating an array of 0s and 1s. We can then look for the first 1 in the array, which will correspond to the first difference between the two strings. The pseudocode for this step is

```
% Compare strings
diff = strings(1,:) ~= strings(2,:)
if sum(diff) == 0
   % Strings match
   result = 0
else
   % Find first difference
   ival = find(diff)
   if strings(1,ival) > strings(2,ival)
      result = 1
   else
      result = -1
   end
end
```

4. **Turn the algorithm into MATLAB statements.**
   The final MATLAB code is

```
function result = c_strcmp(str1,str2)
%C_STRCMP Compare strings like C function "strcmp"
% Function C_STRCMP compares two strings, and returns
% a -1 if str1 < str2, a 0 if str1 == str2, and a
% +1 if str1 > str2.

% Define variables:
%    diff        -- Logical array of string differences
```

```
%     msg        -- Error message
%     result     -- Result of function
%     str1       -- First string to compare
%     str2       -- Second string to compare
%     strings    -- Padded array of strings

%  Record of revisions:
%     Date        Programmer        Description of change
%     ====        ==========        ====================
%  02/25/07    S. J. Chapman        Original code

% Check for a legal number of input arguments.
msg = nargchk(2,2,nargin);
error(msg);

% Check to see if the arguments are strings
if  ~(isstr(str1) & isstr(str2))
   error('Both str1 and str2 must both be strings!')
else

   % Pad strings
   strings = strvcat(str1,str2);

   % Compare strings
   diff = strings(1,:) ~= strings(2,:);
   if sum(diff) == 0

      % Strings match, so return a zero!
      result = 0;
   else
      % Find first difference between strings
      ival = find(diff);
      if strings(1,ival(1)) > strings(2,ival(1))
         result = 1;
      else
         result = -1;
      end
   end
end
```

5. **Test the program.**

Next, we must test the function using various strings.

```
» result = c_strcmp('String 1','String 1')
result =
             0
» result = c_strcmp('String 1','String 1 ')
```

```
    result =
                0
» result = c_strcmp('String 1','String 2')
    result =
                -1
» result = c_strcmp('String 1','String 0')
    result =
                1
» result = c_strcmp('String','str')
    result =
                -1
```

The first test returns a zero, because the two strings are identical. The second test also returns a zero, because the two strings are identical *except for trailing blanks*, and trailing blanks are ignored. The third test returns a -1, because the two strings first differ in position 8, and `'1' < '2'` at that position. The fourth test returns a 1, because the two strings first differ in position 8, and `'1' > '0'` at that position. The fifth test returns a -1, because the two strings first differ in position 1, and `'S' < 's'` in the ASCII collating sequence.

This function appears to be working properly.

◄

Quiz 6.1

This quiz provides a quick check to see if you have understood the concepts introduced in Sections 6.1 through 6.2. If you have trouble with the quiz, reread the section, ask your instructor, or discuss the material with a fellow student. The answers to this quiz are found in the back of the book.

1. What is the value of `result` in the following statements?

(*a*)
```
x = 12 + i*5;
y = 5 - i*13;
result = x > y;
```

(*b*)
```
x = 12 + i*5;
y = 5 - i*13;
result = abs(x) > abs(y);
```

(*c*)
```
x = 12 + i*5;
y = 5 - i*13;
result = real(x) - imag(y);
```

2. If `array` is a complex array, what does the function `plot(array)` do?

3. How can you convert a vector of the `char` data type into a vector of the `double` data type?

For questions 4 through 11, determine whether these statements are correct. If they are, what is produced by each set of statements?

4.
```
str1 = 'This is a test! ';
str2 = 'This line, too.';
res = strcat(str1,str2);
```

5.
```
str1 = 'Line 1';
str2 = 'line 2';
res = strcati(str1,str2);
```

6.
```
str1 = 'This is another test! ';
str2 = 'This line, too.';
res = [str1; str2];
```

7.
```
str1 = 'This is another test! ';
str2 = 'This line, too.';
res = strvcat(str1,str2);
```

8.
```
str1 = 'This is a test! ';
str2 = 'This line, too.';
res = strncmp(str1,str2,5);
```

9.
```
str1 = 'This is a test! ';
res = findstr(str1,'s');
```

10.
```
str1 = 'This is a test! ';
str1(isspace(str1)) = 'x';
```

11.
```
str1 = 'aBcD 1234 !?';
res = isstrprop(str1,'alphanum');
```

12.
```
str1 = 'This is a test! ';
str1(4:7) = upper(str1(4:7));
```

13.
```
str1 = ' 456 '; % Note: Three blanks before & after
str2 = ' abc '; % Note: Three blanks before & after
str3 = [str1 str2];
str4 = [strtrim(str1) strtrim(str2)];
str5 = [deblank(str1) deblank(str2)];
l1 = length(str1);
l2 = length(str2);
l3 = length(str3);
```

```
    l4 = length(str4);
    l5 = length(str4);
14. str1 = 'This way to the egress.';
    str2 = 'This way to the egret.'
    res = strncmp(str1,str2);
```

6.3   Multidimensional Arrays

MATLAB also supports arrays with more than two dimensions. These **multidimensional arrays** are very useful for displaying data that intrinsically has more than two dimensions, or for displaying multiple versions of two-dimensional data sets. For example, measurements of pressure and velocity throughout a three-dimensional volume are very important in such studies as aerodynamics and fluid dynamics. These areas naturally use multidimensional arrays.

Multidimensional arrays are a natural extension of two-dimensional arrays. Each additional dimension is represented by one additional subscript used to address the data.

It is very easy to create multidimensional arrays. They can be created either by assigning values directly in assignment statements or by using the same functions that are used to create one- and two-dimensional arrays. For example, suppose that you have a two-dimensional array created by the assignment statement

```
» a = [1 2 3 4; 5 6 7 8]
a =
          1           2           3           4
          5           6           7           8
```

This is a 2 × 4 array, with each element addressed by two subscripts. The array can be extended to be a three-dimensional 2 × 4 × 3 array with the following assignment statements:

```
» a(:,:,2) = [ 9 10 11 12; 13 14 15 16];
» a(:,:,3) = [ 17 18 19 20; 21 22 23 24]
a(:,:,1) =
          1           2           3           4
          5           6           7           8
a(:,:,2) =
          9          10          11          12
         13          14          15          16
a(:,:,3) =
         17          18          19          20
         21          22          23          24
```

Individual elements in this multidimensional array can be addressed by the array name followed by three subscripts, and subsets of the data can be created using the colon operators. For example, the value of a(2,2,2) is

```
» a(2,2,2)
ans =
          14
```

and the vector a(1,1,:) is

```
» a(1,1,:)
ans(:,:,1) =
          1
ans(:,:,2) =
          9
ans(:,:,3) =
          17
```

Multidimensional arrays can also be created using the same functions as other arrays, for example,

```
» b = ones(4,4,2)
b(:,:,1)  =
          1          1          1          1
          1          1          1          1
          1          1          1          1
          1          1          1          1
b(:,:,2)  =
          1          1          1          1
          1          1          1          1
          1          1          1          1
          1          1          1          1
» c = randn(2,2,3)
c(:,:,1)  =
        -0.4326              0.1253
        -1.6656              0.2877
c(:,:,2)  =
        -1.1465              1.1892
         1.1909             -0.0376
c(:,:,3)  =
         0.3273             -0.1867
         0.1746              0.7258
```

The number of dimensions in a multidimensional array can be found using the ndims function, and the size of the array can be found using the size function.

```
» ndims(c)
ans =
          3
» size(c)
ans =
          2     2     3
```

If you are writing applications that need multidimensional arrays, see the MATLAB Users Guide for more details on the behavior of various MATLAB functions with multidimensional arrays.

✳ Good Programming Practice

Use multidimensional arrays to solve problems that are naturally multivariate in nature, such as aerodynamics and fluid flows.

Also, recall from Chapter 4 that the MATLAB just-in-time compiler cannot compile loops containing arrays with three or more dimensions. If you are working with such arrays, be sure to vectorize your code to increase its speed. Do not rely on the JIT compiler to do the job—it won't.

✳ Good Programming Practice

If you are working with multidimensional arrays, be sure to vectorize your code by hand. The MATLAB JIT compiler cannot handle loops containing multidimensional arrays.

6.4   Additional Data Types

MATLAB also includes a `single` data type and several integer data types. They are briefly discussed in the following sections.

6.4.1   The single Data Type

Variables of type `single` are scalars or arrays of 32-bit *single-precision* floating-point numbers. They can hold real, imaginary, or complex values. Variables of type `single` occupy half the memory of variables of type `double`, but they have lower precision and a more limited range. The real and imaginary components of each `single` variable can be positive or negative numbers in the range 10^{-38} to 10^{38}, with six to seven significant decimal digits of accuracy.

The `single` function creates a variable of type `single`. For example, the following statement creates a variable of type `single` containing the value 3.1:

```
» var = single(3.1)
var =
        3.1000
» whos
      Name      Size      Bytes      Class      Attributes
      var       1x1       4          single
```

Once a `single` variable is created, it can be used in MATLAB operations just like a `double` variable. In MATLAB, an operation performed between a `single` value and a `double` value has a `single` result[3], so the result of the following statements will be of type `single`:

```
» b = 7;
» c = var * b
c =
      21.7000
» whos
  Name        Size      Bytes      Class        Attributes

  b           1x1       8          double
  c           1x1       4          single
  var         1x1       4          single
```

The capability to perform mathematical operations with the `single` data type is a relatively new feature introduced in MATLAB 7.0. Values of type `single` can be used just like values of type `double` in most MATLAB operations. Built-in functions such as `sin`, `cos`, `exp`, and so forth all support the `single` data type, but some M-file functions may not support `single` values yet. (For example, comparisons for near equality between two numbers may be incorrect if the function is expecting `double` values and instead is passed `single` values.) As a practical matter, you will probably never use this data type. Its more limited range and precision make the results more sensitive to cumulative round-off errors or to exceeding the available range. You should consider using this data type only if you have enormous arrays of data that could not fit into your computer memory if they were saved in double precision.

6.4.2 Integer Data Types

MATLAB also includes 8-, 16-, 32-, and 64-bit *signed* and *unsigned* integers. The data types are `int8`, `uint8`, `int16`, `uint16`, `int32`, `uint32`, `int64`, and `uint64`. The difference between a signed and an unsigned integer is the range of numbers represented by the data type. The number of values that can be represented by an integer depends on the number of bits in the integer:

$$\text{number of values} = 2^n \tag{6-11}$$

where n is the number of bits. An 8-bit integer can represent 256 values (2^8), a 16-bit integer can represent 65,536 values (2^{16}), and so forth. Signed integers use half of the available values to represent positive numbers and half for negative numbers, whereas unsigned integers use all of the available values to represent positive numbers. Therefore, the range of values that can be represented in

---

[3]CAUTION: This is unlike the behavior of any other computer language that the author has ever encountered. In every other language (Fortran, C, C++, Java, Basic, etc.), the result of an operation between a `single` and a `double` would be of type `double`.

the int8 data type is −128 to 127 (a total of 256), while the range of values that can be represented in the uint8 data type is 0 to 255 (a total of 256). Similarly, the range of values that can be represented in the int16 data type is −32,768 to 32,767 (a total of 65,536), while the range of values that can be represented in the uint16 data type is 0 to 65,535. The same idea applies to larger integer sizes.

Integer values are created by the int8(), uint8(), int16(), uint16(), int32(), uint32(), int64(), or uint64() functions. For example, the following statement creates a variable of type int8 containing the value 3:

```
» var = int8(3)
var =
        3
» whos
  Name      Size      Bytes     Class     Attributes
  var       1x1       1         int8
```

Integers can be converted to other data types using the double and single functions.

An operation performed between an integer value and a double value has an integer result[4], so the result of the following statements will be of type int8:

```
» b = 7;
» c = var * b
c =
       21
» whos
  Name      Size      Bytes     Class     Attributes
  b         1x1       8         double
  c         1x1       1         int8
  var       1x1       1         int8
```

MATLAB uses *saturating integer arithmetic*. If the result of an integer math operation would be larger than the largest possible value that can be represented in that data type, then the result will be the largest possible value. Similarly, if the result of an integer math operation would be smaller than the smallest possible value that can be represented in that data type, then the result will be the smallest possible value. For example, the largest possible value that can be represented in the int8 data type is 127. The result of the operation int8(100) + int8(50) will be 127, because 150 is larger than 127, the maximum value that can be represented in the data type.

---

[4]CAUTION: This is unlike the behavior of any other computer language that the author has ever encountered. In every other language (Fortran, C, C++, Java, Basic, etc.), the result of an operation between an integer and a double would be of type double.

It is unlikely that you will need to use the integer data type unless you are working with image data. If you do need more information, please consult the MATLAB documentation.

6.4.3 Limitations of the single and Integer Data Types

The `single` data type and integer data types have been around in MATLAB for a while, but they have been mainly used for purposes such as storing image data. Before MATLAB 7.0, it was not possible to perform mathematical operations (+, −, etc.) with these data types. MATLAB is now evolving to make manipulating these data types easier, but the support is still rough in the current release. There are significant gaps. For example, you can add a `single` and a `double`, or an integer and a `double`, but not a `single` and an integer.

```
» a = single(2.1)
a =
          2.1000
» b = int16(4)
b =
            4
» c = a+b
??? Error using ==> plus
Class of operand is not supported.
```

Unless you have some special need to manipulate images, you will probably never need to use either of these data types.

✳ Good Programming Practice

Do not use the `single` or integer data types unless you have a special need, such as image processing.

6.5 Additional Two-Dimensional Plots

In previous chapters, we have learned to create linear, log-log, semilog, and polar plots. MATLAB supports many additional types of plots that you can use to display your data. This section will introduce you to some of these additional plotting options.

6.5.1 Additional Types of Two-Dimensional Plots

In addition to the two-dimensional plots that we have already seen, MATLAB supports *many* other more specialized plots. In fact, the MATLAB help desk lists more than 20 types of two-dimensional plots! Examples include **stem plots**, **stair**

plots, **bar plots**, **pie plots**, and **compass plots**. A *stem plot* is a plot in which each data value is represented by a marker and a line connecting the marker vertically to the *x* axis. A *stair plot* is a plot in which each data point is represented by a horizontal line and successive points are connected by vertical lines, producing a stair-step effect. A *bar plot* is a plot in which each point is represented by a vertical bar or horizontal bar. A *pie plot* is a plot represented by "pie slices" of various sizes. Finally, a *compass plot* is a type of polar plot in which each value is represented by an arrow whose length is proportional to its value. These plots are summarized in Table 6.4, and examples of all of the plots are shown in Figure 6.8.

Stair, stem, vertical bar, horizontal bar, and compass plots are all similar to plot, and they are used in the same manner. For example, the following code produces the stem plot shown in Figure 6.7*a*:

```
x = [1 2 3 4 5 6];
y = [2 6 8 7 8 5];
stem(x,y);
title('\bfExample of a Stem Plot');
xlabel('\bf\itx');
ylabel('\bf\ity');
axis([0 7 0 10]);
```

Table 6.4   Additional Two-Dimensional Plotting Functions

Function	Description
bar(x,y)	This function creates a *vertical* bar plot, with the values in x used to label each bar and the values in y used to determine the height of the bar.
barh(x,y)	This function creates a *horizontal* bar plot, with the values in x used to label each bar and the values in y used to determine the horizontal length of the bar.
compass(x,y)	This function creates a polar plot, with an arrow drawn from the origin to the location of each (x, y) point. Note that the locations of the points to plot are specified in Cartesian coordinates, not polar coordinates.
pie(x) pie(x,explode)	This function creates a pie plot. This function determines the percentage of the total pie corresponding to each value of x, and plots pie slices of that size. The optional array explode controls whether or not individual pie slices are separated from the remainder of the pie.
stairs(x,y)	This function creates a stair plot, with each stair step centered on an (x, y) point.
stem(x,y)	This function creates a stem plot, with a marker at each (x, y) point and a stem drawn vertically from that point to the x axis.

(a)

(b)

Figure 6.8   Additional types of 2D plots: *(a)* stem plot; *(b)* stair plot;

(c)

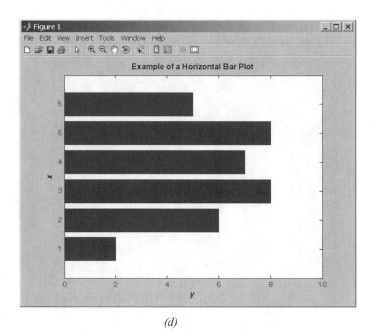

(d)

Figure 6.8   *(continued)* *(c)* vertical bar plot; *(d)* horizontal bar plot;

(e)

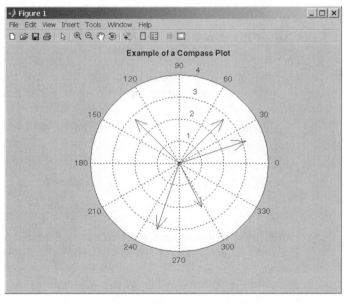

(f)

Figure 6.8   (*continued*) *(e)* pie plot; *(f)* compass plot.

Stair, bar, and compass plots can be created by substituting `stairs`, `bar`, `barh`, or `compass` for `stem` in the preceding code. The details of all of these plots, including any optional parameters, can be found in the MATLAB on-line help system.

Function `pie` behaves differently from the other plots described previously. To create a pie plot, a programmer passes an array `x` containing the data to be plotted, and function `pie` determines the *percentage of the total pie* that each element of `x` represents. For example, if the array `x` is [1 2 3 4], then `pie` will calculate that the first element `x(1)` is 1/10 or 10% of the pie, the second element `x(2)` is 2/10 or 20% of the pie, and so forth. The function then plots those percentages as pie slices.

Function `pie` also supports an optional parameter, `explode`. If present, `explode` is a logical array of 1s and 0s, with an element for each element in array `x`. If a value in `explode` is 1, then the corresponding pie slice is drawn slightly separated from the pie. For example, the code shown below produces the pie plot in Figure 6.7e. Note that the second slice of the pie is "exploded."

```
data = [ 10 37 5 6 6];
explode = [ 0 1 0 0 0];
pie(data,explode);
title('\bfExample of a Pie Plot');
legend('One','Two','Three','Four','Five');
```

6.5.2 Plotting Functions

In all previous plots, we have created arrays of data and passed those arrays to the plotting function. MATLAB also includes two functions that will plot a function directly, without the necessity of creating intermediate data arrays. These functions are `ezplot` and `fplot`.

Function `ezplot` takes one of the following forms:

```
ezplot(fun);
ezplot(fun, [xmin xmax]);
ezplot(fun, [xmin xmax], figure);
```

In each case, `fun` is a *character string* containing the functional expression to be evaluated. The optional parameter [`xmin xmax`] specifies the range of the function to plot. If it is absent, the function will be plotted between -2π and 2π. The optional parameter `figure` specifies the figure number to plot the function on.

For example, the following statements plot the function $f(x) = \sin x/x$ between -4π and 4π. The output of these statements is shown in Figure 6.9.

```
ezplot('sin(x)/x',[-4*pi 4*pi]);
title('Plot of sin x / x');
grid on;
```

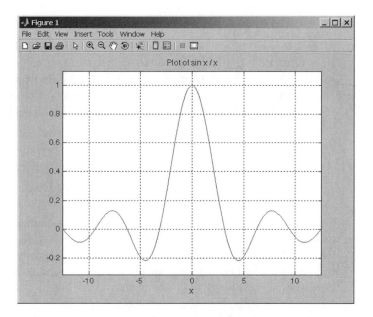

Figure 6.9   The function sin x/x, plotted with function `ezplot`.

Function `fplot` is similar to but more sophisticated than `ezplot`. The first two arguments are the same for both functions, but `fplot` has the following advantages:

1. Function `fplot` is *adaptive*, meaning that it calculates and displays more data points in the regions where the function being plotted is changing most rapidly. The resulting plot is more accurate at locations where a function's behavior changes suddenly.
2. Function `fplot` supports the use of T_EX commands in titles and axis labels, whereas function `ezplot` does not.

In general, you should use `fplot` in preference to `ezplot` whenever you plot functions.

Functions `ezplot` and `fplot` are examples of the "function functions" described in Chapter 5.

✳ Good Programming Practice

Use function `fplot` to plot functions directly without having to create intermediate data arrays.

6.5.3 Histograms

A *histogram* is a plot showing the distribution of values within a data set. To create a histogram, the range of values within the data set is divided into evenly spaced bins, and the number of data values falling into each bin is determined. The resulting count can then be plotted as a function of bin number.

The standard MATLAB histogram function is `hist`. The forms of this function are

```
hist(y)
hist(y,nbins)
hist(y,x);
[n,xout] = hist(y,...)
```

The first form of the function creates and plots a histogram with 10 equally-spaced bins, and the second form creates and plots a histogram with `nbins` equally spaced bins. The third form of the function allows the user to specify the bin centers to use in an array x; the function creates a bin centered around each element in the array. In all three of these cases, the function both creates and plots the histogram. The last form of the function creates a histogram and returns the bin centers in array `xout` and the count in each bin in array n, without actually creating a plot.

For example, the following statements create a data set containing 10,000 Gaussian random values, and generate a histogram of the data using 15 evenly spaced bins. The resulting historam is shown in Figure 6.10.

Figure 6.10 A histogram.

```
y = randn(10000,1);
hist(y,15);
```

MATLAB also includes a function rose to create and plot a histogram on radial axes. It is especially useful for distributions of angular data. You will be asked to use this function in an end-of-chapter exercise.

6.6 Three-Dimensional Plots

MATLAB also includes a rich variety of three-dimensional plots that can be useful for displaying certain types of data. In general, three-dimensional plots are useful for displaying two types of data:

1. Two variables that are functions of the same independent variable, when you wish to emphasize the importance of the independent variable.
2. A single variable that is a function of two independent variables.

6.6.1 Three-Dimensional Line Plots

A three-dimensional line plot can be created with the plot3 function. This function is exactly like the two-dimensional plot function, except that each point is represented by x, y, and z values instead just of x and y values. The simplest form of this function is

```
plot(x,y,z);
```

where x, y, and z are equal-sized arrays containing the locations of data points to plot. Function plot3 supports all the same line size, line style, and color options as plot, and you can use it immediately using the knowledge acquired in previous chapters.

As an example of a three-dimensional line plot, consider the following functions:

$$x(t) = e^{-0.2t} \cos 2t$$
$$y(t) = e^{-0.2t} \sin 2t \qquad (6\text{-}12)$$

These functions might represent the decaying oscillations of a mechanical system in two dimensions, so x and y together represent the location of the system at any given time. Note that x and y are both functions of the *same* independent variable t.

We could create a series of (x, y) points and plot them using the two-dimensional function plot (see Figure 6.11a), but if we do so, the importance of time to the behavior of the system will not be obvious in the graph. The following statements create the two-dimensional plot of the location of the object shown in Figure 6.11a. It is not possible from this plot to tell how rapidly the oscillations are dying out.

```
t = 0:0.1:10;
x = exp(-0.2*t) .* cos(2*t);
```

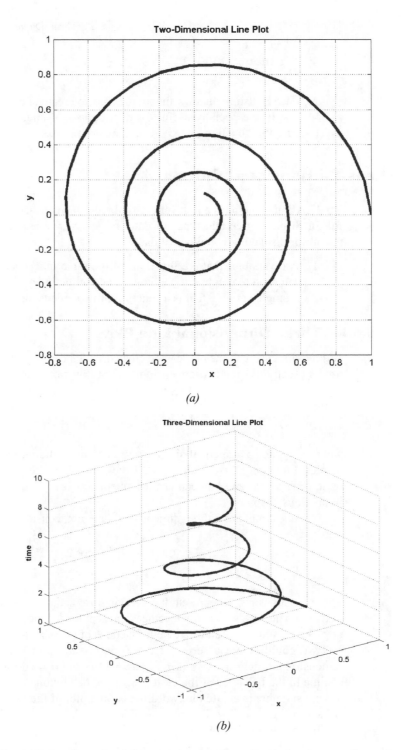

(a)

(b)

Figure 6.11 *(a)* A two-dimensional line plot showing the motion in (x, y) space of a mechanical system. This plot reveals nothing about the time behavior of the system. *(b)* A three-dimensional line plot showing the motion in (x, y) space versus time for the mechanical system. This plot clearly shows the time behavior of the system.

```
y = exp(-0.2*t) .* sin(2*t);
plot(x,y);
title('\bfTwo-Dimensional Line Plot');
xlabel('\bfx');
ylabel('\bfy');
grid on;
```

Instead, we could plot the variables with `plot3` to preserve the time information as well as the two-dimensional position of the object. The following statements will create a three-dimensional plot of Equations (6-12):

```
t = 0:0.1:10;
x = exp(-0.2*t) .* cos(2*t);
y = exp(-0.2*t) .* sin(2*t);
plot3(x,y,t);
title('\bfThree-Dimensional Line Plot');
xlabel('\bfx');
ylabel('\bfy');
zlabel('\bftime');
grid on;
```

The resulting plot is shown in Figure 6.11*b*. Note how this plot emphasizes time-dependence of the two variables *x* and *y*.

6.6.2   Three-Dimensional Surface, Mesh, and Contour Plots

Surface, mesh, and contour plots are convenient ways to represent data that is a function of *two* independent variables. For example, the temperature at a point is a function of both the East-West location (x) and the North-South (y) location of the point. Any value that is a function of two independent variables can be displayed on a three-dimensional surface, mesh, or contour plot. The more common types of plots are summarized in Table 6.5, and examples of each plot are shown in Figure 6.12[5].

To plot data using one of these functions, a user must create three equal-sized arrays. The three arrays must contain the x, y, and z values of every point to be plotted. As a simple example, suppose that we wanted to plot the four points $(-1, -1, 1)$, $(1, -1, 2)$, $(-1, 1, 1)$, and $(1, 1, 0)$. To plot these four points, we must create the arrays $x = \begin{bmatrix} -1 & 1 \\ -1 & 1 \end{bmatrix}$, $y = \begin{bmatrix} -1 & -1 \\ 1 & 1 \end{bmatrix}$, and $z = \begin{bmatrix} 1 & 2 \\ 1 & 0 \end{bmatrix}$. Array

x contains the x values associated with every point to plot, array y contains the y values associated with every point to plot, and array z contains the z values associated with every point to plot. These array are then passed to the plotting function.

---

[5]There are many variations on these basic plot types. Consult the MATLAB Help Browser documentation for a complete description of these variations.

Table 6.5   Selected Mesh, Surface, and Contour Plot Functions

Function	Description
mesh(x,y,z)	This function creates a mesh or wireframe plot, where x is a two-dimensional array containing the x values of every point to display, y is a two-dimensional array containing the y values of every point to display, and z is a two-dimensional array containing the z values of every point to display.
surf(x,y,z)	This function creates a surface plot. Arrays x, y, and z have the same meaning as for a mesh plot.
contour(x,y,z)	This function creates a contour plot. Arrays x, y, and z have the same meaning as for a mesh plot.

The MATLAB function meshgrid makes it easy to create the x and y arrays required for these plots. The form of this function is

```
[x, y] = meshgrid( xstart:xinc:xend, ystart:yinc:yend);
```

where xstart:xinc:xend specifies the x values to include in the grid, and ystart:yinc:yend specifies the y values to be included in the grid.

To create a plot, we use meshgrid to create the arrays of x and y values and then evaluate the function to plot at each of those (x, y) locations. Finally, we call function mesh, surf, or contour to create the plot.

For example, suppose that we wish to create a mesh plot of the function

$$z(x, y) = e^{-0.5[x^2 + 0.5(x-y)^2]} \qquad (6\text{-}13)$$

over the interval $-4 \leq x \leq 4$ and $-4 \leq y \leq 4$. The following statements will create the plot, which is shown in Figure 6.12a:

```
[x,y] = meshgrid(-4:0.2:4);
z = exp(-0.5*(x.^2+y.^2));
mesh(x,y,z);
xlabel('\bfx');
ylabel('\bfy');
zlabel('\bfz');
```

Surface and contour plots may be created by substituting the appropriate function for the mesh function.

6.7 Summary

MATLAB supports complex numbers as an extension of the double data type. They can be defined using the i or j, both of which are predefined as to be $\sqrt{-1}$. Using complex numbers is straightforward, except that the relational

Mesh Plot

(a)

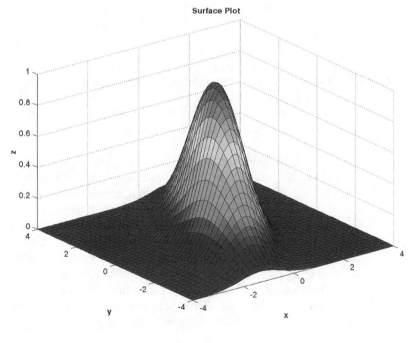

Surface Plot

(b)

Figure 6.12 *(a)* A mesh plot of the function $z(x, y) = e^{-0.5[x^2 + 0.5(x-y)^2]}$. *(b)* A surface plot of the same function.

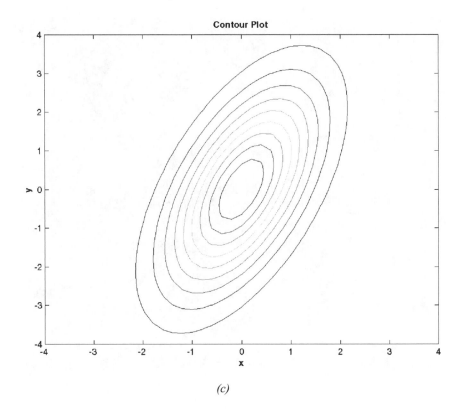

(c)

Figure 6.12 *(continued)* *(c)* A contour plot of the same function.

operators >, >=, <, and <= compare only the *real parts* of complex numbers, not their magnitudes. They must be used with caution when working with complex values.

String functions are functions designed to work with strings, which are arrays of type `char`. These functions allow a user to manipulate strings in a variety of useful ways, including concatenation, comparison, replacement, case conversion, and numeric-to-string and string-to-numeric type conversions.

Multidimensional arrays are arrays with more than two dimensions. They may be created and used in a fashion similar to one- and two-dimensional arrays. Multidimensional arrays appear naturally in certain classes of physical problems.

The `single` data is consists of single-precision floating point numbers. They are created using the `single` function. A mathematical operation between a `single` and a `double` value produces a `single` result.

MATLAB includes signed and unsigned 8-, 16-, 32-, and 64-bit integers. The integer data types are the `int8`, `uint8`, `int16()`, `uint16`, `int32`, `uint32`, `int64`, and `uint64`. Each of these types is created using the corresponding function: `int8()`, `uint8()`, `int16()`, `uint16()`, `int32()`, `uint32()`, `int64()`, or `uint64()`. Mathematical operations (+, −. etc.) can be performed on these data types; the result of an operation between an integer and a `double` has the same type as the integer. If the result of a mathematical operation is too large or too small to be expressed by an integer data type, the result is either the largest or smallest possible integer for that type.

MATLAB includes a rich variety of two- and three-dimensional plots. In this chapter, we introduced stem, stair, bar, compass, mesh, surface, and contour plots.

6.7.1 Summary of Good Programming Practice

The following guidelines should be adhered to:

1. Use the `char` function to create two-dimensional character arrays without worrying about padding each row to the same length.
2. Use function `isstrprop` to determine the characteristics of each character in a string array. This function supersedes the older functions `isletter` and `isspace`, which may be deleted in a future version of MATLAB.
3. Use multidimensional arrays to solve problems that are naturally multivariate in nature, such as aerodynamics and fluid flows.
4. If you are working with multidimensional arrays, be sure to vectorize your code by hand. The MATLAB JIT compiler cannot handle loops containing multidimensional arrays.
5. Do not use the `single` or integer data types unless you have a special need such as image processing.
6. Use function `fplot` to plot functions directly without having to create intermediate data arrays.

6.7.2 MATLAB Summary

The following summary lists all of the MATLAB commands and functions described in this chapter, along with a brief description of each one.

`abs`	Returns absolute value (magnitude) of a number.
`angle`	Returns the angle of a complex number, in radians.
`bar(x,y)`	Create a vertical bar plot.
`barh(x,y)`	Create a horizontal bar plot.
`base2dec`	Convert base B string to decimal integer.
`bin2dec`	Convert binary string to decimal integer.
`blanks`	Create a string of blanks.
`char`	(1) Convert numbers to the corresponding character values. (2) Create a two-dimensional character array from a series of strings.
`compass(x,y)`	Create a compass plot.
`conj`	Compute complex conjugate of a number.
`contour`	Create a contour plot.

deblank	Remove trailing whitespace from a string.
dec2base	Convert decimal integer to base B string.
dec2bin	Convert decimal integer to binary string.
double	Convert characters to the corresponding numeric codes.
find	Find indices and values of nonzero elements in a matrix.
findstr	Find one string within another one.
hex2num	Convert IEEE hexadecimal string to `double`.
hex2dec	Convert hexadecimal string to decimal integer.
hist	Create a histogram of a data set.
full	Convert a sparse matrix into a full matrix.
imag	Returns the imaginary portion of the complex number.
int2str	Convert integer to string.
ischar	Returns true (1) for a character array.
isletter	Returns true (1) for letters of the alphabet.
isreal	Returns true (1) if no element of array has an imaginary component.
isstrprop	Returns true (1) if a character has the specified property.
isspace	Returns true (1) for whitespace.
lower	Convert string to lowercase.
mat2str	Convert matrix to string.
mesh	Create a mesh plot.
meshgrid	Create the (x, y) grid required for mesh, surface, and contour plots.
nnz	Number of nonzero matrix elements.
nonzeros	Return a column vector containing the nonzero elements in a matrix.
num2str	Convert number to string.
nzmax	Amount of storage allocated for nonzero matrix elements.
pie(x)	Create a pie plot.
plot(c)	Plots the real versus the imaginary part of a complex array.
real	Returns the real portion of the complex number.
rose	Create a radial histogram of a data set.
sscanf	Read formatted data from string.
stairs(x,y)	Create a stair plot.
stem(x,y)	Create a stem plot.
str2double	Convert string to `double` value.
str2num	Convert string to number.
strcat	Concatenate strings.
strcmp	Returns true (1) if two strings are identical.

strcmpi	Returns true (1) if two strings are identical ignoring case.
strjust	Justify string.
strncmp	Returns true (1) if first n characters of two strings are identical.
strncmpi	Returns true (1) if first n characters of two strings are identical, ignoring case.
strmatch	Find matches for string.
strtrim	Remove leading and trailing whitespace from a string.
strrep	Replace one string with another.
strtok	Find token in string.
struct	Predefine a structure array.
strvcat	Concatenate strings vertically.
surf	Create a surface plot.
upper	Convert string to uppercase.

6.8   Exercises

6.1   In a sinusoidal steady-state AC circuit, the voltage across a passive element is given by Ohm's Law:

$$\mathbf{V} = \mathbf{I}Z \qquad (6\text{-}14)$$

where $\mathbf{V}$ is the voltage across the element, $\mathbf{I}$ is the current though the element, and Z is the impedance of the element. Note that all three of these values are complex and that these complex numbers are usually specified in the form of a magnitude at a specific phase angle expressed in degrees. For example, the voltage might be $\mathbf{V} = 120\angle30°\text{V}$.

Figure 6.13   The voltage and current relationship on a passive AC circuit element.

Write a program that reads the voltage across an element and the impedance of the element, and calculates the resulting current flow. The input values should be given as magnitudes and angles expressed in degrees, and the resulting answer should be in the same form. Use the function to_complex from Exercise 6.3 to convert the numbers to rectangular for the actual computation of the current, and the function to_polar from Exercise 6.2 to convert the answer into polar form for display.

6.2 Figure 6.14 shows a series *RLC* circuit driven by a sinusoidal ac voltage source whose value is $120\angle 0°$ volts. The impedance of the inductor in this circuit is $Z_L = j2\pi fL$, where j is $\sqrt{-1}$, f is the frequency of the voltage source in hertz and L is the inductance in henrys. The impedance of the capacitor in this circuit is $Z_C = -j\dfrac{1}{2\pi fC}$, where C is the capacitance in farads. Assume that $R = 100\ \Omega$, $L = 0.1$ mH, and $C = 0.25$ nF.

The current **I** flowing in this circuit is given by Kirchhoff's Voltage Law to be

$$\mathbf{I} = \frac{120\angle 0°\ \text{V}}{R + j2\pi fL - j\dfrac{1}{2\pi fC}} \qquad (6\text{-}15)$$

(a) Calculate and plot the magnitude of this current as a function of frequency as the frequency changes from 100 kHz to 10 MHz. Plot this information on both a linear and a log-linear scale. Be sure to include a title and axis labels.

(b) Calculate and plot the phase angle in degrees of this current as a function of frequency as the frequency changes from 100 kHz to 10 MHz. Plot this information on both a linear and a log-linear scale. Be sure to include a title and axis labels.

(c) Plot both the magnitude and phase angle of the current as a function of frequency on two sub-plots of a single figure. Use log-linear scales.

Figure 6.14 A series *RLC* circuit driven by a sinusoidal AC voltage source.

6.3   Write a function `to_polar` that accepts a complex number c, and returns two output arguments containing the magnitude `mag` and angle `theta` of the complex number. The output angle should be in degrees.

6.4   Write a function `to_complex` that accepts two input arguments containing the magnitude `mag` and angle `theta` of the complex number in degrees, and returns the complex number c.

6.5   Write a function that will accept a complex number c, and plot that point on a Cartesian coordinate system with a circular marker. The plot should include both the x and y axes, plus a vector drawn from the origin to the location of c.

6.6   Plot the function $v(t) = 10e^{(-0.2+j\pi)t}$ for $0 \le t \le 10$ using the function `plot(t,v)`. What is displayed on the plot?

6.7   Plot the function $v(t) = 10e^{(-0.2+j\pi)t}$ for $0 \le t \le 10$ using the function `plot(v)`. What is displayed on the plot this time?

6.8   Create a polar plot of the function $v(t) = 10e^{(-0.2+j\pi)t}$ for $0 \le t \le 10$.

6.9   Plot the function $v(t) = 10e^{(-0.2+j\pi)t}$ for $0 \le t \le 10$ using function `plot3`, where the three dimensions to plot are the real part of the function, the imaginary part of the function, and time.

6.10   **Euler's Equation** Euler's equation defines e raised to an imaginary power in terms of sinusoidal functions as follows:

$$e^{j\theta} = \cos\theta + j\sin\theta \tag{6-16}$$

Create a two-dimensional plot of this function as θ varies from 0 to 2π. Create a three-dimensional line plot using function plot3 as θ varies from 0 to 2π (the three dimensions are the real part of the expression, the imaginary part of the expression, and θ).

6.11   Create a mesh, surface plot, and contour plot of the function $z = e^{x+iy}$ for the interval $-1 \le x \le 1$ and $-2\pi \le y \le 2\pi$. In each case, plot the real part of z versus x and y.

6.12   Write a program that accepts an input string from the user and determines the how many times a user-specified character appears within the string. (*Hint:* Look up the `'s'` option of the `input` function using the MATLAB Help Browser.)

6.13   Modify the previous program so that it determines how many times a user-specified character appears within the string without regard to the case of the character.

6.14   Write a program that accepts a string from a user with the `input` function, chops that string into a series of tokens, sorts the tokens into ascending order, and prints them out.

6.15   Write a program that accepts a series of strings from a user with the `input` function, sorts the strings into ascending order, and prints them out.

6.16   Write a program that accepts a series of strings from a user with the `input` function, sorts the strings into ascending order disregarding case, and prints them out.

6.17 MATLAB includes functions `upper` and `lower`, which shift a string to upper case and lower case respectively. Create a new function called `caps`, which capitalizes the first letter in each word and forces all other letters to be lower case. (*Hint:* Take advantage of functions `upper`, `lower`, and `strtok`.)

6.18 Write a function that accepts a character string and returns a `logical` array with true values corresponding to each printable character that is *not* alphanumeric or whitespace (for example, $, %, #, etc.) and false values everywhere else.

6.19 Write a function that accepts a character string and returns a `logical` array with true values corresponding to each vowel and false values everywhere else. Be sure that the function works properly for both lower-case and uppercase characters.

6.20 Plot the function $y = e^{-x} \sin x$ for x between 0 and 2 in steps of 0.1. Create the following plot types: *(a)* stem plot; *(b)* stair plot; *(c)* bar plot; *(d)* compass plot. Be sure to include titles and axis labels on all plots.

6.21 Suppose that George, Sam, Betty, Charlie, and Suzie contributed $15, $5, $10, $5, and $15, respectively, to a colleague's going-away present. Create a pie chart of their contributions. What percentage of the cost was paid by Sam?

6.22 Plot the function $f(x) = 1/\sqrt{x}$ over the range $0.1 \leq x \leq 10.0$ using the function `fplot`. Be sure to label your plot properly.

CHAPTER 7

Advanced Features: Sparse Arrays, Cell Arrays, Structures, and Function Handles

This chapter deals with four very useful features of MATLAB: sparse arrays, cell arrays, structures, and function handles.

Sparse arrays are a special type of array in which memory is allocated only for the nonzero elements in the array. Sparse arrays provide an extremely useful and compact way to represent large arrays containing many zero values.

Cell arrays are very flexible type of array that can hold any sort of data. Each element of a cell array can hold any type of MATLAB data, and different elements within the same array can hold different types of data. They are used extensively in MATLAB graphical user interface (GUI) functions.

Structures are a special type of array with named subcomponents. Each structure can have any number of subcomponents, each with its own name and data type. Structures are the basis of MATLAB objects.

Function handles provide an alternative way to access a function. They are more flexible than simple function names. Function handles make it easy to pass functions to other functions for processing, and also make it easy to save data within a function between calls.

7.1 Sparse Arrays

We learned about ordinary MATLAB arrays in Chapter 2. When an ordinary array is declared, MATLAB creates a memory location for every element in the array. For example, the function $a = eye(10)$ creates 100 elements arranged as a 10×10 structure. In this array, 90 of those elements are zero! This matrix requires

100 elements, but only 10 of them contain nonzero data. This is an example of a **sparse array** or **sparse matrix**. A sparse matrix is a large matrix in which the vast majority of the elements are zero.

```
» a = 2 * eye(10);
a =
     2    0    0    0    0    0    0    0    0    0
     0    2    0    0    0    0    0    0    0    0
     0    0    2    0    0    0    0    0    0    0
     0    0    0    2    0    0    0    0    0    0
     0    0    0    0    2    0    0    0    0    0
     0    0    0    0    0    2    0    0    0    0
     0    0    0    0    0    0    2    0    0    0
     0    0    0    0    0    0    0    2    0    0
     0    0    0    0    0    0    0    0    2    0
     0    0    0    0    0    0    0    0    0    2
```

Now suppose that we create another 10×10 matrix b defined as follows:

```
b =
     1    0    0    0    0    0    0    0    0    0
     0    2    0    0    0    0    0    0    0    0
     0    0    2    0    0    0    0    0    0    0
     0    0    0    1    0    0    0    0    0    0
     0    0    0    0    5    0    0    0    0    0
     0    0    0    0    0    1    0    0    0    0
     0    0    0    0    0    0    1    0    0    0
     0    0    0    0    0    0    0    1    0    0
     0    0    0    0    0    0    0    0    1    0
     0    0    0    0    0    0    0    0    0    1
```

If these two matrices are multiplied together, the result is

```
» c = a * b
c =
     2    0    0    0    0    0    0    0    0    0
     0    4    0    0    0    0    0    0    0    0
     0    0    4    0    0    0    0    0    0    0
     0    0    0    2    0    0    0    0    0    0
     0    0    0    0   10    0    0    0    0    0
     0    0    0    0    0    2    0    0    0    0
     0    0    0    0    0    0    2    0    0    0
     0    0    0    0    0    0    0    2    0    0
     0    0    0    0    0    0    0    0    2    0
     0    0    0    0    0    0    0    0    0    2
```

The process of multiplying these two sparse matrices together requires 1900 multiplications and additions, but because most of the terms being added and multiplied are zeros, it was largely wasted effort.

This problem gets worse rapidly as matrix size increases. For example, suppose that we were to generate two 200×200 sparse matrices a and b as follows:

```
a = 5 * eye(200);
b = 3 * eye(200);
```

Each matrix now contains 20,000 elements, of which 19,800 are zero! Furthermore, multiplying these two matrices together requires **7,980,000** additions and multiplications.

It should be apparent that storing and working with large sparse matrices, most of whose elements are zero, is a serious waste of both computer memory and CPU time. Unfortunately, many real-world problems naturally create sparse matrices, so we need some efficient way to solve problems involving them.

Calculations involving a large electric power system are excellent examples of real-world problems involving sparse matrices. Large electric power systems can have a thousand or more electrical busses at generating plants and transmission and distribution substations. If we wish to know the voltages, currents, and power flows in the system, we must first solve for the voltage at every bus. For a 1000-bus system, this involves the simultaneous solution of 1000 equations in 1000 unknowns, which is equivalent to inverting a matrix with 1,000,000 elements. Solving this matrix requires millions of floating point operations.

However, each bus in the power system is probably connected to an average of only two or three other busses, so 996 of the 1000 terms in each row of the matrix will be zeros, and most of the operations involved in inverting the matrix will be additions and multiplications by zeros. The calculation of the voltages and currents in this power system would be much simpler and more efficient if the zeros could be ignored in the solution process.

7.1.1 The sparse Attribute

MATLAB has a special version of the double data type that is designed to work with sparse arrays. In this special version of the double data type, *only the non-zero elements of an array are allocated memory locations*, and the array is said to have the "sparse" attribute. An array with the sparse attribute actually saves three values for each nonzero element: the value of the element itself and the row and column numbers where the element is located. Even though three values must be saved for each nonzero element, this approach is *much* more memory efficient than allocating full arrays if a matrix has only a few nonzero elements.

To illustrate the use of sparse matrices, we will create a 10×10 identity matrix:

```
» a = eye(10)
a =
     1   0   0   0   0   0   0   0   0   0
     0   1   0   0   0   0   0   0   0   0
     0   0   1   0   0   0   0   0   0   0
     0   0   0   1   0   0   0   0   0   0
     0   0   0   0   1   0   0   0   0   0
     0   0   0   0   0   1   0   0   0   0
     0   0   0   0   0   0   1   0   0   0
     0   0   0   0   0   0   0   1   0   0
     0   0   0   0   0   0   0   0   1   0
     0   0   0   0   0   0   0   0   0   1
```

If this matrix is converted to a sparse matrix using function sparse, the results are

```
» as = sparse(a)
as =
     (1,1)        1
     (2,2)        1
     (3,3)        1
     (4,4)        1
     (5,5)        1
     (6,6)        1
     (7,7)        1
     (8,8)        1
     (9,9)        1
    (10,10)       1
```

Note that the data in the sparse matrix is a list of row and column addresses, followed by the nonzero data value at that point. This is a very efficient way to store data as long as most of the matrix is zero, but if there are many nonzero elements, it can take up even more space than the full matrix because of the need to store the addresses.

If we examine arrays a and as with the whos command, the results are

```
» whos
  Name        Size        Bytes       Class        Attributes
   a          10x10        800        double
   as         10x10        164        double       sparse
```

The a array occupies 800 bytes, because there are 100 elements with 8 bytes of storage each. The as array occupies 164 bytes, because there are 10 nonzero elements with 8 bytes of storage each, plus 20 array indices occupying 4 bytes each, and 4 bytes of overhead. Note that the sparse array occupies much less memory than the full array.

The function issparse can be used to determine whether or not a given array is sparse. If an array is sparse, then issparse(array) returns true (1).

The power of the sparse data type can be seen by considering a 1000×1000 matrix z with an average of 4 nonzero elements per row. If this matrix is stored as a full matrix, it will require 8,000,000 bytes of space. On the other hand, if it is converted to a sparse matrix, the memory usage will drop dramatically.

```
» zs = sparse(z);
» whos
   Name        Size            Bytes          Class
   z         1000x1000       8000000        double array
   zs        1000x1000         51188        sparse array

Grand total is 1003932 elements using 8051188 bytes
```

Generating Sparse Matrices

MATLAB can generate sparse matrices by converting a full matrix into a sparse matrix with the sparse function or by directly generating sparse matrices with the MATLAB functions speye, sprand, and sprandn, which are the sparse equivalents of eye, rand, and randn. For example, the expression a = speye(4) generates a 4×4 sparse matrix.

```
» a = speye(4)
a =
    (1,1)      1
    (2,2)      1
    (3,3)      1
    (4,4)      1
```

The expression b = full(a) converts the sparse matrix into a full matrix.

```
» b = full(a)
b =
     1    0    0    0
     0    1    0    0
     0    0    1    0
     0    0    0    1
```

Working with Sparse Matrices

Once a matrix has been made sparse, individual elements can be added to it or deleted from it using simple assignment statements. For example, the following statement generates a 4×4 sparse matrix and then adds another nonzero element to it:

```
» a = speye(4)
a =
    (1,1)      1
    (2,2)      1
    (3,3)      1
    (4,4)      1
```

```
» a(2,1) = -2
a =
    (1,1)        1
    (2,1)       -2
    (2,2)        1
    (3,3)        1
    (4,4)        1
```

MATLAB allows full and sparse matrices to be freely mixed and used in any combination. The result of an operation between a full matrix and a sparse matrix may be either a full matrix or a sparse matrix depending on which result is the most efficient. Essentially any matrix technique that is supported for full matrices is also available for sparse matrices.

A few of the common sparse matrix functions are listed in Table 7.1.

Table 7.1   Common MATLAB Sparse Matrix Functions

Function	Description
Create Sparse Matrices	
speye	Create a sparse identity matrix.
sprand	Create a sparse uniformly distributed random matrix.
sprandn	Create a sparse normally distributed random matrix.
Full-to-Sparse Conversion Functions	
sparse	Convert a full matrix into a sparse matrix.
full	Convert a sparse matrix into a full matrix.
find	Find indices and values of nonzero elements in a matrix.
Working with Sparse Matrices	
nnz	Number of nonzero matrix elements.
nonzeros	Return a column vector containing the nonzero elements in a matrix.
nzmax	Amount of storage allocated for nonzero matrix elements.
spones	Replace nonzero sparse matrix elements with ones.
spalloc	Allocate space for a sparse matrix.
issparse	Returns 1 (true) for sparse matrix.
spfun	Apply function to nonzero matrix elements.
spy	Visualize sparsity pattern as a plot.

▶

Example 7.1—Solving Simultaneous Equations with Sparse Matrices

To illustrate the ease with which sparse matrices can be used in MATLAB, we will solve the following simultaneous system of equations with both full and sparse matrices.

$$1.0x_1 + 0.0x_2 + 1.0x_3 + 0.0x_4 + 0.0x_5 + 2.0x_6 + 0.0x_7 - 1.0x_8 = 3.0$$
$$0.0x_1 + 1.0x_2 + 0.0x_3 + 0.4x_4 + 0.0x_5 + 0.0x_6 + 0.0x_7 + 0.0x_8 = 2.0$$
$$0.5x_1 + 0.0x_2 + 2.0x_3 + 0.0x_4 + 0.0\,x_5 + 0.0x_6 - 1.0x_7 + 0.0x_8 = -1.5$$
$$0.0x_1 + 0.0x_2 + 0.0x_3 + 2.0x_4 + 0.0x_5 + 1.0x_6 + 0.0x_7 + 0.0x_8 = 1.0$$
$$0.0x_1 + 0.0x_2 + 1.0x_3 + 1.0x_4 + 1.0x_5 + 0.0x_6 + 0.0x_7 + 0.0x_8 = -2.0$$
$$0.0x_1 + 0.0x_2 + 0.0x_3 + 1.0x_4 + 0.0x_5 + 1.0x_6 + 0.0x_7 + 0.0x_8 = 1.0$$
$$0.5x_1 + 0.0x_2 + 0.0x_3 + 0.0x_4 + 0.0x_5 + 0.0x_6 + 1.0x_7 + 0.0x_8 = 1.0$$
$$0.0x_1 + 1.0x_2 + 0.0x_3 + 0.0x_4 + 0.0x_5 + 0.0x_6 + 0.0x_7 + 1.0x_8 = 1.0$$

SOLUTION   To solve this problem, we will create full matrices of the equation coefficients and convert them to sparse form using the sparse function. Then we will solve the equation both ways, comparing the results and the memory required. The script file to perform these calculations is

```
%   Script file: simul.m
%
%   Purpose:
%     This program solves a system of 8 linear equations in 8
%     unknowns (a*x = b), using both full and sparse matrices.
%
%   Record of revisions:
%       Date          Programmer           Description of change
%       ====          ==========           =====================
%     03/03/07      S. J. Chapman          Original code
%
% Define variables:
%   a               -- Coefficients of x (full matrix)
%   as              -- Coefficients of x (sparse matrix)
%   b               -- Constant coefficients (full matrix)
%   bs              -- Constant coefficients (sparse matrix)
%   x               -- Solution (full matrix)
%   xs              -- Solution (sparse matrix)

% Define coefficients of the equation a*x = b for
% the full matrix solution.
a = [ 1.0    0.0    1.0    0.0    0.0    2.0    0.0  -1.0; ...
      0.0    1.0    0.0    0.4    0.0    0.0    0.0   0.0; ...
      0.5    0.0    2.0    0.0    0.0    0.0   -1.0   0.0; ...
      0.0    0.0    0.0    2.0    0.0    1.0    0.0   0.0; ...
      0.0    0.0    1.0    1.0    1.0    0.0    0.0   0.0; ...
      0.0    0.0    0.0    1.0    0.0    1.0    0.0   0.0; ...
      0.5    0.0    0.0    0.0    0.0    0.0    1.0   0.0; ...
      0.0    1.0    0.0    0.0    0.0    0.0    0.0   1.0];

b = [ 3.0    2.0   -1.5    1.0   -2.0    1.0    1.0   1.0]';
```

```
% Define coefficients of the equation a*x = b for
% the sparse matrix solution.
as = sparse(a);
bs = sparse(b);

% Solve the system both ways
disp ('Full matrix solution:');
x = a\b

disp ('Sparse matrix solution:');
xs = as\bs

% Show workspace
disp('Workspace contents after the solutions:')
whos
```

When this program is executed, the results are

```
» simul
Full matrix solution:
x =
      0.5000
      2.0000
     -0.5000
     -0.0000
     -1.5000
      1.0000
      0.7500
     -1.0000
Sparse matrix solution:
xs =
    (1,1)        0.5000
    (2,1)        2.0000
    (3,1)       -0.5000
    (5,1)       -1.5000
    (6,1)        1.0000
    (7,1)        0.7500
    (8,1)       -1.0000
Workspace contents after the solutions:
  Name      Size         Bytes      Class         Attributes

    a        8x8          512       double
    as       8x8          276       double        sparse
    b        8x1           64       double
    bs       8x1          104       double        sparse
    x        8x1           64       double
    xs       8x1           92       double        sparse
```

The answers are the same for both solutions. Note that the sparse solution does not contain a solution for x_4, because that value is zero and zeros aren't carried in a sparse matrix! Also, note that the sparse form of matrix b actually takes up more space than the full form. This happens because the sparse representation must store the indices as well as the values in the arrays, so it is less efficient if most of the elements in an array are nonzero.

◀

7.2   Cell Arrays

A **cell array** is a special MATLAB array whose elements are *cells*, containers that can hold other MATLAB arrays. For example, one cell of a cell array might contain an array of real numbers, another an array of strings, and yet another a vector of complex numbers (see Figure 7.1).

In programming terms, each element of a cell array is a *pointer* to another data structure, and those data structures can be of different types. Figure 7.2 illustrates this concept. Cell arrays provide a great way to collect information about a problem, because all of the information can be kept together and accessed by a single name.

Cell arrays use braces "{}" instead of parentheses "()" for selecting and displaying the contents of cells. This difference is due to the fact that *cell arrays contain data structures instead of data*. Suppose that the cell array a is defined

cell 1,1	**cell 1,2**
$\begin{bmatrix} 1 & 3 & -7 \\ 2 & 0 & 6 \\ 0 & 5 & 1 \end{bmatrix}$	'This is a text string.'
cell 2,1	**cell 2,2**
$\begin{bmatrix} 3+i4 & -5 \\ -i10 & 3-i4 \end{bmatrix}$	[ ]

Figure 7.1   The individual elements of a cell array may point to real arrays, complex arrays, string, other cell arrays, or even empty arrays.

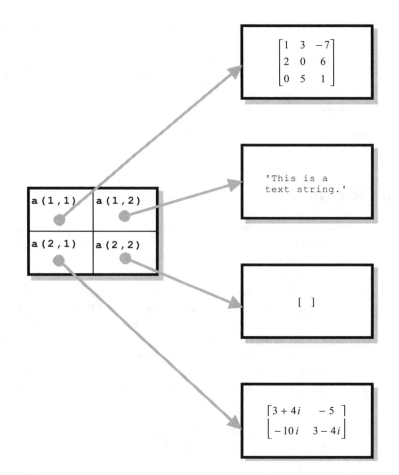

Figure 7.2 Each element of a cell array holds a *pointer* to another data structure, and different cells in the same cell array can point to different types of data structures.

as shown in Figure 7.2. Then the contents of element a(1,1) is a data structure containing a 3 × 3 array of numeric data, and a reference to a(1,1) displays the *contents* of the cell, which is the data structure.

```
» a(1,1)
ans =
    [3x3 double]
```

By contrast, a reference to a{1,1} displays *the contents of the data item contained in the cell*.

```
» a{1,1}
ans =
        1    3   -7
        2    0    6
        0    5    1
```

In summary, the notation a(1,1) refers to the contents of cell a(1,1) (which is a data structure), while the notation a{1,1} refers to the contents of the data structure within the cell.

💣 Programming Pitfalls

Be careful not to confuse () with {} when addressing cell arrays. They are very different operations!

7.2.1 Creating Cell Arrays

Cell arrays can be created in two ways:

- By using assignment statements
- By preallocating a cell array using the `cell` function

The simplest way to create a cell array is to directly assign data to individual cells, one cell at a time. However, preallocating cell arrays is more efficient, so you should preallocate really large cell arrays.

Allocating Cell Arrays Using Assignment Statements

You can assign values to cell arrays one cell at a time using assignment statements. There are two ways to assign data to cells, known as **content indexing** and **cell indexing**.

Content indexing involves placing braces "{}" around the cell subscripts, together with cell contents in ordinary notation. For example, the following statement creates the 2 × 2 cell array in Figure 7.2:

```
a{1,1} = [1 3 -7; 2 0 6; 0 5 1];
a{1,2} = 'This is a text string.';
a{2,1} = [3+4*i -5; -10*i 3 - 4*i];
a{2,2} = [];
```

This type of indexing defines the *contents of the data structure contained in a cell.*

Cell indexing involves placing braces "{}" around the data to be stored in a cell, together with cell subscripts in ordinary subscript notation. For example, the following statement creates the 2 × 2 cell array in Figure 7.2:

```
a(1,1) = {[1 3 -7; 2 0 6; 0 5 1]};
a(1,2) = {'This is a text string.'};
a(2,1) = {[3+4*i -5; -10*i 3 - 4*i]};
a(2,2) = {[]};
```

This type of indexing *creates a data structure containing the specified data and then assigns that data structure to a cell.*

These two forms of indexing are completely equivalent, and they may be freely mixed in any program.

Programming Pitfalls

Do not attempt to create a cell array with the same name as an existing numeric array. If you do this, MATLAB will assume that you are trying to assign cell contents to an ordinary array, and it will generate an error message. Be sure to clear the numeric array before trying to create a cell array with the same name.

Preallocating Cell Arrays with the `cell` Function

The `cell` function allows you to preallocate empty cell arrays of the specified size. For example, the following statement creates an empty 2×2 cell array:

```
a = cell(2,2);
```

Once a cell array has been created, you can use assignment statements to fill values in the cells.

7.2.2   Using Braces `{ }` as Cell Constructors

It is possible to define many cells at once by placing all of the cell contents between a single set of braces. Individual cells on a row are separated by commas, and rows are separated by semicolons. For example, the following statement creates a 2×3 cell array:

```
b = {[1 2], 17, [2;4]; 3-4*i, 'Hello', eye(3)}
```

7.2.3   Viewing the Contents of Cell Arrays

MATLAB displays the data structures in each element of a cell array in a condensed form that limits each data structure to a single line. If the entire data structure can be displayed on the single line, it will be. Otherwise, a summary would be displayed. For example, cell arrays a and b would be displayed as:

```
» a
a =
    [3x3 double]     [1x22 char]
    [2x2 double]             []
» b
b =
           [1x2 double]   [    17] [2x1 double]
    [3.0000- 4.0000i]    'Hello' [3x3 double]
```

Note that MATLAB *is displaying the data structures*, complete with brackets or apostrophes, not the entire contents of the data structures.

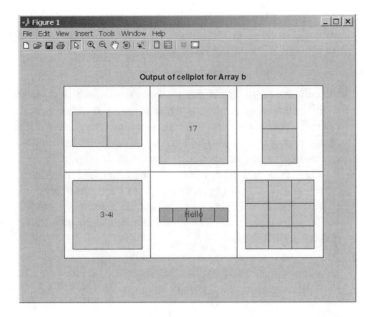

Figure 7.3   The structure of cell array b is displayed as a nested series of boxes by function cellplot.

If you would like to see the full contents of a cell array, use the celldisp function. This function displays *the contents of the data structures in each cell.*

```
» celldisp(a)
a{1,1} =
      1    3    -7
      2    0     6
      0    5     1
a{2,1} =
   3.0000 + 4.0000i   -5.0000
        0 -10.0000i    3.0000 - 4.0000i
a{1,2} =
This is a text string.
a{2,2} =
      []
```

For a high-level graphical display of the structure of a cell array, use function cellplot. For example, the function cellplot(b) produces the plot shown in Figure 7.3.

7.2.4   Extending Cell Arrays

If a value is assigned to a cell array element that does not currently exist, the element will be automatically created, and any additional cells necessary to preserve the shape of the array will be automatically created. For example, suppose that array a has been defined to be a 2 × 2 cell array as shown in Figure 7.1. If the following statement is executed

```
a{3,3} = 5
```

the cell array will be automatically extended to 3 × 3, as shown in Figure 7.4.

Preallocating cell arrays with the `cell` function is much more efficient than extending the arrays one element at a time using assignment statements. When a new element is added to an existing array as we did previously, MATLAB must create a new array large enough to include this new element, copy the old data into the new array, add the new value to the array, and then delete the old array. This is a very time-consuming process. Instead, you should always allocate the cell array to be the largest size that you can, and then add values to it one element at a time. If you do that, only the new element needs to be added; the rest of the array can remain undisturbed.

cell 1,1	cell 1,2	cell 1,3
$\begin{bmatrix} 1 & 3 & -7 \\ 2 & 0 & 6 \\ 0 & 5 & 1 \end{bmatrix}$	'This is a text string.'	[ ]
cell 2,1	**cell 2,2**	**cell 2,3**
$\begin{bmatrix} 3+i4 & -5 \\ -i10 & 3-i4 \end{bmatrix}$	[ ]	[ ]
cell 3,1	**cell 3,2**	**cell 3,3**
[ ]	[ ]	[5]

Figure 7.4 The result of assigning a value to a{3,3}. Note that four other empty cells were created to preserve the shape of the cell array.

The program that follows illustrates the advantages of preallocation. It creates a cell array containing 50,000 strings added one at a time, with and without preallocation.

```
%  Script file: test_preallocate.m
%
%  Purpose:
%    This program tests the creation of cell arrays with
%    and without preallocation.
%
%  Record of revisions:
%      Date         Programmer          Description of change
%      ====         ==========          =====================
%    03/04/07     S. J. Chapman         Original code
%
% Define variables:
%    a                -- Cell array
%    maxvals          -- Maximum values in cell array

% Create array without preallocation
clear all
maxvals = 50000;
tic
for ii = 1:maxvals
   a{ii} = ['Element ' int2str(ii)];
end
disp( ['Elapsed time without preallocation = ' num2str(toc)] );

% Create array with preallocation
clear all
maxvals = 50000;
tic
a = cell(1,maxvals);
for ii = 1:maxvals
   a{ii} = ['Element ' int2str(ii)];
end
disp( ['Elapsed time with preallocation = ' num2str(toc)] );
```

When this program is executed using MATLAB 7.4 on a 1.8 GHz Pentium Core 2 Duo computer, the results are as shown below. The advantages of preallocation are obvious.

```
» test_preallocate
Elapsed time without preallocation = 7.6484
Elapsed time with preallocation    = 2.6934
```

❋ **Good Programming Practice**

Always preallocate all cell arrays before assigning values to the elements of the array. This practice greatly increases the execution speed of a program.

7.2.5 Deleting Cells in Arrays

To delete an entire cell array, use the `clear` command. Subsets of cells may be deleted by assigning an empty array to them. For example, assume that a is the 3×3 cell array defined previously.

```
» a
a =
    [3x3 double]    [1x22 char]    []
    [2x2 double]             []    []
            []              []    [5]
```

It is possible to delete the entire third row with the statement

```
» a(3,:) = []
a =
    [3x3 double]    [1x22 char]    []
    [2x2 double]             []    []
```

7.2.6 Using Data in Cell Arrays

The data stored inside the data structures within a cell array may be used at any time, with either content indexing or cell indexing. For example, suppose that a cell array c is defined as

```
c = {[1 2;3 4], 'dogs'; 'cats', i}
```

The contents of the array stored in cell c(1,1) can be accessed as follows:

```
» c{1,1}
ans =
     1     2
     3     4
```

and the contents of the array in cell c(2,1) can be accessed as follows:

```
» c{2,1}
ans =
cats
```

Subsets of a cell's contents can be obtained by concatenating the two sets of subscripts. For example, suppose that we would like to get the element (1, 2) from the array stored in cell `c(1,1)` of cell array `c`. To do this, we would use the expression `c{1,1}(1,2)`, which says "select element (1, 2) from the contents of the data structure contained in cell `c(1,1)`."

```
» c{1,1}(1,2)
ans =
    2
```

7.2.7   Cell Arrays of Strings

It is often convenient to store groups of strings in a cell array instead of storing them in rows of a standard character array, because each string in a cell array can have a different length, whereas every row of a standard character array must have an identical length. This fact means that *strings in cell arrays do not have to be padded with blanks*. Many MATLAB graphical user interface functions use cell arrays for precisely this reason, as we will see in Chapter 10.

Cell arrays of strings can be created in one of two ways. Either the individual strings can be inserted into the array with brackets, or else function `cellstr` can be used to convert a two-dimensional string array into a cell array of strings.

The following example creates a cell array of strings by inserting the strings into the cell array one at a time, and displays the resulting cell array. Note that the individual strings can be of different lengths.

```
» cellstring{1} = 'Stephen J. Chapman';
» cellstring{2} = 'Male';
» cellstring{3} = 'SSN 999-99-9999';
» cellstring
    'Stephen J. Chapman'    'Male'    'SSN 999-99-9999'
```

Function `cellstr` creates a cell array of strings from a two-dimensional string array. Consider the character array

```
» data = ['Line 1         ';'Additional Line']
data =
Line 1
Additional Line
```

This 2 × 15 character array can be converted into a cell array of strings with the function `cellstr` as follows:

```
» c = cellstr(data)
c =
    'Line 1'
    'Additional Line'
```

and it can be converted back to a standard character array using function `char`:

```
» newdata = char(c)
newdata =
Line 1
Additional Line
```

7.2.8 The Significance of Cell Arrays

Cell arrays are extremely flexible, since any amount of any type of data can be stored in each cell. As a result, cell arrays are used in many internal MATLAB data structures. We must understand them in order to use many features of the MATLAB graphical user interface, which we will study in Chapter 10.

In addition, the flexibility of cell arrays makes them regular features of functions with variable numbers of input arguments and output arguments. A special input argument, `varargin`, is available within user-defined MATLAB functions to support variable numbers of input arguments. This argument appears as the last item in an input argument list, and it returns a cell array; thus, *a single dummy input argument can support any number of actual arguments*. Each actual argument becomes one element of the cell array returned by `varargin`. If it is used, `varargin` must be the *last* input argument in a function, following all of the required input arguments.

For example, suppose that we are writing a function that may have any number of input arguments. This function could be implemented as shown here:

```
function test1(varargin)
disp(['There are ' int2str(nargin) ' arguments.']);
disp('The input arguments are:');
disp(varargin);
```

When this function is executed with varying numbers of arguments, the results are

```
» test1
There are 0 arguments.
The input arguments are:
» test1(6)
There are 1 arguments.
The input arguments are:
    [6]
» test1(1,'test 1',[1 2;3 4])
There are 3 arguments.
The input arguments are:
    [1]    'test 1'    [2x2 double]
```

As you can see, the arguments become a cell array within the function.

A sample function making use of variable numbers of arguments is shown at the end of this paragraph. Function `plotline` accepts an arbitrary number of

1×2 row vectors, with each vector containing the (x, y) position of one point to plot. The function plots a line connecting all of the (x, y) values together. Note that this function also accepts an optional line specification string and passes that specification on to the plot function.

```
function plotline(varargin)
%PLOTLINE Plot points specified by [x,y] pairs.
% Function PLOTLINE accepts an arbitrary number of
% [x,y] points and plots a line connecting them.
% In addition, it can accept a line specification
% string, and pass that string on to function plot.

% Define variables:
%   ii          -- Index variable
%   jj          -- Index variable
%   linespec    -- String defining plot characteristics
%   msg         -- Error message
%   varargin    -- Cell array containing input arguments
%   x           -- x values to plot
%   y           -- y values to plot

% Record of revisions:
%     Date      Programmer          Description of change
%     ====      ==========          =====================
%   03/05/07    S. J. Chapman       Original code

% Check for a legal number of input arguments.
% We need at least 2 points to plot a line...
msg = nargchk(2,Inf,nargin);
error(msg);

% Initialize values
jj = 0;
linespec = '';

% Get the x and y values, making sure to save the line
% specification string, if one exists.
for ii = 1:nargin

  % Is this argument an [x,y] pair or the line
  % specification?
  if ischar(varargin{ii})

     % Save line specification
     linespec = varargin{ii};

  else

     % This is an [x,y] pair. Recover the values.
     jj = jj + 1;
```

```
        x(jj) = varargin{ii}(1);
        y(jj) = varargin{ii}(2);
   end
end

% Plot function.
if isempty(linespec)
   plot(x,y);
else
   plot(x,y,linespec);
end
```

When this function is called with the following arguments, the resulting plot is shown in Figure 7.5. Try the function with different numbers of arguments and see for yourself how it behaves.

```
plotline([0 0],[1 1],[2 4],[3 9],'k--');
```

There is also a special output argument, varargout, to support variable numbers of output arguments. This argument appears as the last item in an output argument list, and it returns a cell array; thus, *a single dummy output argument can*

Figure 7.5 The plot produced by function plotline.

support any number of actual arguments. Each actual argument becomes one element of the cell array stored in `varargout`.

If it is used, `varargout` must be the *last* output argument in a function, following all of the required input arguments. The number of values to be stored in `varargout` can be determined from function `nargout`, which specifies the number of actual output arguments for any given function call.

A sample function `test2` is shown below. This function detects the number of output arguments expected by the calling program, using the function `nargout`. It returns the number of random values in the first output argument and then fills the remaining output arguments with random numbers taken from a Gaussian distribution. Note that the function uses `varargout` to hold the random numbers, so that there can be an arbitrary number of output values.

```
function [nvals,varargout] = test2(mult)
% nvals is the number of random values returned
% varargout contains the random values returned
nvals = nargout - 1;
for ii = 1:nargout-1
    varargout{ii} = randn * mult;
end
```

When this function is executed, the results are as follows:

```
» test2(4)
ans =
    -1
» [a b c d] = test2(4)
a =
    3
b =
    -1.7303
c =
    -6.6623
d =
    0.5013
```

✳ Good Programming Practice

Use cell array arguments `varargin` and `varargout` to create functions that support varying numbers of input and output arguments.

Table 7.2   Common MATLAB Cell Functions

Function	Description
`cell`	Predefine a cell array structure.
`celldisp`	Display contents of a cell array.
`cellplot`	Plot structure of a cell array.
`cellstr`	Convert a two-dimensional character array to a cell array of strings.
`char`	Convert a cell array of strings into a two-dimensional character array.

7.2.9   Summary of `cell` Functions

The common MATLAB cell functions are summarized in Table 7.2.

7.3   Structure Arrays

An *array* is a data type in which there is a name for the whole data structure, but individual elements within the array are known only by number. Thus the fifth element in the array named `arr` would be accessed as `arr(5)`. All of the individual elements in an array must be of the *same* type.

A *cell array* is a data type in which there is a name for the whole data structure, but individual elements within the array are known only by number. However, the individual elements in the cell array may be of *different* types.

In contrast, a **structure** is a data type in which each individual element has a name. The individual elements of a structure are known as **fields**, and each field in a structure may have a different type. The individual fields are addressed by combining the name of the structure with the name of the field, separated by a period.

Figure 7.6 shows a sample structure named `student`. This structure has five fields; `name,addr1,city,state`, and `zip`. The field "name" would be addressed as `student.name`.

A **structure array** is an array of structures. Each structure in the array will have identically the same fields, but the data stored in each field can differ. For example, a class could be described by an array of the structure `student`. The first student's name would be addressed as `student(1).name`, the second student's city would be addressed as `student(2).city`, and so forth.

7.3.1   Creating Structure Arrays

Structure arrays can be created in two ways:

- A field at a time, using assignment statements
- All at once, using the `struct` function

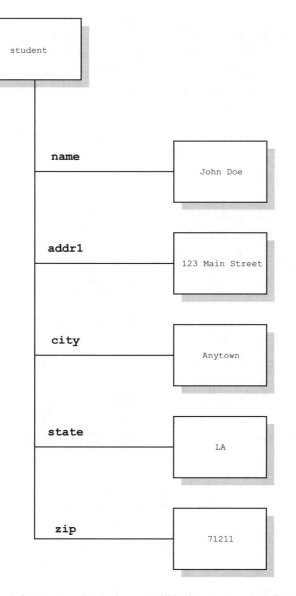

Figure 7.6   A sample structure. Each element within the structure is called a field, and each field is addressed by name.

Building a Structure with Assignment Statements

You can build a structure one field at a time using assignment statements. Each time data is assigned to a field, that field is automatically created. For example, the structure shown in Figure 7.6 can be created with the following statements:

```
» student.name = 'John Doe';
» student.addr1 = '123 Main Street';
» student.city = 'Anytown';
» student.state = 'LA';
» student.zip = '71211'
student =
    name:    'John Doe'
    addr1:   '123 Main Street'
    city:    'Anytown'
    state:   'LA'
    zip:     '71211'
```

A second student can be added to the structure by adding a subscript to the structure name (*before* the period).

```
» student(2).name = 'Jane Q. Public'
student =
1x2 struct array with fields:
    name
    addr1
    city
    state
    zip
```

`student` is now a 1 × 2 array. Note that when a structure array has more than one element, only the field names are listed, not their contents. The contents of each element can be listed by typing the element separately in the Command Window:

```
» student(1)
ans =
      name:  'John Doe'
     addr1:  '123 Main Street'
      city:  'Anytown'
     state:  'LA'
       zip:  '71211'
» student(2)
ans =
      name:   'Jane Q. Public'
     addr1:   []
      city:   []
     state:   []
       zip:   []
```

Note that *all of the fields of a structure are created for each array element whenever that element is defined,* even if they are not initialized. The uninitialized fields will contain empty arrays, which can be initialized with assignment statements at a later time.

The field names used in a structure can be recovered at any time using the `fieldnames` function. This function returns a list of the field names in a cell array of strings and is very useful for working with structure arrays within a program.

Creating Structures with the `struct` Function

The `struct` function allows you to preallocate a structure or an array of structures. The basic form of this function is

```
str_array = struct('field1',val1,'field2',val2, ...)
```

where the arguments are field names and their initial values. With this syntax, function `struct` initializes every field to the specified value.

To preallocate an entire array with the `struct` function, simply assign the output of the `struct` function to the *last value* in the array. All of the values before that will be automatically created at the same time. For example, the statements that follow create an array containing 1000 structures of type `student`.

```
student(1000) = struct('name',[].'addr1',[], ...
                       'city',[],'state',[],'zip',[])

student =
1x1000 struct array with fields:
    name
    addr1
    city
    state
    zip
```

All of the elements of the structure are preallocated, which will speed up any program using the structure.

There is another version of the `struct` function that will preallocate an array and at the same time assign initial values to all of its fields. You will be asked to do this in an end-of-chapter exercise.

7.3.2 Adding Fields to Structures

If a new field name is defined for any element in a structure array, the field is automatically added to all of the elements in the array. For example, suppose that we add some exam scores to Jane Public's record:

```
» student(2).exams = [90 82 88]
student =
1x2 struct array with fields:
    name
    addr1
    city
    state
    zip
    exams
```

There is now a field called exams in every record of the array, as shown at the end of this paragraph. This field will be initialized for student(2) and will be an empty array for all other students until appropriate assignment statements have been issued.

```
» student(1)
ans =
        name: 'John Doe'
       addr1: '123 Main Street'
        city: 'Anytown'
       state: 'LA'
         zip: '71211'
       exams: []
» student(2)
ans =
        name: 'Jane Q. Public'
       addr1: []
        city: []
       state: []
         zip: []
       exams: [90 82 88]
```

7.3.3 Removing Fields from Structures

A field may be removed from a structure array using the rmfield function. The form of this function is

```
struct2 = rmfield(str_array,'field')
```

where str_array is a structure array, 'field' is the field to remove, and struct2 is the name of a new structure with that field removed. For example, we can remove the field 'zip' from structure array student with the following statement:

```
» stu2 = rmfield(student,'zip')
stu2 =
1x2 struct array with fields:
    name
    addr1
    city
    state
    exams
```

7.3.4 Using Data in Structure Arrays

Now let's assume that structure array student has been extended to include 3 students, and all data has been filled in, as shown in Figure 7.7. How do we use the data in this structure array?

To access the information in any field of any array element, just name the array element followed by a period and the field name:

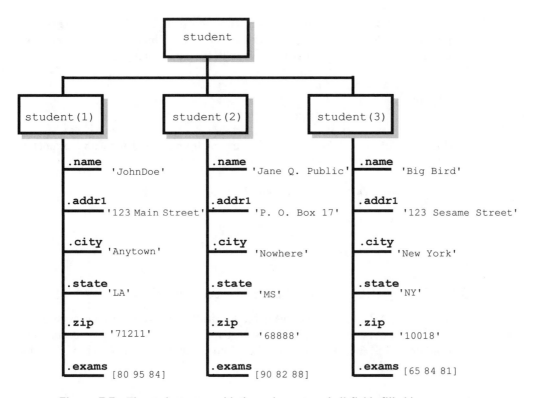

Figure 7.7 The student array with three elements and all fields filled in.

```
» student(2).addr1
ans =
P. O. Box 17
» student(3).exams
ans =
     65     84     81
```

To access an individual item within a field, add a subscript after the field name. For example, the second exam of the third student is

```
» student(3).exams(2)
ans =
     84
```

The fields in a structure array can be used as arguments in any function that supports that type of data. For example, to calculate student(2)'s exam average, we could use the function

```
» mean(student(2).exams)
ans =
     86.6667
```

To extract the values from a given field across multiple array elements, simply place the structure and field name inside a set of brackets. For example, we can get access to an array of zip codes with the expression [student.zip]:

```
» [student.zip]
ans =
      71211     68888     10018
```

Similarly, we can get the average of *all* exams from *all* students with the function mean([student.exams]).

```
» mean([student.exams])
ans =
      83.2222
      71211     68888     10018
```

7.3.5   The getfield and setfield Functions

Two MATLAB functions are available to make structure arrays easier to use in programs. Function getfield gets the current value stored in a field, and function setfield inserts a new value into a field. The structure of function getfield is

```
f = getfield(array,{array_index},'field',{field_index})
```

where the field_index is optional, and array_index is optional for a 1-by-1 structure array. The function call corresponds to the statement

```
f = array(array_index).field(field_index);
```

but it can be used even if the programmer doesn't know the names of the fields in the structure array at the time the program is written.

For example, suppose that we needed to write a function to read and manipulate the data in an unknown structure array. This function could determine the field names in the structure using a call to fieldnames, and then could read the data using function getfield. To read the zip code of the second student, the function would be

```
» zip = getfield(student,{2},'zip')
zip =
      68888
```

Similarly, a program could modify values in the structure using function setfield. The structure of function setfield is

```
f = setfield(array,{array_index},'field',{field_index},value)
```

where f is the output structure array, the `field_index` is optional, and `array_index` is optional for a 1-by-1 structure array. The function call corresponds to the statement

```
array(array_index).field(field_index) = value;
```

7.3.6 Dynamic Field Names

Beginning with MATLAB 7.0, there is an alternative way to access the elements of a structure: **dynamic field names**. A dynamic field name is a string enclosed in parentheses at a location where a field name is expected. For example, the name of student 1 can be retrieved with either static or dynamic field names as follows:

```
» student(1).name           % Static field name
ans =
John Doe
» student(1).('name')       % Dynamic field name
ans =
John Doe
```

Dynamic field names perform the same function as static field names, but *dynamic field names can be changed during program execution*. This allows a user to access different information in the same function within a program.

For example, the following function accepts a structure array and a field name and calculates the average of the values in the specified field for all elements in the structure array. It returns that average (and optionally the number of values averaged) to the calling program.

```
function [ave, nvals] = calc_average(structure,field)
%CALC_AVERAGE Calculate the average of values in a field.
% Function CALC_AVERAGE calculates the average value
% of the elements in a particular field of a structure
% array. It returns the average value and (optionally)
% the number of items averaged.

% Define variables:
%   arr       -- Array of values to average
%   ave       -- Average of arr
%   ii        -- Index variable
%
% Record of revisions:
%     Date       Programmer        Description of change
%     ====       ==========        =====================
%   03/04/07    S. J. Chapman      Original code
%
```

```
% Check for a legal number of input arguments.
msg = nargchk(2,2,nargin);
error(msg);

% Create an array of values from the field
arr = [];
for ii = 1:length(structure)
   arr = [arr structure(ii).(field)];
end

% Calculate average
ave = mean(arr);

% Return number of values averaged
if nargout == 2
   nvals = length(arr);
end
```

A program can average the values in different fields by simply calling this function multiple times with different structure names and different field names. For example, we can calculate the average values in fields exams and zip as follows:

```
» [ave,nvals] = calc_average(student,'exams')
ave =
   83.2222
nvals =
      9
» ave = calc_average(student,'zip')
ave =
      50039
```

7.3.7 Using the size Function with Structure Arrays

When the size function is used with a structure array, it returns the size of the structure array itself. When the size function is used with a *field* from a particular element in a structure array, it returns the size of that field instead of the size of the whole array. For example,

```
» size(student)
ans =
      1      3
```

```
» size(student(1).name)
ans =
     1    8
```

7.3.8   Nesting Structure Arrays

Each field of a structure array can be of any data type, including a cell array or a structure array. For example, the following statements define a new structure array as a field under array `student` to carry information about each class that the student is enrolled in.

```
student(1).class(1).name = 'COSC 2021'
student(1).class(2).name = 'PHYS 1001'
student(1).class(1).instructor = 'Mr. Jones'
student(1).class(2).instructor = 'Mrs. Smith'
```

After these statements have been issued, `student(1)` contains the following data. Note the technique used to access the data in the nested structures.

```
» student(1)
ans =
        name:   'John Doe'
       addr1:   '123 Main Street'
        city:   'Anytown'
       state:   'LA'
         zip:   '71211'
       exams:   [80 95 84]
       class:   [1x2 struct]
» student(1).class
ans =
1x2 struct array with fields:
    name
    instructor
» student(1).class(1)
ans =
          name:   'COSC 2021'
    instructor:   'Mr. Jones'
» student(1).class(2)
ans =
          name:   'PHYS 1001'
    instructor:   'Mrs. Smith'
» student(1).class(2).name
ans =
PHYS 1001
```

Table 7.3   Common MATLAB Structure Functions

Function	Description
fieldnames	Return a list of field names in a cell array of strings.
getfield	Get current value from a field.
rmfield	Remove a field from a structure array.
setfield	Set new value into a field.
struct	Pre-define a structure array.

7.3.9   Summary of `structure` Functions

The common MATLAB structure functions are summarized in Table 7.3.

7.4   Function Handles

A **function handle** is a MATLAB data type that holds information to be used in referencing a function. When you create a function handle, MATLAB captures all the information about the function that it needs to execute it later on. Once the handle is created, it can be used to execute the function at any time.

As we will see in Chapter 10, function handles are key to the operation of MATLAB graphical user interfaces. We will learn about them here, and we will apply that knowledge in Chapter 10.

7.4.1   Creating and Using Function Handles

A function handle can be created either of two possible ways: the @ operator or the str2func function. To create a function handle with the @ operator, just place it in front of the function name. To create a function handle with the str2func function, call the function with the function name in a string. For example, suppose that function my_func is defined as follows:

```
function res = my_func(x)
res = x.^2 - 2*x + 1;
```

Then either of the following lines will create a function handle for function my_func:

```
hndl = @my_func
hndl = str2func('my_func');
```

Once a function handle has been created, the function can be executed by naming the function handle followed by any calling parameters. The result will be exactly the same as if the function itself were named.

```
» hndl = @my_func
hndl =
    @my_func
» hndl(4)
ans =
    9
» my_func(4)
ans =
    9
```

If a function has no calling parameters, then the function handle must be followed by empty parentheses when it is used to call the function.

```
» h1 = @randn;
» h1()
ans =
  -0.4326
```

After a function handle has been created, it appears in the current workspace with the data type "function handle":

```
» whos
Name       Size       Bytes      Class            Attributes
ans        1x1            8       double
h1         1x1           16       function_handle
hndl       1x1           16       function_handle
```

A function handle can also be executed using the `feval` function. This provides a convenient way to execute function handles within a MATLAB program.

```
» feval(hndl,4)
ans =
    9
```

It is possible to recover the function name from a function handle using the `func2str` function.

```
» func2str(hndl)
ans =
my_func
```

This feature is useful when we want to create descriptive messages, error messages, or labels inside a function that accepts and evaluates function handles.

For example, the function that follows accepts a function handle in the first argument and plots the function at the points specified in the second argument. It also prints out a title containing the name of the function being plotted.

```
function plotfunc(fun,points)
% PLOTFUNC Plots a function between the specified points.
% Function PLOTFUNC accepts a function handle, and
% plots the function at the points specified.

% Define variables:
%    fun          -- Function handle
%    msg          -- Error message
%
% Record of revisions:
%     Date          Programmer           Description of change
%     ====          ==========           =====================
%   03/05/07      S. J. Chapman         Original code

% Check for a legal number of input arguments.
msg = nargchk(2,2,nargin);
error(msg);

% Get function name
fname = func2str(fun);

% Plot the data and label the plot
plot(points,fun(points));
title(['\bfPlot of ' fname '(x) vs x']);
xlabel('\bfx');
ylabel(['\bf' fname '(x)']);
grid on;
```

For example, this function can be used to plot the function $\sin x$ from -2π to 2π with the following statement:

```
plotfunc(@sin, [-2*pi:pi/10:2*pi])
```

The resulting function is shown in Figure 7.8.

Some common MATLAB functions used with function handles are summarized in Table 7.4.

7.4.2 The Significance of Function Handles

Either function names or function handles can be used to execute most functions. However, function handles have certain advantages over function names. These advantages include:

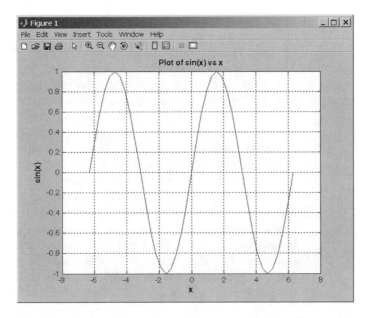

Figure 7.8 Plot of function sin x from -2π to 2π, created using function plotfunc.

1. **Passing Function Access Information to Other Functions.** As we saw in the previous section, you can pass a function handle as an argument in a call to another function. The function handle enables the receiving function to call the function attached to the handle. You can execute a function handle from within another function *even if the handle's function is not in the scope of the evaluating function*. This is because the function handle has a complete description of the function to execute; the calling function does not have to search for it.

2. **Improved Performance in Repeated Operations.** MATLAB performs a search for a function at the time that you create a function handle and then

Table 7.4 MATLAB Functions that Manipulate Function Handles

Function	Description
@	Create a function handle.
feval	Evaluate a function using a function handle.
func2str	Recover the function name associated with a given function handle.
functions	Recover miscellaneous information from a function handle. The data is returned in a structure.
str2func	Create a function handle from a specified string.

stores this access information in the handle itself. Once it has been defined, you can use this handle over and over, without having to look it up again. This makes function execution faster.

3. **Allow Wider Access to Subfunctions and Private Functions.** All MATLAB functions have a certain scope. They are visible to other MATLAB entities within that scope but not visible outside of it. You can call a function directly from another function that is within its scope but *not* from a function outside that scope. Subfunctions, private functions, and nested functions are limited in their visibility to other MATLAB functions. You can invoke a subfunction only from another function that is defined within the same M-file. You can invoke a private function only from a function in the directory immediately above the `private` subdirectory. You can invoke a nested function only from within the host function or another nested function at the same level. However, when you create a handle to a function that has limited scope, the function handle stores all the information MATLAB needs to evaluate the function from *any* location in the MATLAB environment. If you create a handle to a subfunction within the M-file that defines the subfunction, you can then pass the handle to code that resides outside of that M-file and evaluate the subfunction from beyond its usual scope. The same holds true for private functions and nested functions.

4. **Include More Functions per M-File for Easier File Management.** You can use function handles to help reduce the number of M-files required to contain your functions. The problem with grouping a number of functions in one M-file has been that this defines them as subfunctions and thus reduces their scope in MATLAB. Using function handles to access these subfunctions removes this limitation. This enables you to group functions as you want and reduce the number of files you have to manage.

7.4.3 Function Handles and Nested Functions

When MATLAB invokes an ordinary function, a special workspace is created to contain the function's variables. The function executes to completion, and then the workspace is destroyed. All the data in the function workspace is lost, except for any values labeled `persistent`. If the function is executed again, a completely new workspace is created for the new execution.

By contrast, when a host function creates a handle for a nested function and returns that handle to a calling program, the host function's workspace is created and *remains in existence for as long as the function handle remains in existence*. Since the nested function has access to the host function's variables, MATLAB has to preserve the host's function's data as long as there is any chance that the nested function will be used. This means that *we can save data in a function between uses*.

This idea is illustrated in the function that follows. When function `count_calls` is executed, it initializes a local variable `current_count` to a user-specified initial count and then creates and returns a handle to the nested function `increment_count`. When `increment_count` is called using that function handle, the count is increased by one and the new value is returned.

```
function fhandle = count_calls(initial_value)

% Save initial value in a local variable
% in the host function.
current_count = initial_value;

% Create and return a function handle to the
% nested function below.
fhandle = @increment_count;

    % Define a nested function to increment counter
    function count = increment_count
    current_count = current_count + 1;
    count = current_count;
    end % function increment_count

end % function count_calls
```

When this program is executed, the results are as shown here. Each call to the function handle increments the count by one.

```
» fh = count_calls(4);
» fh()
ans =
     5
» fh()
ans =
     6
» fh()
ans =
     7
```

Even more importantly, *each function handle created for a function has its own independent workspace*. If we create two different handles for this function, each one will have its own local data, and they will be independent of each other. As you can see, we can increment either counter independently by calling the function with the proper handle.

```
» fh1 = count_calls(4);
» fh2 = count_calls(20);
» fh1()
ans =
     5
» fh1()
ans =
     6
```

```
» fh2()
ans =
        21
» fh1()
ans =
         7
```

You can use this feature to run multiple counters and so forth within a program without their interfering with each other.

7.4.4 An Example Application: Solving Ordinary Differential Equations

One very important application of function handles occurs in the MATLAB functions designed to solve ordinary differential equations. MATLAB includes a plethora of functions to solve differential equations under various conditions, but the most useful of them is ode45. This function solves ordinary differential equations of the form

$$y' = (t, y) \tag{7-1}$$

using a Runge-Kutta (4,5) integration algorithm, and it works well for many types of equations with many different input conditions.

The calling sequence for this function is

```
[t,y] = ode45(odefun_handle,tspan,y0,options)
```

where the calling parameters are as follows:

odefun_handle	A *handle* to a function $f(t, y)$ that calculates the derivative y' of the differential equation.
tspan	A vector containing the times to integrate. If this is a two-element array [t0 tend], then the values are interpreted as the starting and ending times to integrate. The integrator applies the initial conditions at time t0 and integrates the equation until time tend. If the array has more than two elements, then the integrator returns the values of the differential equation at exactly the specified times.
y0	The initial conditions for the variable at time t0.
options	A structure of optional parameters that change the default integration properties. (We will not use this parameter in this book.)

and the results are

---

t   A column vector of time points at which the differential equation was solved.

y   The solution array. Each row of y contains the solutions to all variables at the time specified in the same row of t.

This function also works well for systems of simultaneous first-order differential equations, where there are vectors of dependent variables y_1, y_2, and so on.

We will try a few example differential equations to get a better understanding of this function. First, consider the simple first-order linear time-invariant differential equation

$$\frac{dy}{dt} + 2y = 0 \tag{7-2}$$

with the initial condition $y(0) = 1$. The function that would specify the derivative of the differential equation is

$$\frac{dy}{dt} = -2y \tag{7-3}$$

This function could be programmed in MATLAB as follows:

```
function yprime = fun1(t,y)
yprime = -2 * y;
```

Function ode45 could be used to solve Equation (7-2) for $y(t)$

```
% Script file: ode45_test1.m
%
% Purpose:
%   This program solves a differential equation of the
%   form dy/dt + 2 * y = 0, with the initial condition
%   y(0) = 1.
%
% Record of revisions:
%     Date         Programmer          Description of change
%     ====         ==========          =====================
%   03/15/07     S. J. Chapman         Original code
%
% Define variables:
%   odefun_handle -- Handle to function that defines the
derivative
%   tspan          -- Duration to solve equation for
%   yo             -- Initial condition for equation
```

```
%   t               -- Array of solution times
%   y               -- Array of solution values

% Get a handle to the function that defines the
% derivative.
odefun_handle = @fun1;

% Solve the equation over the period 0 to 5 seconds
tspan = [0 5];

% Set the initial conditions
y0 = 1;

% Call the differential equation solver.
[t,y] = ode45(odefun_handle,tspan,y0);

% Plot the result
figure(1);
plot(t,y,'b-','LineWidth',2);
grid on;
title('\bfSolution of Differential Equation');
xlabel('\bfTime (s)');
ylabel('\bf\ity''');
```

When this script file is executed, the resulting output is shown in Figure 7.9. This sort of exponential decay is exactly what would be expected for a first-order linear differential equation.

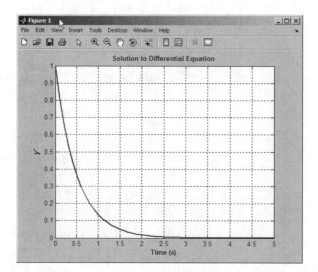

Figure 7.9   Solution to the differential equation $dy/dt + 2y = 0$ with the initial condition $y(0) = 1$.

▶

Example 7.1—Radioactive Decay Chains

The radioactive isotope thorium 227 decays into radium 223 with a half life of 18.68 days, and radium 223 in turn decays into radon 219 with a half life of 11.43 days. The radioactive decay constant for thorium 227 is λ_{th} = 0.03710638/day, and the radioactive decay constant for radon is λ_{ra} = 0.0606428/day. Assume that initially we have 1 million atoms of thorium 227, and calculate and plot the amount of thorium 227 and radium 223 that will be present as a function of time.

SOLUTION   The rate of decrease in thorium 227 is equal to the amount of thorium 227 present at a given moment times the decay constant for the material.

$$\frac{dn_{th}}{dt} = -\lambda_{th}n_{th} \tag{7-4}$$

where n_{th} is the amount of thorium 227 and λ_{th} is the decay rate per day. The rate of decrease in radium 223 is equal to the amount of radium 223 present at a given moment times the decay constant for the material. However, the amount of radium 223 is *increased* by the number of atoms of thorium 227 that have decayed, so the total change in the amount of radium 223 is

$$\frac{dn_{ra}}{dt} = -\lambda_{ra}n_{ra} - \frac{dn_{th}}{dt}$$

$$\frac{dn_{ra}}{dt} = -\lambda_{ra}n_{ra} + \lambda_{th}n_{th} \tag{7-5}$$

where n_{ra} is the amount of radon 219 and λ_{ra} is the decay rate per day. Equations (7-4) and (7-5) must be solved simultaneously to determine the amount of thorium 227 and radium 223 present at any given time.

1. **State the problem.**
    Calculate and plot the amount of thorium 227 and radium 223 present as a function of time, given that there were initially 1,000,000 atoms of thorium 227 and no radium 223.

2. **Define the inputs and outputs.**
    There are no inputs to this program. The outputs from this program are the plots of thorium 227 and radium 223 as a function of time.

3. **Describe the algorithm.**
    This program can be broken down into three major steps:

```
Create a function to describe the derivatives of
    Thorium 227 and Radium 223
Solve the differential equations using ode45
Plot the resulting data
```

The first major step is to create a function that calculates the rate of change of thorium 227 and radium 223. This is just a direct implementation of Equations (7-4) and (7-5). The detailed pseudocode is

```
function yprime = decay1(t,y)
yprime(1) = -lambda_th * y(1);
yprime(2) = -lambda_ra * y(2) + lambda_th * y(1);
```

Next we have to solve the differential equation. To do this, we need to set the initial conditions and the duration, and then call ode45. The detailed pseudocode is

```
% Get a function handle.
odefun_handle = @decay1;

% Solve the equation over the period 0 to 100 days
tspan = [0 100];

% Set the initial conditions
y0(1) = 1000000;     % Atoms of Thorium 227
y0(2) = 0;           % Atoms of Radium 223

% Call the differential equation solver.
[t,y] = ode45(odefun_handle,tspan,y0);
```

The final step is plotting the results. Each result appears in its own column of the output array y, so y(:,1) will contain the amount of thorium 227 and y(:,2) will contain the amount of radium 223.

4. **Turn the algorithm into MATLAB statements.**
The MATLAB code for the selection sort function is

```
% Script file: calc_decay.m
%
% Purpose:
%   This program calculates the amount of Thorium 227 and
%   Radium 223 left as a function of time, given an inital
%   concentration of 1 gram of Thorium 227 and no grams of
%   Radium 223.
%
% Record of revisions:
%    Date          Programmer         Description of change
%    ====          ==========         =====================
%   03/15/07       S. J. Chapman      Original code
%
% Define variables:
%   odefun_handle -- Handle to function that defines the
derivative
```

```
%    tspan             -- Duration to solve equation for
%    yo                -- Initial condition for equation
%    t                 -- Array of solution times
%    y                 -- Array of solution values

% Get a handle to the function that defines the derivative.
odefun_handle = @decay1;

% Solve the equation over the period 0 to 100 days
tspan = [0 100];

% Set the initial conditions
y0(1) = 1000000;      % Atoms of Thorium 227
y0(2) = 0;            % Atoms of Radium 223

% Call the differential equation solver.
[t,y] = ode45(odefun_handle,tspan,y0);

% Plot the result
figure(1);
plot(t,y(:,1),'b-','LineWidth',2);
hold on;
plot(t,y(:,2),'k--','LineWidth',2);
title('\bfAmount of Thorium 227 and Radium 223 vs Time');
xlabel('\bfTime (days)');
ylabel('\bfNumber of Atoms');
legend('Thorium 227','Radium 223');
grid on;
hold off;
```

The function to calculate the derivatives is

```
function yprime = decay1(t,y)
%DECAY1 Calculates the decay rates of Thorium 227 and
  Radium 223.
% Function DECAY1 Calculates the rates of change of
  Thorium 227
% and Radium 223 (yprime) for a given current
  concentration y.

% Define variables:
%    t                 -- Time (in days)
%    y                 -- Vector of current concentrations
%
```

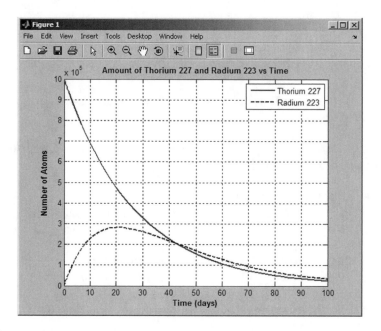

Figure 7.10 Plot of radioactive decay of Thorium 227 and Radium 223 vs time.

```
%  Record of revisions:
%     Date           Programmer            Description of change
%     ====           ==========            =====================
%   03/15/07      S. J. Chapman             Original code

% Set decay constants.
lambda_th = 0.03710636;
lambda_ra = 0.0606428;

% Calculate rates of decay
yprime = zeros(2,1);
yprime(1) = -lambda_th * y(1);
        yprime(2) = -lambda_ra * y(2) + lambda_th * y(1);
```

5. **Test the program.**

   When this program is executed, the results are as shown in Figure 7.10. These results look reasonable. The initial amount of thorium 227 starts high and decreases exponentially with a half life of about 18 days. The initial amount of radium 223 starts a zero and rises rapidly due to the decay of thorium 227, and then starts decaying as the amount increase from the decay of thorium 227 slows. ◄

Quiz 7.1

This quiz provides a quick check to see if you have understood the concepts introduced in Sections 7.1 through 7.4. If you have trouble with the quiz, reread the section, ask your instructor, or discuss the material with a fellow student. The answers to this quiz are found in the back of the book.

1. What is a sparse array? How does it differ from a full array? How can you convert from a sparse array to a full array, and vice versa?

2. What is a cell array? How does it differ from an ordinary array?

3. What is the difference between content indexing and cell indexing?

4. What is a structure? How does it differ from ordinary arrays and cell arrays?

5. What is the purpose of `varargin`? How does it work?

6. What is a function handle? How do you create a function handle? How do you call a function using a function handle?

7. Given the definition of array *a* shown here, what will be produced by each of the following sets of statements? (*Note:* Some of these statements may be illegal. If a statement is illegal, explain why.)

```
a{1,1} = [1 2 3; 4 5 6; 7 8 9];
a(1,2) = {'Comment line'};
a{2,1} = j;
a{2,2} = a{1,1} - a{1,1}(2,2);
```

(*a*) `a(1,1)`
(*b*) `a{1,1}`
(*c*) `2*a(1,1)`
(*d*) `2*a{1,1}`
(*e*) `a{2,2}`
(*f*) `a(2,3) = {[-17; 17]}`
(*g*) `a{2,2}(2,2)`

8. Given the definition of structure array b shown here, what will be produced by each of the following sets of statements? (*Note:* Some of these statements may be illegal. If a statement is illegal, explain why.)

```
b(1).a = -2*eye(3);
b(1).b = 'Element 1';
b(1).c = [1 2 3];
b(2).a = [b(1).c' [-1; -2; -3] b(1).c'];
b(2).b = 'Element 2';
b(2).c = [1 0 -1];
```

(*a*) `b(1).a - b(2).a`
(*b*) `strncmp(b(1).b,b(2).b,6)`
(*c*) `mean(b(1).c)`

(*d*) `mean(b.c)`

(*e*) `b`

(*f*) `b(1).('b')`

(*g*) `b(1)`

9. What will be returned by the following function, if it is called with the expression `myfun(@cosh)`?

```
function res = myfun(x)
res = func2str(x);
end % function myfun
```

7.5 Summary

Sparse arrays are special arrays in which memory is allocated only for nonzero elements. Three values are saved for each nonzero element—a row number, a column number, and the value itself. This form of storage is much more efficient than for arrays for the situation where only a tiny fraction of the elements are nonzero. MATLAB includes functions and intrinsic calculations for sparse arrays, so they can be freely and transparently mixed with full arrays.

Cell arrays are arrays whose elements are *cells*, containers that can hold other MATLAB arrays. Any sort of data may be stored in a cell, including structure arrays and other cell arrays. They are a very flexible way to store data, and are used in many internal MATLAB graphical user interface functions.

Structure arrays are a data type in which each individual element is given a name. The individual elements of a structure are known as fields, and each field in a structure may have a different type. The individual fields are addressed by combining the name of the structure with the name of the field, separated by a period. Structure arrays are useful for grouping together all of the data related to a particular person or thing into a single location.

Function handles are a special data type containing all the information required to invoke a function. Function handles are created with the @ operator or the `str2func` function, and they are used to call the corresponding function by naming the handle following by parentheses and the required calling arguments. If a function handle is created for a nested function, the workspace of the host function will be preserved between calls to the nested function using the function handle.

7.5.1 Summary of Good Programming Practice

The following guidelines should be adhered to:

1. Always preallocate all cell arrays before assigning values to the elements of the array. This practice greatly increases the execution speed of a program.
2. Use cell array arguments `varargin` and `varargout` to create functions that support varying numbers of input and output arguments.

7.5.2   MATLAB Summary

The following summary lists all of the MATLAB commands and functions described in this chapter, along with a brief description of each one.

`@`	Create a function handle.
`cell`	Predefine a cell array structure.
`celldisp`	Display contents of a cell array.
`cellplot`	Plot structure of a cell array.
`cellstr`	Convert a two-dimensional character array to a cell array of strings.
`feval`	Evaluate a function using a function handle.
`fieldnames`	Return a list of field names in a cell array of strings.
`func2str`	Get the name of the function pointed to by the specified function handle.
`functions`	Recover miscellaneous information from a function handle in a structure.
`getfield`	Get current value from a field.
`full`	Convert a sparse matrix into a full matrix.
`nnz`	Number of nonzero matrix elements.
`nonzeros`	Return a column vector containing the nonzero elements in a matrix.
`nzmax`	Amount of storage allocated for nonzero matrix elements.
`ode45`	Function to solve ordinary differential equations using a Runge-Kutta (4,5) technique.
`rmfield`	Remove a field from a structure array.
`setfield`	Set new value into a field.
`spalloc`	Allocate space for a sparse matrix.
`sparse`	Convert a full matrix into a sparse matrix.
`speye`	Create a sparse identity matrix.
`spfun`	Apply function to nonzero matrix elements.
`spones`	Replace nonzero sparse matrix elements with ones.
`sprand`	Create a sparse uniformly distributed random matrix.
`sprandn`	Create a sparse normally distributed random matrix.
`sprintf`	Write formatted data to string.
`spy`	Visualize sparsity pattern as a plot.
`str2func`	Create a function handle for the function named in a string argument.

7.6 Exercises

7.1 Write a MATLAB function that will accept a cell array of strings and sort them into ascending order according to the lexicographic order of the ASCII character set. (You may use function `c_strcmp` from Chapter 6 for the comparisons if you wish.)

7.2 Write a MATLAB function that will accept a cell array of strings and sort them into ascending order according to *alphabetical order*. (This implies that you must treat 'A' and 'a' as the same letter.)

7.3 Create a sparse 100 × 100 array a in which about 5% of the elements contain normally distributed random values, and all of the other elements are zero (use function `sprandn` to generate these values). Next, set all of the diagonal elements in the array to 1. Next, define a 100-element sparse column array b, and initialize that array with 100 uniformly distributed values produced by function `rand`. Answer the following questions about these arrays:

(*a*) Create a full array a_full from the sparse array a. Compare the memory required to store the full array and the sparse array. Which is more efficient?

(*b*) Plot the distribution of values in array a using function `spy`.

(*c*) Create a full array b_full from the sparse array b. Compare the memory required to store the full array and the sparse array. Which is more efficient?

(*d*) Solve the system of equations a * x = b using both the full arrays and the sparse arrays. How do the two sets of answers compare? Time the two solutions. Which one is faster?

7.4 Create a function that accepts any number of numeric input arguments and sums up all of individual elements in the arguments. Test your function by

passing it the four arguments $a = 10$, $b = \begin{bmatrix} 4 \\ -2 \\ 2 \end{bmatrix}$, $c = \begin{bmatrix} 1 & 0 & 3 \\ -5 & 1 & 2 \\ 1 & 2 & 0 \end{bmatrix}$, and

$d = \begin{bmatrix} 1 & 5 & -2 \end{bmatrix}$.

7.5 Modify the function of the previous exercise so that it can accept either ordinary numeric arrays or cell arrays containing numeric values. Test your

function by passing it the two arguments a and b, where $a = \begin{bmatrix} 1 & 4 \\ -2 & 3 \end{bmatrix}$,

$b\{1\} = \begin{bmatrix} 1 & 5 & 2 \end{bmatrix}$, and $b\{2\} = \begin{bmatrix} 1 & -2 \\ 2 & 1 \end{bmatrix}$.

7.6 Create a structure array containing all of the information needed to plot a data set. At a minimum, the structure array should have the following fields:

- x_data   *x*-data (one or more data sets in separate cells)
- y_data   *y*-data (one or more data sets in separate cells)

- `type`   linear, semilogx, and so forth
- `plot_title`   plot title
- `x_label`   *x*-axis label
- `y_label`   *y*-axis label
- `x_range`   *x*-axis range to plot
- `y_range`   *y*-axis range to plot

You may add additional fields that would enhance your control of the final plot.

    After this structure array has been created, create a MATLAB function that accepts an array of this structure and produces one plot for each structure in the array. The function should apply intelligent defaults if some data fields are missing. For example, if the `plot_title` field is an empty matrix, then the function should not place a title on the graph. Think carefully about the proper defaults before starting to write your function!

    To test your function, create a structure array containing the data for three plots of three different types and pass that structure array to your function. The function should correctly plot all three data sets in three different figure windows.

7.7   Define a structure `point` containing two fields `x` and `y`. The `x` field will contain the *x*-position of the point, and the `y` field will contain the *y*-position of the point. Then write a function `dist3` that accepts two points and returns the distance between the two points on the Cartesian plane. Be sure to check the number of input arguments in your function.

7.8   Write a function that will accept a structure as an argument and return two cell arrays containing the names of the fields of that structure and the data types of each field. Be sure to check that the input argument is a structure and generate an error message if it is not.

7.9   Write a function that will accept a structure array of `student` as defined in this chapter, and calculate the final average of each one, assuming that all exams have equal weighting. Add a new field to each array to contain the final average for that student, and return the updated structure to the calling program. Also, calculate and return the final class average.

7.10   Write a function that will accept two arguments—the first a structure array, and the second a field name stored in a string. Check to make sure that these input arguments are valid. If they are not valid, print out an error message. If they are valid and the designated field is a string, concatenate all of the strings in the specified field of each element in the array, and return the resulting string to the calling program.

7.11   **Calculating Directory Sizes.** Function `dir` returns the contents of a specified directory. The `dir` command returns a structure array with four fields, as shown here:

```
» d = dir('chap7')
d =
```

```
36x1 struct array with fields:
    name
    date
    bytes
    isdir
```

The field name contains the names of each file, date contains the last modification date for the file, bytes contains the size of the file in bytes, and isdir is 0 for conventional files and 1 for directories. Write a function that accepts a directory name and path and returns the total size of all files in the directory in bytes.

7.12 **Recursion.** A function is said to be *recursive* if the function calls itself. Modify the function created in Problem 7.11 so that it calls itself when it finds a subdirectory and sums up the size of all file in the current directory plus all subdirectories.

7.13 **Function Generators.** Write a nested function that evaluates a polynomial of the form $y = ax^2 + bx + c$. The host function gen_func should have three calling arguments—a, b, and c—to initialize the coefficients of the polynomial. It should also create and return a function handle for the nested function eval_func. The nested function eval_func(x) should calculate a value of y for a given value of x, using the values of a, b, and c stored in the host function. This is effectively a function generator, since each combination of a, b, and c values produces a function handle that evaluates a unique polynomial. Then perform the following steps:

(a) Call gen_func(1,2,1) and save the resulting function handle in variable h1. This handle now evaluates the function $y = x^2 + 2x + 1$.
(b) Call gen_func(1,4,3) and save the resulting function handle in variable h2. This handle now evaluates the function $y = x^2 + 4x + 3$.
(c) Write a function that accepts a function handle and plots the specified function between two specified limits.
(d) Use this function to plot the two polynomials generated in parts (a) and (b).

7.14 **Function Generators.** Generalize the function generator of the previous problem to handle polynomial of arbitrary dimension. Test it by creating function handles and plots the same way that you did in the previous problem. [*Hint:* Use varagrin.]

7.15 Look up function struct in the MATLAB Help Brower, and learn how to preallocate a structure and simultaneously initialize all of the elements in the structure array to the same value. Then create a 2000-element array of type student, with the values in every array element initialized with the fields shown here:

```
 name: 'John Doe'
addr1: '123 Main Street'
 city: 'Anytown'
state: 'LA'
  zip: '71211'
```

7.16   RC Circuits.   Figure 7.11a shows a simple series RC circuit with the output voltage taken across the capacitor. Assume that there is no voltage or power in this circuit before time $t = 0$, and that the voltage v_{in} (t) is applied at time $t \geq 0$. Calculate and plot the output voltage of this circuit for time $0 \leq t \leq 10$ s. [Hint: The output voltage from this circuit can be found by writing a Kirchoff's Voltage Law (KVL) equation at the output, and solving for v_{out} (t). The KVL equation is

$$\frac{v_{out}(t) - v_{in}(t)}{R} + C\frac{dv_{out}\ (t)}{dt} = 0 \tag{7-6}$$

Collecting terms in this equation produces the result

$$\frac{dv_{out}\ (t)}{dt} + \frac{1}{RC}\,v_{out}\ (t) = \frac{1}{RC}\,v_{in}(t) \tag{7-7}$$

Solve this equation for v_{out} (t).]
7.17   Calculate and plot the output v of the following differential equation:

$$\frac{dv(t)}{dt} + v(t) = \begin{cases} t & 0 \leq t \leq 5 \\ 0 & \text{elsewhere} \end{cases} \tag{7-8}$$

$$R = 1\,M\Omega \qquad\qquad C = 1\,\mu F$$

(a)

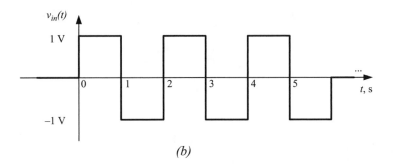

(b)

Figure 7.11   (a) A simple series RC circuit. (b) The input voltage to this circuit as a function of time. Note that the voltage is 0 for all times before zero and all times after $t = 6$ s.

CHAPTER **8**

Input/Output Functions

In Chapter 2, we learned how to load and save MATLAB data using the `load` and `save` commands, and how to write out formatted data using the `fprintf` function. In this chapter we will learn more about MATLAB's input/output capabilities. First, we will learn about `textread`, a very useful function for reading ASCII data from a file. Then, we will spend a bit more time examining the `load` and `save` commands. Finally, we will look at the other file I/O options available in MATLAB.

Those readers familiar with C will find much of this material very familiar. However, be careful—there are subtle differences between MATLAB and C functions that can trip you up.

8.1 The `textread` Function

The `textread` function reads ASCII files that are formatted into columns of data where each column can be of a different type, and stores the contents of each column in a separate output array. This function is *very* useful for importing tables of data printed out by other applications.

The form of the `textread` function is

```
[a,b,c,...] = textread(filename,format,n)
```

where `filename` is the name of the file to open, `format` is a string containing a description of the type of data in each column, and `n` is the number of lines to read. (If `n` is missing, the function reads to the end of the file.) The format string contains the same types of format descriptors as function `fprintf`. Note that the number of output arguments must match the number of columns that you are reading.

For example, suppose that file `test_input.dat` contains the following data:

```
James    Jones    O+    3.51    22    Yes
Sally    Smith    A+    3.28    23    No
```

This data could be read into a series of arrays with the following function:

```
[first,last,blood,gpa,age,answer] = ...
        textread('test_input.dat','%s %s %s %f %d %s')
```

When this command is executed, the results are

```
» [first,last,blood,gpa,age,answer] = ...
        textread('test_input.dat','%s %s %s %f %d %s')

first =
    'James'
    'Sally'
last =
    'Jones'
    'Smith'
blood =
    'O+'
    'A+'
gpa =
    3.5100
    3.2800
age =
    42
    28
answer =
    'Yes'
    'No'
```

This function can also skip selected columns by adding an asterisk to the corresponding format descriptor (for example, `%*s`). The following statement reads only the `first`, `last`, and `gpa` from the file:

```
» [first,last,gpa] = ...
        textread('test_input.dat','%s %s %*s %f %*d %*s')
first =
    'James'
    'Sally'
last =
    'Jones'
    'Smith'
gpa =
    3.5100
    3.2800
```

Function `textread` is much more useful and flexible than the `load` command. The `load` command assumes that all of the data in the input file is of a single type—it cannot support different types of data in different columns. In addition, it stores all of the data into a single array. In contrast, the `textread` function allows each column to go into a separate variable, which is *much* more convenient when working with columns of mixed data.

Function `textread` has a number of additional options that increase its flexibility. Consult the MATLAB on-line documentation for details of these options.

8.2   More about the `load` and `save` Commands

The `save` command saves MATLAB workspace data to disk, and the `load` command load data from disk into the workspace. The `save` command can save data either in a special binary format called a MAT-file or in an ordinary ASCII file. The form of the `save` command is

```
save filename [content] [options]
```

The command `save` all by itself saves all of the data in the current workspace to a file named `matlab.mat` in the current directory. If a filename is included, the data will be saved in file "`filename.mat`". If a list of variables is included at the `content` position, then only those particular variables will be saved.

For example, suppose that a workspace contains a 1000-element double array x and a character string `str`. We can save these two variables to a MAT-file with the following command:

```
save test_matfile x str
```

This command creates a MAT-file with the name `test_matfile.mat`. The contents of this file can be examined with `-file` option of the `whos` command:

```
» whos -file test_matfile.mat
  Name      Size        Bytes    Class       Attributes
  str       1x11           22    char
  x         1x1000       8000    double
```

The content to be saved can be specified in several ways, as described in Table 8.1.

The more important options supported by the save command are shown in Table 8.2; a complete list can be found in the MATLAB on-line documentation.

The `load` command can load data from MAT-files or from ordinary ASCII files. The form of the `load` command is

```
load filename [options] [content]
```

The command `load` all by itself loads all of the data in file `matlab.mat` into the current workspace. If a filename is included, the data will be loaded from that

Table 8.1 Ways of Specifying save Command Content

Values for content	Description
`<nothing>`	Save all data in current workspace.
`varlist`	Save only the values in the variable list.
`-regexp exprlist`	Save all variables that match any of the regular expressions in the expression list.
`-struct s`	Save as individual variables all fields of the scalar structure `s`.
`-struct s fieldlist`	Save as individual variables only the specified fields of structure `s`.

Table 8.2 Selected save Command Options

Option	Description
`-mat`	Save data in MAT-file format (default).
`-ascii`	Save data in space-separated ASCII format.
`-append`	Adds the specified variables to an existing MAT file.
`-v4`	Save the MAT-file in a format readable by MATLAB version 4 or later.
`-v6`	Save the MAT-file in a format readable by MATLAB versions 5 and 6 or later.
`-v7`	Save the MAT-file in a format readable by MATLAB versions 7 through 7.2 or later.
`-v7.3`	Save the MAT-file in a format readable by MATLAB versions 7.3 or later.

filename. If specific variables are included in the content list, then only those variables will be loaded from the file. For example,

```
load                  % Loads entire content of matlab.mat
load mydat.mat        % Loads entire content of mydat.mat
load mydat.mat a b c  % Loads only a, b, and c from mydat.mat
```

The options supported by the `load` command are shown in Table 8.3.

Although it is not immediately obvious, the `save` and `load` commands are the most powerful and useful I/O commands in MATLAB. Their advantages include the following:

1. These commands are very easy to use.
2. MAT-files are *platform independent*. A MAT-file written on any type of computer that supports MATLAB can be read on any other computer

Table 8.3   `load` Command Options

Option	Description
-mat	Treat file as a MAT-file (default if file extent is `mat`).
-ascii	Treat file as a space-separated ASCII file (default if file extent is *not* `mat`).

   supporting MATLAB. This format transfers freely among PCs, Macs, and many different versions of Linux and Unix. Also, the Unicode character encoding ensures that character strings will be preserved properly across platforms.

3. MAT-files are efficient users of disk space, using only the amount of memory required for each data type. They store the full precision of every variable—no precision is lost due to conversion to and from ASCII format. MAT-files can also be compressed to save even more disk space.

4. MAT-files preserve all of the information about each variable in the workspace, including its class, name, and whether or not it is global. All of this information is lost in other types of I/O. For example, suppose that the workspace contains the following information:

```
» whos
  Name       Size      Bytes     Class       Attributes

  a          10x10     800       double
  b          10x10     800       double
  c          2x2       32        double
  string     1x14      28        char
  student    1x3       888       struct
```

If this workspace is saved with the command `save workspace.mat`, a file named `workspace.mat` will be created. When this file is loaded, all of the information will be restored, including the type of each item and whether or not it is global.

   A disadvantage of these commands is that the MAT-file format is unique to MATLAB and cannot be used to share data with other programs. The −ascii option can be used if you wish to share data with other programs, but it has serious limitations.

✳ Good Programming Practice

Unless you must exchange data with non-MATLAB programs, always use the `load` and `save` commands to save data sets in MAT-file format. This format is efficient and transportable across MATLAB implementations, and it preserves all details of all MATLAB data types.

The `save —ascii` command will not save cell or structure array data at all, and it converts string data to numbers before saving it. The `load —ascii` command will load only space- or tab-separated data with an equal number of elements on each row, and it will place all of the data into a single variable with the same name as the input file. If you need anything more elaborate (e.g., saving and loading strings, cells, structure arrays, etc. in formats suitable for exchanging with other programs), then it will be necessary to use the other file I/O commands described in this chapter.

If the filename and the names of the variables to be loaded or saved are in strings, then you should use the function forms of the `load` and `save` commands. For example, the following fragment of code asks the user for a filename, and saves the workspace in that file.

```
filename = input('Enter save file name: ','s');
save (filename,'-mat');
```

8.3 An Introduction to **MATLAB** File Processing

To use files within a MATLAB program, we need some way to select the desired file and to read from or write to it. MATLAB has a very flexible method to read and write files, whether they are on disk, memory stick, or some other device attached to the computer. This mechanism is known as the **file id** (sometimes known as **fid**). The file id is a number assigned to a file when it is opened and is used for all reading, writing, and control operations on that file. The file id is a positive integer. Two file id's are always open—file id 1 is the standard output device (`stdout`), and file id 2 is the standard error (`stderr`) device for the computer on which MATLAB is executing. Additional file id's are assigned as files are opened and released as files are closed.

Several MATLAB functions may be used to control disk file input and output. The file I/O functions are summarized in Table 8.4.

File id's are assigned to disk files or devices using the `fopen` statement, and detached from them using the `fclose` statement. Once a file is attached to a file id using the `fopen` statement, we can read and write to that file using MATLAB file input and output statements. When we are through with the file, the `fclose` statement closes the file and makes the file id invalid. The `frewind` and `fseek` statements may be used to change the current reading or writing position in a file while it is open.

Data can be written to and read from files in two possible ways: as binary data or as formatted character data. Binary data consists of the actual bit patterns that are used to store the data in computer memory. Reading and writing binary data is very efficient, but a user cannot read the data stored in the file. Data in formatted files is translated into characters that can be read directly by a user. However, formatted I/O operations are slower and less efficient than binary I/O operations. We will discuss both types of I/O operations later in this chapter.

Table 8.4   MATLAB Input/Output Functions

Category	Function	Description
Load/Save Workspace	load	Load workspace.
	save	Save workspace.
File Opening and Closing	fopen	Open file.
	fclose	Close file.
Binary I/O	fread	Read binary data from file.
	fwrite	Write binary data to file.
Formatted I/O	fscanf	Read formatted data from file.
	fprintf	Write formatted data to file.
	fgetl	Read line from file, discard newline character.
	fgets	Read line from file, keep newline character.
File Positioning, Status, and Miscellaneous	delete	Delete file.
	exist	Check for the existence of a file.
	ferror	Inquire file I/O error status.
	feof	Test for end-of-file.
	fseek	Set file position.
	ftell	Check file position.
	frewind	Rewind file.
Temporary Files	tempdir	Get temporary directory name.
	tempname	Get temporary file name.

8.4   File Opening and Closing

The file opening and closing functions, fopen and fclose, are described in the following subsections.

8.4.1   The fopen Function

The fopen function opens a file and returns a file id number for use with the file. The basic forms of this statement are

```
fid = fopen(filename,permission)
[fid, message] = fopen(filename,permission)
[fid, message] = fopen(filename,permission,format)
```

where *filename* is a string specifying the name of the file to open, *permission* is a character string specifying the mode in which the file is opened, and

Table 8.5 fopen File Permissions

File Permission	Meaning
`'r'`	Open an existing file for reading only (default).
`'r+'`	Open an existing file for reading and writing.
`'w'`	Delete the contents of an existing file (or create a new file) and open it for writing only.
`'w+'`	Delete the contents of an existing file (or create a new file) and open it for reading and writing.
`'a'`	Open an existing file (or create a new file) and open it for writing only, appending to the end of the file.
`'a+'`	Open an existing file (or create a new file) and open it for reading and writing, appending to the end of the file.
`'W'`	Write without automatic flushing (special command for tape drives).
`'A'`	Append without automatic flushing (special command for tape drives).

format is an optional string specifying the numeric format of the data in the file. If the open is successful, `fid` will contain a positive integer after this statement is executed, and `message` will be an empty string. If the open fails, `fid` will contain a -1 after this statement is executed, and `message` will be a string explaining the error. If a file is opened for reading and it is not in the current directory, MATLAB will search for it along the MATLAB search path.

The possible permission strings are shown in Table 8.5.

On some platforms such as PCs, it is important to distinguish between text files and binary files. If a file is to be opened in text mode, then a `t` should be added to the permissions string (for example, `'rt'` or `'rt+'`). If a file is to be opened in binary mode, a `b` may be added to the permissions string (for example, `'rb'`), but this is not actually required, since files are opened in binary mode by default. This distinction between text and binary files does not exist on Unix or Linux computers, so the `t` or `b` is never needed on those systems.

The *format* string in the `fopen` function specifies the numeric format of the data stored in the file. This string is needed only when transferring files between computers with incompatible numeric data formats, so it is rarely used. A few of the possible numeric formats are shown in Table 8.6; see the MATLAB Language Reference Manual for a complete list of possible numeric formats.

There are also two forms of this function that provide information rather than open files. The function

```
fids = fopen('all')
```

returns a row vector containing a list of all file id's for currently open files (except for `stdout` and `stderr`). The number of elements in this vector is equal to the number of open files. The function

```
[filename, permission, format] = fopen(fid)
```

Table 8.6 fopen Format Strings

File Permission	Meaning
'native' or 'n'	Numeric format for the machine MATLAB is executing on (default)
'ieee-le' or 'l'	IEEE floating point with little-endian byte ordering
'ieee-be' or 'b'	IEEE floating point with big-endian byte ordering
'ieee-le.l64' or 'a'	IEEE floating point with little-endian byte ordering and 64-bit long data type
'ieee-le.b64' or 's'	IEEE floating point with big-endian byte ordering and 64-bit long data type

returns the file name, permission string, and numeric format for an open file specified by file id.

Some examples of correct fopen functions are shown below.

Case 1: Opening a Binary File for Input

The function that follows opens a file named example.dat for binary input only.

```
fid = fopen('example.dat','r')
```

The permission string is 'r', indicating that the file is to be opened for reading only. The string could have been 'rb', but this is not required, because binary access is the default case.

Case 2: Opening a File for Text Output

The functions that follow open a file named outdat for text output only.

```
fid = fopen('outdat','wt')
```

or

```
fid = fopen('outdat','at')
```

The 'wt' permissions string specifies that the file is a new text file; if it already exists, then the old file will be deleted and a new empty file will be opened for writing. This is the proper form of the fopen function for an *output file* if we want replace preexisting data.

The 'at' permissions string specifies that we want to append to an existing text file. If it already exists, then it will be opened and new data will be appended to the currently existing information. This is the proper form of the fopen function for an *output file* if we don't want to replace preexisting data.

Case 3: Opening a Binary File for Read/Write Access

The function that follows opens a file named junk for binary input and output.

```
fid = fopen('junk','r+')
```

The function that follows also opens the file for binary input and output.

```
fid = fopen('junk','w+')
```

The difference between the first and the second statements is that the first statement requires the file to exist before it is opened, whereas the second statement will delete any preexisting file.

✳ Good Programming Practice

Always be careful to specify the proper permissions in `fopen` statements, depending on whether you are reading from or writing to a file. This practice will help prevent errors such as accidentally overwriting data files that you want to keep.

It is important to check for errors after you attempt to open a file. If the `fid` is −1, then the file failed to open. You should report this problem to the user, and allow him or her to either select another file or else quit the program.

✳ Good Programming Practice

Always check the status after a file open operation to make sure that it is successful. If the file open fails, tell the user and provide a way to recover from the problem.

8.4.2 The `fclose` Function

The `fclose` function closes a file. Its form is

```
status = fclose(fid)
status = fclose('all')
```

where `fid` is a file id and `status` is the result of the operation. If the operation is successful, `status` will be 0; if it is unsuccessful, `status` will be −1.

The form `status = fclose('all')` closes all open files except for stdout (`fid = 1`) and stderr (`fid = 2`). It returns a status of 0 if all files close successfully, and −1 otherwise.

8.5 Binary I/O Functions

The binary I/O functions, `fwrite` and `fread`, are described in the following subsections.

8.5.1 The `fwrite` Function

The `fwrite` function writes binary data in a user-specified format to a file. Its form is

```
count = fwrite(fid,array,precision)
count = fwrite(fid,array,precision,skip)
```

where `fid` is the file id of a file opened with the `fopen` function, `array` is the array of values to write out, and `count` is the number of values written to the file.

MATLAB writes out data in *column order*, which means that the entire first column is written out, followed by the entire second column, and so on. For example, if

$$\text{array} = \begin{bmatrix} 1 & 2 \\ 3 & 4 \\ 5 & 6 \end{bmatrix}$$, the data will be written out in the order 1, 3, 5, 2, 4, 6.

The optional *precision* string specifies the format in which the data will be output. MATLAB supports both platform-independent precision strings, which are the same for all computers that MATLAB runs on, and platform-dependent precision strings, which vary among different types of computers. *You should use only the platform-independent strings*, and those are the only forms presented in this book.

For convenience, MATLAB accepts some C and Fortran data type equivalents for the MATLAB precision strings. If you are a C or Fortran programmer, you may find it more convenient to use the names of the data types in the language that you are most familiar with.

The possible platform-independent precisions are presented in Table 8.7. All of these precisions work in units of bytes, except for `'bitN'` or `'ubitN'`, which work in units of bits.

Table 8.7   Selected MATLAB Precision Strings

MATLAB Precision String	C/Fortran Equivalent	Meaning
`'char'`	`'char*1'`	8-bit characters
`'schar'`	`'signed char'`	8-bit signed character
`'uchar'`	`'unsigned char'`	8-bit unsigned character
`'int8'`	`'integer*1'`	8-bit integer
`'int16'`	`'integer*2'`	16-bit integer
`'int32'`	`'integer*4'`	32-bit integer
`'int64'`	`'integer*8'`	64-bit integer
`'uint8'`	`'integer*1'`	8-bit unsigned integer
`'uint16'`	`'integer*2'`	16-bit unsigned integer
`'uint32'`	`'integer*4'`	32-bit unsigned integer
`'uint64'`	`'integer*8'`	64-bit unsigned integer
`'float32'`	`'real*4'`	32-bit floating point
`'float64'`	`'real*8'`	64-bit floating point
`'bitN'`		N-bit signed integer, $1 \leq N \leq 64$
`'ubitN'`		N-bit unsigned integer, $1 \leq N \leq 64$

The optional argument *skip* specifies the number of bytes to skip in the output file before each write. This option is useful for placing values at certain points in fixed-length records. Note that if *precision* is a bit format like `'bitN'` or `'ubitN'`, skip is specified in bits instead of bytes.

8.5.2 The `fread` Function

The `fread` function reads binary data in a user-specified format from a file and returns the data in a (possibly different) user-specified format. Its form is

```
[array,count] = fread(fid,size,precision)
[array,count] = fread(fid,size,precision,skip)
```

where `fid` is the file id of a file opened with the `fopen` function, `size` is the number of values to read, `array` is the array to contain the data, and `count` is the number of values read from the file.

The optional argument *size* specifies the amount of data to be read from the file. There are three versions of this argument:

- n—Read exactly n values. After this statement, `array` will be a column vector containing n values read from the file.
- `Inf`—Read until the end of the file. After this statement, `array` will be a column vector containing all of the data until the end of the file.
- [n m]—Read exactly n × m values and format the data as an n × m array.

If `fread` reaches the end of the file and the input stream does not contain enough bits to write out a complete array element of the specified precision, `fread` pads the last byte or element with zero bits until the full value is obtained. If an error occurs, reading is done up to the last full value.

The *precision* argument specifies both the format of the data on the disk and the format of the data array to be returned to the calling program. The general form of the precision string is

```
'disk_precision => array_precision'
```

where `disk_precision` and `array_precision` are both one of the precision strings found in Table 8.6. The `array_precision` value can be defaulted. If it is missing, then the data is returned in a `double` array. There is also a shortcut form of this expression if the disk precision and the array precision are the same: `'*disk_precision'`.

A few examples of `precision` strings are shown as follows:

`'single'`	Read data in single precision format from disk and return it in a `double` array.
`'single=>single'`	Read data in single precision format from disk and return it in a `single` array.

`'*single'`	Read data in single precision format from disk and return it in a `single` array (a shorthand version of the previous string).
`'double=>real*4'`	Read data in double precision format from disk and return it in a `single` array.

▶

Example 8.1—Writing and Reading Binary Data

The example script file that follows creates an array containing 10,000 random values, opens a user-specified file for writing only, writes the array to disk in 64-bit floating-point format, and closes the file. It then opens the file for reading and reads the data back into a 100×100 array. It illustrates the use of binary I/O operations.

```
% Script file: binary_io.m
%
% Purpose:
%   To illustrate the use of binary i/o functions.
%
% Record of revisions:
%   Date          Programmer        Description of change
%   ====          ==========        =====================
%   03/21/07      S. J. Chapman     Original code
%
% Define variables:
%   count        -- Number of values read / written
%   fid          -- File id
%   filename     -- File name
%   in_array     -- Input array
%   msg          -- Open error message
%   out_array    -- Output array
%   status       -- Operation status

% Prompt for file name
filename = input('Enter file name: ','s');

% Generate the data array
out_array = randn(1,10000);

% Open the output file for writing.
[fid,msg] = fopen(filename,'w');

% Was the open successful?
if fid > 0

    % Write the output data.
    count = fwrite(fid,out_array,'float64');
```

```
    % Tell user
    disp([int2str(count) ' values written...']);

    % Close the file
    status = fclose(fid);

else

    % Output file open failed. Display message.
    disp(msg);

end

% Now try to recover the data. Open the
% file for reading.
[fid,msg] = fopen(filename,'r');

% Was the open successful?
if fid > 0

    % Write the output data.
    [in_array, count] = fread(fid,[100 100],'float64');

    % Tell user
    disp([int2str(count) ' values read...']);

    % Close the file
    status = fclose(fid);

else

    % Input file open failed. Display message.
    disp(msg);

end
```

When this program is executed, the result are

```
» binary_io
Enter file name: testfile
10000 values written...
10000 values read...
```

An 80,000-byte file named testfile was created in the current directory. This file is 80,000 bytes long because it contains 10,000 64-bit values and each value occupies 8 bytes.

◀

Quiz 8.1

This quiz provides a quick check to see if you have understood the concepts introduced in Sections 8.1 through 8.5. If you have trouble with the quiz, reread the section, ask your instructor, or discuss the material with a fellow student. The answers to this quiz are found in the back of the book.

1. Why is the `textread` function especially useful for reading data created by programs written in other languages?

2. What are the advantages and disadvantages of saving data in a MAT-file?

3. What MATLAB functions are used to open and close files? What is the difference between opening a binary file and opening a text file?

4. Write the MATLAB statement to open a preexisting file named `myinput.dat` for appending new text data.

5. Write the MATLAB statements required to open an unformatted input file for reading only. Check to see if the file exists, and generate an appropriate error message if it doesn't.

For questions 6 and 7, determine whether the MATLAB statements are correct or not. If they are in error, specify what is wrong with them.

6. ```
fid = fopen('file1','rt');
array = fread(fid,Inf)
fclose(fid);
```

7. ```
fid = fopen('file1','w');
x = 1:10;
count = fwrite(fid,x);
fclose(fid);
fid = fopen('file1','r');
array = fread(fid,[2 Inf])
fclose(fid);
```

8.6 Formatted I/O Functions

The formatted I/O functions are described in the following subsections.

8.6.1 The `fprintf` Function

The `fprintf` function writes formatted data in a user-specified format to a file. Its form is

```
count = fprintf(fid,format,val1,val2,...)
fprint(format,val1,val2,...)
```

where `fid` is the file id of a file to which the data will be written and `format` is the format string controlling the appearance of the data. If `fid` is missing, the data is written to the standard output device (the Command Window). This is the form of `fprintf` that we have been using since Chapter 2.

The Components of a Format Specifier

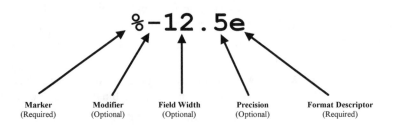

| Marker | Modifier | Field Width | Precision | Format Descriptor |
| (Required) | (Optional) | (Optional) | (Optional) | (Required) |

Figure 8.1 The structure of a typical format specifier.

The format string specifies the alignment, significant digits, field width, and other aspects of output format. It can contain ordinary alphanumeric characters along with special sequences of characters that specify the exact format in which the output data will be displayed. The structure of a typical format is shown in Figure 8.1. A single % character always marks the beginning of a format—if an ordinary % sign is to be printed out, then it must appear in the format string as %%. After the % character, the format can have a flag, a field width and precision specifier, and a conversion specifier. The % character and the conversion specifier are always required in any format, whereas the field and field width and precision specifier are optional.

The possible conversion specifiers are listed in Table 8.8, and the possible flags are listed in Table 8.9. If a field width and precision are specified in a format, then the number before the decimal point is the field width, which is the number of characters used to display the number. The number after the decimal

Table 8.8 Format Conversion Specifiers for fprintf

Specifier	Description
%c	Single character
%d	Decimal notation (signed)
%e	Exponential notation (using a lowercase e as in 3.1416e+00)
%E	Exponential notation (using an uppercase E as in 3.1416E+00)
%f	Fixed-point notation
%g	The more compact of %e or %f. Insignificant zeros do not print.
%G	Same as %g, but using an uppercase E
%o	Octal notation (unsigned)
%s	String of characters
%u	Decimal notation (unsigned)
%x	Hexadecimal notation (using lowercase letters a–f)
%X	Hexadecimal notation (using uppercase letters A–F)

Table 8.9   Format Flags

Flag	Description
Minus sign ($-$)	Left-justifies the converted argument in its field (example: `%-5.2d`). If this flag is not present, the argument is right justified.
+	Always print a + or – sign (example: `%+5.2d`).
0	Pad argument with leading zeros instead of blanks (example: `%05.2d`).

Table 8.10   Escape Characters in Format Strings

Escape Sequences	Description
`\n`	New line.
`\t`	Horizontal tab.
`\b`	Backspace.
`\r`	Carriage return.
`\f`	Form feed.
`\\`	Print an ordinary backslash (\) symbol.
`\''` or `''`	Print an apostrophe or single quote.
`%%`	Print an ordinary percent (%) symbol.

point is the precision, which is the minimum number of significant digits to display after the decimal point.

In addition to ordinary characters and formats, certain special escape characters can be used in a format string. These special characters are listed in Table 8.10.

8.6.2   Understanding Format Conversion Specifiers

Because the best way to understand the wide variety of format conversion specifiers is by example, we will now present several examples along with their results.

Case 1: Displaying Decimal Data

Decimal (integer) data is displayed with the `%d` format conversion specifier. The d may be preceded by a flag and a field-width and precision specifier, if desired. If used, the precision specifier will set a minimum number of digits to display. If there are not enough digits, leading zeros will be added to the number.

If a nondecimal number is displayed with the `%d` conversion specifier, the specifier will be ignored, and the number will be displayed in exponential format. For example,

```
fprintf('%6d\n',123.4)
```

produces the result `1.234000e+002`.

Function	Result	Comment
`fprintf('%d\n',123)`	----|----| 123	Display the number using as many characters as required. For the number 123, three characters are required.
`fprintf('%6d\n',123)`	----|----| 123	Display the number in a 6-character-wide field. By default the number is *right justified* in the field.
`fprintf('%6.4d\n',123)`	----|----| 0123	Display the number in a 6-character-wide field using a minimum of 4 characters. By default the number is *right justified* in the field.
`fprintf('%-6.4d\n',123)`	----|----| 0123	Display the number in a 6-character-wide field using a minimum of 4 characters. The number is *left justified* in the field.
`fprintf('%+6.4d\n',123)`	----|----| +0123	Display the number in a 6-character-wide field using a minimum of 4 characters plus a sign character. By default the number is *right justified* in the field.

Case 2: Displaying Floating-Point Data

Floating-point data can be displayed with the `%e`, `%f`, or `%g` format conversion specifiers. They may be preceded by a flag and a field width and precision specifier, if desired. If the specified field width is too small to display the number, it is ignored. Otherwise, the specified field width is used.

Function	Result	Comment
`fprintf('%f\n',123.4`	----|----| 123.400000	Display the number using as many characters as required. The default case for `%f` is to display 6 digits after the decimal place.
`fprintf('%8.2f\n',123.4)`	----|----| 123.40	Display the number in an 8-character wide field, with two places after the decimal point. The number is *right justified* in the field.
`fprintf('%4.2f\n',123.4)`	----|----| 123.40	Display the number in a 6-character-wide field. The width specification was ignored because it was too small to display the number.

```
fprintf('%10.2e\n',123.4)   ----|----|        Display the number in
                            1.23e+002          exponential format in a
                                               10-character-wide field using
                                               2 decimal places. By default
                                               the number is *right justified*
                                               in the field.

fprintf('%10.2E\n',123.4)   ----|----|        The same but with a capital E for
                            1.23E+002          the exponent.
```

Case 3: Displaying Character Data

Character data may be displayed with the %c or %s format conversion specifiers. They may be preceded by field width specifier, if desired. If the specified field with is too small to display the number, it is ignored. Otherwise, the specified field width is used.

Function	Result	Comment
fprintf('%c\n','s')	----|----| s	Display a single character.
fprintf('%s\n','string')	----|----| string	Display the character string.
fprintf('%8s\n','string')	----|----| string	Display the character string in an 8-character-wide field. By default the string is *right justified* in the field.
fprintf('%-8s\n','string')	----|----| string	Display the character string in an 8-character-wide field. The string is *left justified* in the field.

8.6.3 How Format Strings Are Used

The fprintf function contains a format string followed by zero or more values to print out. When the fprintf function is executed, the list of output values associated with the fprintf function is processed together with the format string. The function begins at the left end of the variable list and the left end of the format string, and scans from left to right, associating the first value in the output list with the first format descriptor in the format string, and so forth. The variables in the output list must be of the same type and in the same order as the format descriptors in the format, or unexpected results may be produced. For example, if we attempt to display a floating-point number such as 123.4 with a %c or %d descriptor, the descriptor is ignored totally and the number is printed in exponential notation.

🎇 Programming Pitfalls

Make sure that there is a one-to-one correspondence between the types of the data in an `fprintf` function and the types of the format conversion specifiers in the associated format string, or your program will produce unexpected results.

As the program moves from left to right through the variable list of an `fprintf` function, it also scans from left to right through the associated format string. Format strings are scanned according to the following rules:

1. *Format strings are scanned in order from left to right.* The first format conversion specifier in the format string is associated with the first value in the output list of the `fprintf` function, and so forth. The type of each format conversion specifier must match the type of the data being output. In the example shown below, specifier `%d` is associated with variable `a`, `%f` with variable `b`, and `%s` with variable `c`. Note that the specifier types match the data types.

   ```
   a = 10; b = pi; c = 'Hello';
   fprintf('Output: %d %f %s\n',a,b,c);
   ```

2. If the scan reaches the end of the format string before the `fprintf` function runs out of values, the program starts over *at the beginning of the format string.* For example, the statements

   ```
   a = [10 20 30 40];
   fprintf('Output = %4d %4d\n',a);
   ```

   will produce the output

   ```
   ----|----|----|----|
   Output = 10 20
   Output = 30 40
   ```

   When the function reaches the end of the format string after printing `a(2)`, it starts over at the beginning of the string to print `a(3)` and `a(4)`.

3. If the `fprintf` function runs out of variables before the end of the format string, *the use of the format string stops at the first format conversion specifier without a corresponding variable, or at the end of the format string, whichever comes first.* For example, the statements

   ```
   a = 10; b = 15; c = 20;
   fprintf('Output = %4d\nOutput = %4.1f\n',a,b,c);
   ```

   will produce the output

   ```
   Output = 10
   Output = 15.0
   Output = 20
   Output = »
   ```

The use of the format string stops at %4.1f, which is the first unmatched format conversion specifier. On the other hand, the statements

```
voltage = 20;
fprintf('Voltage = %6.2f kV.\n',voltage);
```

will produce the output

```
Voltage = 20.00 kV.
```

since there are no unmatched format conversion specifiers, and the use of the format stops at the end of the format string.

8.6.4 The `sprintf` Function

The `sprintf` function is exactly like `fprintf`, except that it writes formatted data to a character string instead of a file. Its form is

```
string = sprint(format,val1,val2,...)
```

where `format` is the format string controlling the appearance of the data. This function is very useful for creating formatted data that can be displayed within a program.

▶
━━━━━━━━━━━━━━━━━━━━━━━━━━━━━━━━━━━━━━━

Example 8.2—Generating a Table of Information

A good way to illustrate the use of `fprintf` functions is to generate and print out a table of data. The example script file that follows generates the square roots, squares, and cubes of all integers between 1 and 10 and presents the data in a table with appropriate headings.

```
% Script file: table.m
%
% Purpose:
%   To create a table of square roots, squares, and
%   cubes.
%
% Record of revisions:
%     Date          Programmer           Description of change
%     ====          ==========           =====================
%   03/22/07    S. J. Chapman         Original code
%
% Define variables:
%   cube            -- Cubes
%   ii              -- Index variable
%   square          -- Squares
%   square_roots    -- Square roots
%   out             -- Output array

% Print the title of the table.
fprintf(' Table of Square Roots, Squares, and Cubes\n\n');
```

```
% Print column headings
fprintf(' Number     Square Root      Square    Cube\n');
fprintf(' ======     ===========      ======    ====\n');

% Generate the required data
ii = 1:10;
square_root = sqrt(ii);
square = ii.^2;
cube = ii.^3;

% Create the output array
out = [ii' square_root' square' cube'];

% Print the data
for ii = 1:10
   fprintf (' %2d     %11.4f     %6d     %8d\n',out(ii,:));
end
```

When this program is executed, the result is

```
» table
Table of Square Roots, Squares, and Cubes

Number      Square Root      Square    Cube
======      ===========      ======    ====
    1          1.0000            1        1
    2          1.4142            4        8
    3          1.7321            9       27
    4          2.0000           16       64
    5          2.2361           25      125
    6          2.4495           36      216
    7          2.6458           49      343
    8          2.8284           64      512
    9          3.0000           81      729
   10          3.1623          100     1000
```

◄

8.6.5   The fscanf Function

The fscanf function reads formatted data in a user-specified format from a file. Its form is

```
array = fscanf(fid,format)
[array, count] = fscanf(fid,format,size)
```

where fid is the file id of a file from which the data will be read, format is the format string controlling how the data is read, and array is the array that receives the data. The output argument count returns the number of values read from the file.

The optional argument *size* specifies the amount of data to be read from the file. There are three versions of this argument:

- n—Read exactly n values. After this statement, `array` will be a column vector containing n values read from the file.
- `Inf`—Read until the end of the file. After this statement, `array` will be a column vector containing all of the data until the end of the file.
- [n m]—Read exactly n × m values and format the data as an n × m array.

The format string specifies the format of the data to be read. It can contain ordinary characters along with format conversion specifiers. The `fscanf` function compares the data in the file with the format conversion specifiers in the format string. As long as the two match, `fscanf` converts the value and stores it in the output array. This process continues until the end of the file or until the amount of data in *size* has been read, whichever comes first.

If the data in the file does not match the format conversion specifiers, the operation of `fscanf` stops immediately.

The format conversion specifiers for `fscanf` are basically the same as those for `fprintf`. The most common specifiers are shown in Table 8.11.

To illustrate the use of `fscanf`, we will attempt to read a file called `x.dat` containing the following values on two lines:

```
10.00   20.00
30.00   40.00
```

1. If the file is read with the statement

```
[z, count] = fscanf(fid,'%f');
```

variable z will be the column vector $\begin{bmatrix} 10 \\ 20 \\ 30 \\ 40 \end{bmatrix}$ and `count` will be 4.

Table 8.11   Format Conversion Specifiers for `fscanf`

Specifier	Description
%c	Read a single character. This specifier reads any character including blanks, new lines, etc.
%Nc	Read *N* characters.
%d	Read a decimal number (ignores blanks).
%e %f %g	Read a floating-point number (ignores blanks).
%i	Read a signed integer (ignores blanks).
%s	Read a string of characters. The string is terminated by blanks or other special characters such as new lines.

2. If the file is read with the statement

```
[z, count] = fscanf(fid,'%f',[2 2]);
```

variable z will be the array $\begin{bmatrix} 10 & 30 \\ 20 & 40 \end{bmatrix}$ and count will be 4.

3. Next, let's try to read this file as decimal values. If the file is read with the statement

```
[z, count] = fscanf(fid,'%d',Inf);
```

variable z will be the single value 10 and count will be 1. This happens because the decimal point in the 10.00 does not match the format conversion specifier, and fscanf stops at the first mismatch.

4. If the file is read with the statement

```
[z, count] = fscanf(fid,'%d.%d',[1 Inf]);
```

variable z will be the row vector [10 0 20 0 30 0 40 0] and count will be 8. This happens because the decimal point is now matched in the format conversion specifier and the numbers on either side of the decimal point are interpreted as separate integers.

5. Now let's try to read the file as individual characters. If the file is read with the statement

```
[z, count] = fscanf(fid,'%c');
```

variable z will be a row vector containing every character in the file, including all spaces and newline characters! Variable count will be equal to the number of characters in the file.

6. Finally, let's try to read the file as a character string. If the file is read with the statement

```
[z, count] = fscanf(fid,'%s');
```

variable z will be a row vector containing the 20 characters 10.0020.0030.0040.00, and count will be 4. This happens because the string specifier ignores white space, and the function found four separate strings in the file.

8.6.6 The `fgetl` Function

The `fgetl` function reads the next line *excluding the end-of-line characters* from a file as a character string. It form is

```
line = fgetl(fid)
```

where `fid` is the file id of a file from which the data will be read, and `line` is the character array that receives the data. If `fgetl` encounters the end of a file, the value of `line` is set to −1.

8.6.7 The `fgets` Function

The `fgets` function reads the next line *including the end-of-line characters* from a file as a character string. Its form is

```
line = fgets(fid)
```

where `fid` is the file id of a file from which the data will be read and `line` is the character array that receives the data. If `fgets` encounters the end of a file, the value of `line` is set to −1.

8.7 Comparing Formatted and Binary I/O Functions

Formatted I/O operations produce formatted files. A **formatted file** contains recognizable characters, numbers, and so forth. stored as ASCII text. These files are easy to distinguish, because we can see the characters and numbers in the file when we display them on the screen or print them on a printer. However, to use data in a formatted file, a MATLAB program must translate the characters in the file into the internal data format used by the computer. Format conversion specifiers provide the instructions for this translation.

Formatted files have the advantages that we can readily see what sort of data they contain and it is easy to exchange data between different types of programs using them. However, they also have disadvantages. A program must do a good deal of work to convert a number between the computer's internal representation and the characters contained in the file. All of this work is just wasted effort if we are going to be reading the data back into another MATLAB program. Also, the internal representation of a number usually requires much less space than the corresponding representation of the number found in a formatted file. For example, the internal representation of a 64-bit floating-point value requires 8 bytes of space. The character representation of the same value would be $\pm$d.ddddddddddddddE$\pm$ee, which requires 21 bytes of space (one byte per character). Thus, storing data in character format is inefficient and wasteful of disk space.

Unformatted files (or **binary files**) overcome these disadvantages by copying the information from the computer's memory directly to the disk file with no conversions at all. Since no conversions occur, no computer time is wasted formatting the data. In MATLAB, binary I/O operations are *much* faster than formatted I/O operations because there is no conversion. Furthermore, the data occupies a much smaller amount of disk space. On the other hand, unformatted data cannot be examined and interpreted directly by humans. In addition, it usually cannot be moved between different types of computers, because different types of computers have different internal ways to represent integers and floating-point values.

Formatted and unformatted files are compared in Table 8.12. In general, formatted files are best for data that people must examine or for data that may have to be moved between different programs on different computers. Unformatted files

Table 8.12 Comparison of Formatted and Unformatted Files

Formatted Files	Unformatted Files
Can display data on output devices.	Cannot display data on output devices.
Can easily transport data between different computers.	Cannot easily transport data between computers with different internal data representations.
Require a relatively large amount of disk space.	Require relatively little disk space.
Slow—require a lot of computer time.	Fast—require little computer time.
Truncation or rounding errors possible in formatting.	No truncation or rounding errors.

are best for storing information that will not need to be examined by human beings and that will be created and used on the same type of computer. Under those circumstances, unformatted files are faster and occupy less disk space.

☀ Good Programming Practice

Use formatted files to create data that must be readable by humans, or that must be transferable between programs on computers of different types. Use unformatted files to efficiently store large quantities of data that do not have to be directly examined and that will remain on only one type of computer. In addition, use unformatted files when I/O speed is critical.

Example 8.3—Comparing Formatted and Binary I/O

The program that follows compares the time required to read and write a 10,000-element array using both formatted and binary I/O operations. Note that each operation is repeated 10 times and the average time is reported.

```
% Script file: compare.m
%
% Purpose:
%   To compare binary and formatted I/O operations.
%   This program generates an array of 10,000 random
%   values and writes it to disk both as a binary and
%   as a formatted file.
%
```

```
% Record of revisions:
%     Date         Programmer        Description of change
%     ====         ==========        =====================
%   03/22/07     S. J. Chapman       Original code
%
% Define variables:
%    count         -- Number of values read / written
%    fid           -- File id
%    in_array      -- Input array
%    msg           -- Open error message
%    out_array     -- Output array
%    status        -- Operation status
%    time          -- Elapsed time in seconds

%%%%%%%%%%%%%%%%%%%%%%%%%%%%%%%%%%%%%%%%%%%%%%%%%%%%%
% Generate the data array.
%%%%%%%%%%%%%%%%%%%%%%%%%%%%%%%%%%%%%%%%%%%%%%%%%%%%%
out_array = randn(1,10000);

%%%%%%%%%%%%%%%%%%%%%%%%%%%%%%%%%%%%%%%%%%%%%%%%%%%%%
% First, time the binary output operation.
%%%%%%%%%%%%%%%%%%%%%%%%%%%%%%%%%%%%%%%%%%%%%%%%%%%%%
% Reset timer
tic;

% Loop for 10 times
for ii = 1:10

    % Open the binary output file for writing.
    [fid,msg] = fopen('unformatted.dat','w');

    % Write the data
    count = fwrite(fid,out_array,'float64');

    % Close the file
    status = fclose(fid);

end

% Get the average time
time = toc / 10;
fprintf ('Write time for unformatted file = %6.3f\n',time);

%%%%%%%%%%%%%%%%%%%%%%%%%%%%%%%%%%%%%%%%%%%%%%%%%%%%%
% Next, time the formatted output operation.
%%%%%%%%%%%%%%%%%%%%%%%%%%%%%%%%%%%%%%%%%%%%%%%%%%%%%
```

```
% Reset timer
tic;

% Loop for 10 times
for ii = 1:10

    % Open the formatted output file for writing.
    [fid,msg] = fopen('formatted.dat','wt');

    % Write the data
    count = fprintf(fid,'%23.15e\n',out_array);

    % Close the file
    status = fclose(fid);

end

% Get the average time
time = toc / 10;
fprintf ('Write time for formatted file = %6.3f\n',time);


%%%%%%%%%%%%%%%%%%%%%%%%%%%%%%%%%%%%%%%%%%%%%%%%%%%%%
% Time the binary input operation.
%%%%%%%%%%%%%%%%%%%%%%%%%%%%%%%%%%%%%%%%%%%%%%%%%%%%

% Reset timer
tic;

% Loop for 10 times
for ii = 1:10

    % Open the binary file for reading.
    [fid,msg] = fopen('unformatted.dat','r');

    % Read the data
    [in_array, count] = fread(fid,Inf,'float64');

    % Close the file
    status = fclose(fid);

end

% Get the average time
time = toc / 10;
fprintf ('Read time for unformatted file = %6.3f\n', time);
```

```
%%%%%%%%%%%%%%%%%%%%%%%%%%%%%%%%%%%%%%%%%%%%%%%
% Time the formatted input operation.
%%%%%%%%%%%%%%%%%%%%%%%%%%%%%%%%%%%%%%%%%%%%%%%

% Reset timer
tic;

% Loop for 10 times
for ii = 1:10

    % Open the formatted file for reading.
    [fid,msg] = fopen('formatted.dat','rt');

    % Read the data
    [in_array, count] = fscanf(fid,'%f',Inf);

    % Close the file
    status = fclose(fid);

end

% Get the average time
time = toc / 10;
fprintf ('Read time for formatted file = %6.3f\n', time);
```

When this program is executed in MATLAB 7.4 on a 1.8 GHz Intel Core 2 Duo computer running Windows XP Professional, the results are

```
» compare
Write time for unformatted file = 0.002
Write time for formatted file =   0.033
Read time for unformatted file =  0.001
Read time for formatted file =    0.023
```

The files written to disk are

```
C:\book\matlab\chap8>dir *.dat
Volume in drive C is SYSTEM
Volume Serial Number is 0866-1AC5

Directory of C:\book\matlab\chap8

23/03/2007  04:24p  <DIR>                         .
23/03/2007  04:24p  <DIR>                         ..
23/03/2007  05:26p                 250,000 formatted.dat
23/03/2007  05:26p                  80,000 unformatted.dat
              4 File(s)            330,000 bytes
              2 Dir(s)     30,364,545,024 bytes free
```

Note that the write time for the formatted file was more than 15 times slower than the write time for the unformatted file, and the read time for the formatted file was more than 20 times slower than the read time for the unformatted file. Furthermore, the formatted file was 3 times larger than the unformatted file.

It is clear from these results that unless you *really* need formatted data, binary I/O operations are the preferred way to save data in MATLAB.

◀

Quiz 8.2

This quiz provides a quick check to see if you have understood the concepts introduced in Sections 8.6 and 8.7. If you have trouble with the quiz, reread the section, ask your instructor, or discuss the material with a fellow student. The answers to this quiz are found in the back of the book.

1. What is the difference between unformatted (binary) and formatted I/O operations?
2. When should formatted I/O be used? When should unformatted I/O be used?
3. Write the MATLAB statements required to create a table contains the sine and cosine of x for $x = 0, 0.1\pi, \ldots, \pi$. Be sure to include a title and label on the table.

For questions 4 and 5, determine whether the MATLAB statements are correct or not. If they are in error, specify what is wrong with them.

4.
```
a = 2*pi;
b = 6;
c = 'hello';
fprintf(fid,'%s %d %g\n',a,b,c);
```
5.
```
data1 = 1:20;
data2 = 1:20;
fid = fopen('xxx','w+');
fwrite(fid,data1);
fprintf(fid,'%g\n',data2);
```

8.8 **File Positioning and Status Functions**

As was stated previously, MATLAB files are sequential—they are read in order from the first record in the file to the last record in the file. However, we sometimes need to read a piece of data more than once, or to process a whole file more than once during a program. How can we skip around within a sequential file?

The MATLAB function `exist` can determine whether or not a file exists before it is opened. There are two functions to tell us where we are within a file once it is opened: `feof` and `ftell`. In addition, there are two functions to help us move around within the file: `frewind` and `fseek`.

Finally, MATLAB includes a function `ferror` that provides a detailed description of cause of I/O errors when they occur. We will now explore these five functions, looking at `ferror` first, since it can be used with all of the other functions.

8.8.1 The `exist` Function

The MATLAB function `exist` checks for the existence of a variable in a work-space, a built-in function, or a file in the MATLAB search path. The forms of the `ferror` function are

```
ident = exist('item');
ident = exist('item','kind');
```

If `'item'` exists, this function returns a value based on its type. The possible results are shown in Table 8.13.

The second form of the `exist` function restricts the search for an item to a specified kind. The legal types are `'var'`, `'file'`, `'builtin'`, and `'dir'`.

The `exist` function is very important, because we can use it to check for the existence of a file before it is overwritten by `fopen`. The permissions `'w'` and `'w+'` delete the contents of an existing file when they open it. Before a programmer allows `fopen` to delete an existing file, he or she should check with the user to confirm that the file really should be deleted.

Table 8.13 Values Returned by the `exist` Function

Value	Meaning
0	Item not found.
1	Item is a variable in the current workspace.
2	Item is an M-file or a file of unknown type.
3	Item is a MEX file.
4	Item is a MDL file.
5	Item is a built-in function.
6	Item is a P file.
7	Item is a directory.
8	Item is a Java class.

▶

Example 8.4—Opening an Output File

The program that follows gets an output file name from the user and checks to see whether it exists. If it exists, the program checks to see whether the user wants to delete the existing file or to append the new data to it. If the file does not exist, then the program simply opens the output file.

```
% Script file: output.m
%
% Purpose:
%   To demonstrate opening an output file properly.
%   This program checks for the existence of an output
%   file. If it exists, the program checks to see if
%   the old file should be deleted, or if the new data
%   should be appended to the old file.
%
% Record of revisions:
%     Date          Programmer           Description of change
%     ====          ==========           =====================
%   03/24/07       S. J. Chapman         Original code
%
% Define variables:
%   fid              -- File id
%   out_filename     -- Output file name
%   yn               -- Yes/No response

% Get the output file name.
out_filename = input('Enter output filename: ','s');

% Check to see if the file exists.
if exist(out_filename,'file')

   % The file exists
   disp('Output file already exists.');
   yn = input('Keep existing file? (y/n) ','s');

   if yn == 'n'
      fid = fopen(out_filename,'wt');
   else
      fid = fopen(out_filename,'at');
   end

else
```

```
    % File doesn't exist
    fid = fopen(out_filename,'wt');

end

% Output data
fprintf(fid,'%s\n',date);

% Close file
fclose(fid);
```

When this program is executed, the results are

```
» output
Enter output filename: xxx          (Nonexistent file)
» type xxx

23-Mar-2007

» output
Enter output filename: xxx
Output file already exists.
Keep existing file? (y/n) y        (Keep current file)
» type xxx

23-Mar-2007
23-Mar-2007                        (Note new data added)

» output
Enter output filename: xxx
Output file already exists.
Keep existing file? (y/n) n        (Replace current file)
» type xxx

23-Mar-2007
```

The program appears to be functioning correctly in all three cases. ◀

✳ Good Programming Practice

Do not overwrite an output file without confirming that the user would like to delete the preexisting information.

8.8.2 The `ferror` Function

The MATLAB I/O system has several internal variables, including a special error indicator that is associated with each open file. This error indicator is updated by every I/O operation. The `ferror` function gets the error indicator and translates it into an easy to understand character message. The forms of the `ferror` function are

```
message = ferror(fid)
message = ferror(fid,'clear')
[message,errnum] = ferror(fid)
```

This function returns the most recent error message (and optionally error number) associated with the file attached to `fid`. It may be called at any time after any I/O operation to obtain a more detailed description of what went wrong. If this function is called after a successful operation, the message will be '..' and the error number will be 0.

The 'clear' argument clears the error indicator for a particular file id.

8.8.3 The `feof` Function

The `feof` function tests to see whether the current file position is at the end of the file. The form of the `feof` function is

```
eofstat = feof(fid)
```

This function returns a logical true (1) if the current file position is at the end of the file, and a logical false (0) otherwise.

8.8.4 The `ftell` Function

The `ftell` function returns the current location of the file position indicator for the file specified by `fid`. The position is a nonnegative integer specified in bytes from the beginning of the file. A returned value of −1 for position indicates that the query was unsuccessful. If this happens, use `ferror` to determine why the request failed. The form of the `ftell` function is

```
position = ftell(fid)
```

8.8.5 The `frewind` Function

The `frewind` function allows a programmer to reset a file's position indicator to the beginning of the file. The form of the `frewind` function is

```
frewind(fid)
```

This function does not return status information.

8.8.6   The `fseek` Function

The `fseek` function allows a programmer to set a file's position indicator to an arbitrary location within a file. The form of the `fseek` function is

```
status = fseek(fid,offset,origin)
```

This function repositions the file position indicator in the file with the given `fid` to the byte with the specified `offset` relative to `origin`. The `offset` is measured in bytes, with a positive number indicating motion towards the end of the file and a negative number indicating motion towards the head of the file. The `origin` is a string that can have one of three possible values:

- `'bof'`—This is the beginning of the file.
- `'cof'`—This is the current position within the file.
- `'eof'`—This is the end of the file.

The returned `status` is zero if the operation is successful and -1 if the operation fails. If the returned status is -1, use `ferror` to determine why the request failed.

As an example of using `fseek` and `ferror` together, consider the following statements:

```
[fid,msg] = fopen('x','r');
status = fseek(fid,-10,'bof');
if status ~= 0
   msg = ferror(fid);
   disp(msg);
end
```

These commands open a file and attempt to set the file pointer 10 bytes before the beginning of the file. Since this is impossible, `fseek` returns a status of -1, and `ferror` gets an appropriate error message. When these statements are executed, the result is an informative error message:

```
Offset is bad - before beginning-of-file.
```

▶
━━━

Example 8.5—Fitting a Line to a Set of Noisy Measurements

In Example 4.6, we learned how to perform a fit of a noisy set of measurements (x, y) to a line of the form

$$y = mx + b \tag{8-1}$$

The standard method for finding the regression coefficients m and b is the method of least squares. This method is named "least squares" because it produces the line $y = mx + b$ for which the sum of the squares of the differences between the observed y values and the predicted y values is as small as possible. The slope of the least squares line is given by

$$m = \frac{(\sum xy) - (\sum x)\bar{y}}{(\sum x^2) - (\sum x)\bar{x}} \qquad (8\text{-}2)$$

and the intercept of the least squares line is given by

$$b = \bar{y} - m\bar{x} \qquad (8\text{-}3)$$

where

$\sum x$ is the sum of the x values
$\sum x^2$ is the sum of the squares of the x values
$\sum xy$ is the sum of the products of the corresponding x and y values
$\bar{x}$ is the mean (average) of the x values
$\bar{y}$ is the mean (average) of the y values

Write a program that will calculate the least-squares slope m and y-axis intercept b for a given set of noisy measured data points (x, y) *that are to be found in an input data file.*

SOLUTION

1. **State the problem.**
   Calculate the slope m and intercept b of a least-squares line that best fits an input data set consisting of an arbitrary number of (x, y) pairs. The input (x, y) data resides in a user-specified input file.

2. **Define the inputs and outputs.**
   The inputs required by this program are pairs of points (x, y), where x and y are real quantities. Each pair of points will be located on a separate line in the input disk file. The number of points in the disk file is not known in advance.
   The outputs from this program are the slope and intercept of the least-squares fitted line, plus the number of points going into the fit.

3. **Describe the algorithm.**
   This program can be broken down into four major steps:

   ```
   Get the name of the input file and open it
   Accumulate the input statistics
   Calculate the slope and intercept
   Write out the slope and intercept
   ```

   The first major step of the program is to get the name of the input file and to open the file. To do this, we will have to prompt the user to enter the name of the input file. After the file has been opened, we must check to see that the open was successful. Next, we must read the file and keep track of the number of values entered, plus the sums Σx, Σy, Σx^2, and Σxy. The pseudocode for these steps is

```
Initialize n, sum_x, sum_x2, sum_y, and sum_xy to 0
Prompt user for input file name
Open file 'filename'
Check for error on open
if no error
   Read x, y from file 'filename'
   while not at end-of-file
      n <- n + 1
      sum_x <- sum_x + x
      sum_y <- sum_y + y
      sum_x2 <- sum_x2 + x^2
      sum_xy <- sum_xy + x*y
      Read x, y from file 'filename'
   end
 (further processing)
end
```

Next, we must calculate the slope and intercept of the least-squares line. The pseudocode for this step is just the MATLAB versions of Equations 8-2 and 8-3.

```
x_bar <- sum_x / n
y_bar <- sum_y / n
slope <- (sum_xy - sum_x*y_bar) / (sum_x2 - sum_x*x_bar)
y_int <- y_bar - slope * x_bar
```

Finally, we must write out the results.

```
Write out slope 'slope' and intercept 'y_int'.
```

4. **Turn the algorithm into MATLAB statements.**
   The final MATLAB program is

```
% Script file: lsqfit.m
%
% Purpose:
%   To perform a least-squares fit of an input data set
%   to a straight line, and print out the resulting slope
%   and intercept values. The input data for this fit
%   comes from a user-specified input data file.
%
% Record of revisions:
%     Date         Programmer          Description of change
%     ====         ==========          =====================
%   03/24/07     S. J. Chapman        Original code
```

```
%
% Define variables:
%   count        -- number of values read
%   filename     -- Input file name
%   fid          -- File id
%   msg          -- Open error message
%   n            -- Number of input data pairs (x,y)
%   slope        -- Slope of the line
%   sum_x        -- Sum of all input X values
%   sum_x2       -- Sum of all input X values squared
%   sum_xy       -- Sum of all input X*Y values
%   sum_y        -- Sum of all input Y values
%   x            -- An input X value
%   x_bar        -- Average X value
%   y            -- An input Y value
%   y_bar        -- Average Y value
%   y_int        -- Y-axis intercept of the line

% Initialize sums
n = 0; sum_x = 0; sum_y = 0; sum_x2 = 0; sum_xy = 0;

% Prompt user and get the name of the input file.
disp('This program performs a least-squares fit of an');
disp('input data set to a straight line. Enter the name');
disp('of the file containing the input (x,y) pairs: ' );
filename = input(' ','s');

% Open the input file
[fid,msg] = fopen(filename,'rt');

% Check to see if the open failed.
if fid < 0

   % There was an error--tell user.
   disp(msg);

else

   % File opened successfully. Read the (x,y) pairs from
   % the input file. Get first (x,y) pair before the
   % loop starts.
   [in,count] = fscanf(fid,'%g %g',2);

   while ~feof(fid)
      x = in(1);
      y = in(2);
      n      = n + 1;                  %
      sum_x  = sum_x + x;              % Calculate
      sum_y  = sum_y + y;              % statistics
```

```
      sum_x2 = sum_x2 + x.^2;             %
      sum_xy = sum_xy + x * y;             %

      % Get next (x,y) pair
      [in,count] = fscanf(fid,'%f',[1 2]);

  end

  % Close the file
  fclose(fid);

  % Now calculate the slope and intercept.
  x_bar = sum_x / n;
  y_bar = sum_y / n;
  slope = (sum_xy - sum_x*y_bar) / (sum_x2 - sum_x*x_bar);
  y_int = y_bar - slope * x_bar;

  % Tell user.
  fprintf('Regression coefficients for the least-squares line:\n');
  fprintf(' Slope (m)     = %12.3f\n',slope);
  fprintf(' Intercept (b) = %12.3f\n',y_int);
  fprintf(' No of points  = %12d\n',n);
end
```

5. **Test the program.**

   To test this program, we will try a simple data set. For example, if every point in the input data set actually falls along a line, then the resulting slope and intercept should be exactly the slope and intercept of that line. Thus the data set

1.1	1.1
2.2	2.2
3.3	3.3
4.4	4.4
5.5	5.5
6.6	6.6
7.7	7.7

   should produce a slope of 1.0 and an intercept of 0.0. If we place these values in a file called input1, and run the program, the results are

   ```
   » lsqfit
   This program performs a least-squares fit of an
   input data set to a straight line. Enter the name
   of the file containing the input (x,y) pairs:
   input1
   Regression coefficients for the least-squares line:
        Slope (m)     = 1.000
        Intercept (b) = 0.000
        No of points  =        7
   ```

Now let's add some noise to the measurements. The data set becomes

```
1.1   1.01
2.2   2.30
3.3   3.05
4.4   4.28
5.5   5.75
6.6   6.48
7.7   7.84
```

If these values are placed in a file called `input2`, and the program is run on that file, the results are

```
» lsqfit
This program performs a least-squares fit of an
input data set to a straight line. Enter the name
of the file containing the input (x,y) pairs:
input2
Regression coefficients for the least-squares line:
   Slope (m)      =    1.024
   Intercept (b)  =  -0.120
   No of points   =       7
```

If we calculate the answer by hand, it is easy to show that the program gives the correct answers for our two test data sets. The noisy input data set and the resulting least-squares fitted line are shown in Figure 8.2.

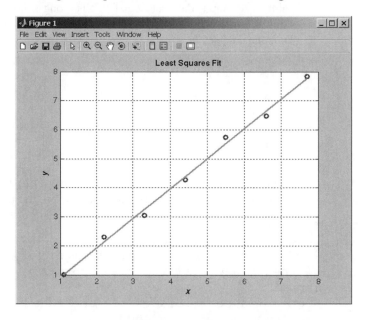

Figure 8.2 A noisy input data set and the resulting least-squares fitted line.

8.9   The textscan Function

The textscan function reads ASCII files that are formatted into columns of data, where each column can be of a different type, and stores the contents into the columns of a cell array. This function is *very* useful for importing tables of data printed out by other applications. It is new as of MATLAB 7.0. It is basically similar to textread, except that it is faster and more flexible.

The form of the textscan function is

```
a = textscan(fid, 'format')
a = textscan(fid, 'format', N)
a = textscan(fid, 'format', param, value, ...)
a = textscan(fid, 'format', N, param, value, ...)
```

where fid is the file id of a file that has already been opened with fopen, format is a string containing a description of the type of data in each column, and n is the number of times to use the format specifier. (If n is -1 or is missing, the function reads to the end of the file.) The format string contains the same types of format descriptors as function fprintf. Note that there is only one output argument, with all of the values returned in a cell array. The cell array will contain a number of elements equal to the number of format descriptors to read.

For example, suppose that file test_input1.dat contains the following data:

```
James     Jones     O+     3.51     22     Yes
Sally     Smith     A+     3.28     23     No
Hans      Carter    B-     2.84     19     Yes
Sam       Spade     A+     3.12     21     Yes
```

This data could be read into a cell array with the following function:

```
fid = fopen('test_input1.dat','rt');
a = textscan(fid,'%s %s %s %f %d %s',-1);
fclose(fid);
```

When this command is executed, the results are

```
» fid = fopen('test_input1.dat','rt');
» a = textscan(fid,'%s %s %s %f %d %s',-1)

a =
  {4x1 cell}    {4x1 cell}    {4x1 cell}    [4x1 double]
  [4x1 int32]   {4x1 cell}
» a{1}
ans =
    'James'
    'Sally'
    'Hans'
    'Sam'
```

```
» a{2}
ans =
    'Jones'
    'Smith'
    'Carter'
    'Spade'
» a{3}
ans =
    'O+'
    'A+'
    'B-'
    'A+'
» a{4}
ans =
    3.5100
    3.2800
    2.8400
    3.1200
» fclose(fid);
```

This function can also skip selected columns by adding an asterisk to the corresponding format descriptor (e.g., %*s). For example, the following statements read only the first name, last name, and gpa from the file:

```
fid = fopen('test_input1.dat','rt');
a = textscan(fid,'%s %s %*s %f %*d %*s',-1);
fclose(fid);
```

Function textscan is similar to function textread, but it is more flexible and faster. The advantages of textscan include the following:

1. The textscan function offers better performance than textread, making it a better choice when reading large files.
2. With textscan, you can start reading at any point in the file. When the file is opened with fopen, you can move to any position in the file with fseek and begin the textscan at that point. The textread function requires that you start reading from the beginning of the file.
3. Subsequent textscan operations start reading the file at a point where the last textscan left off. The textread function always begins at the start of the file, regardless of any prior textread operations.
4. Function textscan returns a single cell array regardless of how many fields you read. With textscan, you don't need to match the number of output arguments with the number of fields being read, as you would with textread.
5. Function textscan offers more choices in how the data being read is converted.

Function `textscan` has a number of additional options that increase its flexibility. Consult the MATLAB on-line documentation for details of these options.

✴ **Good Programming Practice**

Use function `textscan` in preference to `textread` to import ASCII data in column format from programs written in other languages or exported from applications such as spreadsheets.

8.10   Function `uiimport`

Function `uiimport` is a GUI-based way to import data from a file or from the clipboard. This command takes the forms

```
uiimport
structure = uiimport;
```

In the first case, the imported data is inserted directly into the current MATLAB workspace. In the second case, the data is converted into a structure and saved in variable `structure`.

When the command `uiimport` is typed, the Import Wizard is displayed in a window (see Figure 8.3 for the PC version of this window). The user can then select the file that he or she would like to import from and the specific data within that file. Many different formats are supported; a partial list is given in Table 8.13. In addition, data can be imported from almost *any* application by saving the data on the clipboard. This flexibility can be very useful when you are trying to get data into MATLAB for analysis.

8.11   Summary

In Chapter 8, we have presented an introduction to file I/O operations. Many MATLAB I/O functions are quite similar to C functions, but there are differences in some details.

The `textread` and `textscan` functions can be used to import ASCII data in column format from programs written in other languages or exported from applications such as spreadsheets. Of these two functions, `textscan` is preferred because it is more flexible and faster than `textread`.

The `load` and `save` commands using MAT-files are very efficient; are transportable across MATLAB implementations; and preserve full precision, data types, and global status for all variables. MAT-files should be used as the first-choice method of I/O, unless data must be shared with other applications or must be readable by humans.

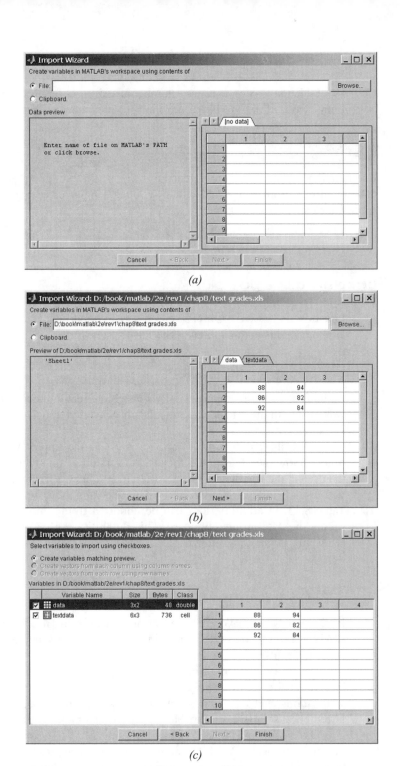

Figure 8.3 Using `uiimport`: *(a)* The Import Wizard after it is started. *(b)* After a data file has been selected, one or more data arrays are created, and their contents can be examined. *(c)* Next, the user can select which of the data arrays will be imported into MATLAB.

There are two types of I/O statements in MATLAB: binary and formatted. Binary I/O statements store or read data in unformatted files, and formatted I/O statements store or read data in formatted files.

MATLAB files are opened with the `fopen` function and closed with the `fclose` function. Binary reads and writes are performed with the `fread` and `fwrite` functions, while formatted reads and writes are performed with the `fscanf` and `fprintf` functions. Functions `fgets` and `fgetl` simply transfer a line of text from a formatted file into a character string.

The `exist` function can be used to determine whether a file exists before it is opened. This is useful to ensure that existing data is not accidentally overwritten.

It is possible to move around within a disk file using the `frewind` and `fseek` functions. The `frewind` function moves the current file position to the beginning of the file, while the `fseek` function move the current file position to a point a

Table 8.13   Selected File Formats Supported by `uiimport`

File Extents	Meaning
*.gif	Image files
*.jpg	
*.jpeg	
*.ico	
*.png	
*.pcx	
*.tif	
*.tiff	
*.bmp	
*.cur	Cursor format
*.hdf	Hierarchical Data Format file
*.au	Sound files
*.snd	
*.wav	
*.avi	Movie file
*.csv	Spreadsheet files
*.xls	
*.wk1	
*.txt	Text files
*.dat	
*.dlm	
*.tab	

specified number of bytes ahead or behind a reference point. The reference point may be the current file position, the beginning of the file, or the end of the file.

8.11.1 Summary of Good Programming Practice

The following guidelines should be adhered to when working with MATLAB I/O functions.

1. Unless you must exchange data with non-MATLAB programs, always use the `load` and `save` commands to save data sets in MAT-file format. This format is efficient and transportable across MATLAB implementations, and it preserves all details of all MATLAB data types.
2. Always be careful to specify the proper permissions in `fopen` statements, depending on whether you are reading from or writing to a file. This practice will help prevent errors such as accidentally overwriting data files that you want to keep.
3. Always check the status after a file-open operation to make sure that it is successful. If the file open fails, tell the user and provide a way to recover from the problem.
4. Use formatted files to create data that must be readable by humans or that must be transferable between programs on computers of different types. Use unformatted files to efficiently store large quantities of data that do not have to be directly examined and that will remain on only one type of computer. Also, use unformatted files when I/O speed is critical.
5. Do not overwrite an output file without confirming that the user would like to delete the preexisting information.
6. Use function `textscan` in preference to `textread` to import ASCII data in column format from programs written in other languages or exported from applications such as spreadsheets.

8.11.2 MATLAB Summary

The following summary lists all of the MATLAB commands and functions described in this chapter, along with a brief description of each one.

`exist`	Check for the existence of a file.
`fclose`	Close file.
`feof`	Test for end-of-file.
`ferror`	Inquire file I/O error status.
`fgetl`	Read a line from file, and discard newline character.
`fgets`	Read a line from file, and keep newline character.
`fopen`	Open file.
`fprintf`	Write formatted data to file.

fread	Read binary data from file.
frewind	Rewind file.
fscanf	Read formatted data from file.
fseek	Set file position.
ftell	Check file position.
fwrite	Write binary data to a file.
sprintf	Write formatted data to a character string.
textread	Read data of various types organized in column format from an ASCII file and store the data in each column in separate variables.
textscan	Read data of various types organized in column format from an ASCII file, and store the data in a cell array.
uiimport	Starts a GUI tool for importing data.

8.12 Exercises

8.1 What is the difference between binary and formatted I/O? Which MATLAB functions perform each type of I/O?

8.2 **Table of Logarithms** Write a MATLAB program to generate a table of the base-10 logarithms between 1 and 10 in steps of 0.1. The table should start in a new page, and it should include a title describing the table and row and column headings. This table should be organized as shown here:

	X.0	X.1	X.2	X.3	X.4	X.5	X.6	X.7	X.8	X.9
1.0	0.000	0.041	0.079	0.114	...					
2.0	0.301	0.322	0.342	0.362	...					
3.0	...									
4.0	...									
5.0	...									
6.0	...									
7.0	...									
8.0	...									
9.0	...									
10.0	...									

8.3 Write a MATLAB program that reads in a time in seconds since the start of the day (this value will be somewhere between 0. and 86400.), and prints a character string containing time in the form HH:MM:SS using the 24-hour clock convention. Use the proper format converter to ensure that leading zeros are preserved in the MM and SS fields. Also, be sure to check the input number of seconds for validity, and write an appropriate error message if an invalid number is entered.

8.4 **Gravitational Acceleration** The acceleration due to the Earth's gravity at any height h above the surface of the Earth is given by the equation

$$g = -G\frac{M}{(R + h)^2} \tag{8-4}$$

where G is the gravitational constant (6.672×10^{-11} N m^2/ kg^2), M is the mass of the earth (5.98×10^{24} kg), R is the mean radius of the Earth (6371 km), and h is the height above the Earth's surface. If M is measured in kg and R and h in meters, then the resulting acceleration will be in units of meters per second squared. Write a program to calculate the acceleration due to the Earth's gravity in 500 km increments at heights from 0 km to 40,000 km above the surface of the Earth. Print out the results in a table of height versus acceleration with appropriate labels, including the units of the output values. Plot the data as well.

8.5 The program in Example 8.5 illustrated the use of formatted I/O commands to read (x, y) pairs of data from disk. However, this could also be done with the load –ascii function. Rewrite this program to use load instead of the formatted I/O functions. Test your rewritten program to confirm that it gives the same answers as Example 8.5.

8.6 Rewrite the program in Example 8.5 to use the textread function instead of the formatted I/O functions.

8.7 Rewrite the program in Example 8.5 to use the textscan function instead of the formatted I/O functions. How difficult was it to use textscan, compared to using textread, load –ascii, or the formatted I/O functions?

8.8 Write a program that reads an arbitrary number of real values from a user-specified input data file, rounds the values to the nearest integer, and writes the integers out to a user-specified output file. Make sure that the input file exists, and if not, tell the user and ask for another input file. If the output file exists, ask the user whether or not to delete it. If not, prompt for a different output file name.

8.9 **Table of Sines and Cosines** Write a program to generate a table containing the sine and cosine of θ for θ between 0° and 90°, in 1° increments. The program should properly label each of the column in the table.

8.10 File int.dat (available at this book's Web site) contains 25 integer values in 'int8' format. Write a program that reads these values into a single array using function fread.

8.11 **Interest Calculations** Suppose that you have a sum of money P in an interest-bearing account at a local bank (P stands for *present value*). If the bank pays you interest on the money at a rate of i percent per year and compounds the interest monthly, the amount of money that you will have in the bank after n months is given by the equation

$$F = P\left(1 + \frac{i}{1200}\right)^n \tag{8-5}$$

where F is the future value of the account and $\dfrac{i}{12}$ is the monthly percentage interest rate (the extra factor of 100 in the denominator converts the interest rate from percentages to fractional amounts). Write a MATLAB program that will read an initial amount of money P and an annual interest rate i, and will calculate and write out a table showing the future value of the account every month for the next five years. The table should be written to an output file called 'interest'. Be sure to properly label the columns of your table.

8.12   Write a program to read a set of integers from an input data file, and locate the largest and smallest values within the data file. Print out the largest and smallest values, together with the lines on which they were found. Assume that you do not know the number of values in the file before the file is read.

8.13   Create a 400×400 element double array x, and fill it with random data using function rand. Save this array to a MAT file x1.dat, and then save it again to a second MAT file x2.dat using the –compress option. How do the sizes of the two files compare?

8.14   **Means** In Exercise 4.27, we wrote a MATLAB program that calculated the arithmetic mean (average), rms average, geometric mean, and harmonic mean for a set of numbers. Modify that program to read an arbitrary number of values from an input data file, and calculate the means of those numbers. To test the program, place the following values into an input data file and run the program on that file: 1.0, 2.0, 5.0, 4.0, 3.0, 2.1, 4.7, 3.0.

8.15   **Converting Radians to Degrees/Minutes/Seconds** Angles are often measured in degrees (°), minutes ('), and seconds ("), with 360 degrees in a circle, 60 minutes in a degree, and 60 seconds in a minute. Write a program that reads angles in radians from an input disk file and converts them into degrees, minutes, and seconds. Test your program by placing the following four angles expressed in radians into an input file and reading that file into the program: 0.0, 1.0, 3.141593, 6.0.

8.16   Create a data set in some other program on your computer, such as Microsoft Word, Microsoft Excel, a text editor, and so forth. Copy the data set to the clipboard using the Windows or Unix copy function, and then use function uiimport to load the data set into MATLAB.

CHAPTER 9

Handle Graphics

Handle graphics is the name of a set of low-level graphics functions that control the characteristics of graphics objects generated by MATLAB. These functions are normally hidden inside M-files, but they are very important to the programmer, since they allow him or her to have fine control of the appearance of the plots and graphs that they generate. For example, it is possible to use handle graphics to turn on a grid on the *x*-axis only, or to choose a line color such as orange, which is not supported by the standard `LineSpec` option of the `plot` command. In addition, handle graphics enable a programmer to create graphical user interfaces (GUIs) for programs, as we will see in Chapter 10.

This chapter introduces the structure of the MATLAB graphics system and explains how to control the properties of graphical objects to create a desired display.

9.1 The MATLAB Graphics System

The MATLAB graphics system is based on a hierarchical system of core **graphics objects**, each of which is known by a unique number called a **handle**. Each graphics object has special data known as **properties** associated with it, and modifying those properties will modify the behavior of the object. For example, a **line** is one type of graphics object. The properties associated with a line object include: *x*-data, *y*-data, color, line style, line width, marker type, and so forth. Modifying any of these properties will change the way the line is displayed in a Figure Window.

Every component of a MATLAB graph is a graphical object. For example, each line, axes, and text string is a separate object with its own unique identifying number (handle) and characteristics. All graphical objects are arranged in a hierarchy

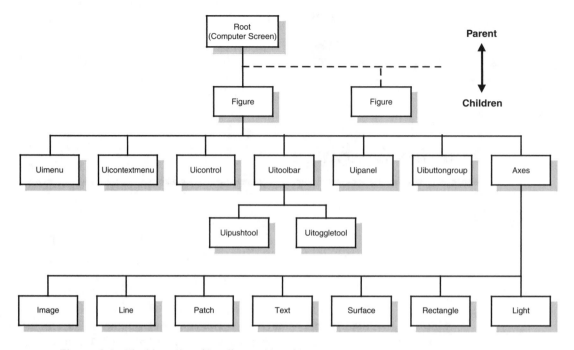

Figure 9.1 The hierarchy of handle graphics objects.

with **parent objects** and **child objects**, as shown in Figure 9.1. When a child object is created, it inherits many of its properties from its parent.

The highest-level graphics object in MATLAB is the **root**, which can be thought of as the entire computer screen. The handle of the root object is always 0. It is created automatically when MATLAB starts up, and it is always present until the program is shut down. The properties associated with the root object are the defaults that apply to all MATLAB windows.

Under the root there can be one or more Figure Windows, or just **figures**. Each figure is a separate window on the computer screen that can display graphical data, and each figure has its own properties. The properties associated with a figure include color, color map, paper size, paper orientation, and pointer type, among others.

Each figure can contain seven types of objects: uimenus, uicontextmenus, uicontrols, uitoolbars, uipanels, uibuttongroups, and axes. Uimenus, uicontextmenus, uicontrols, uitoolbars, uipanels, and uibuttongroups are special graphics objects used to create graphical user interfaces—-they are described in Chapter 10. Axes are regions within a figure where data is actually plotted. There can be more than one set of axes in a single figure.

Each set of axes can contain as many `lines`, `text` strings, `patches`, etc. as necessary to create the plot of interest.

9.2   Object Handles

Each graphics object has a unique name called a **handle**. The handle is a unique integer or real number that is used by MATLAB to identify the object. A handle is automatically returned by any function that creates a graphics object. For example, the function call

> » **hndl = figure;**

creates a new figure and returns the handle of that figure in variable `hndl`. The handle of the root object is always 0, and the handle of each figure is normally a small positive integer, such as 1, 2, 3, ... The handles of all other graphics objects are arbitrary floating-point numbers.

There are MATLAB functions available to get the handles of figures, axes, and other objects. For example, the function `gcf` returns the handle of the currently selected figure, `gca` returns the handle of the currently selected axes within the currently selected figure, and `gco` returns the handle of the currently selected object. These functions are discussed in more detail further on in this chapter.

By convention, handles are usually stored in variables that begin with the letter `h`. This practice helps us to recognize handles in MATLAB programs.

9.3   Examining and Changing Object Properties

Object properties are special values associated with an object that control some aspect of how that object behaves. Each property has a **property name** and an associated value. The property names are strings that are typically displayed in mixed case with the first letter of each word capitalized, but MATLAB recognizes a property name regardless of the case in which it is written.

9.3.1   Changing Object Properties at Creation Time

When an object is created, all of its properties are automatically initialized to default values. These default values can be overridden at creation time by including `'PropertyName'`, `value` pairs in the object creation function[1].

---

[1]Examples of object creation functions include `figure`, which creates a new figure; `axes`, which creates a new set of axes within a figure; and `line`, which creates a line within a set of axes. High-level functions such as `plot` are also object creation functions.

For example, we saw in Chapter 2 that the width of a line could be modified in the `plot` command as follows:

```
plot(x,y,'LineWidth',2);
```

This function overrides the default `LineWidth` property with the value 2 at the time the line object is created.

9.3.2 Changing Object Properties after Creation Time

The properties of any object can be examined at any time using the `get` function, and modified using the `set` function. The most common forms of the `get` function are

```
value = get(handle,'PropertyName');
value = get(handle);
```

where `value` is the value contained in the specified property of the object whose handle is supplied. If only the handle is included in the function call, then the function returns a structure array in which the field names are all of the properties of the object and the field values are the property values.

The most common form of the `set` function is

```
set(handle,'PropertyName1',value1,...);
```

where there can be any number of `'PropertyName'`, `value` pairs in a single function.

For example, suppose that we plotted the function $y(x) = x^2$ from 0 to 2 with the following statements:

```
x = 0:0.1:2;
y = x.^2;
hndl = plot(x,y);
```

The resulting plot is shown in Figure 9.2*a*. The handle of the plotted line is stored in `hndl`, and we can use it to examine or modify the properties of the line. The function `get(hndl)` will return all of the properties of this line in a structure, with each property name being an element of the structure.

```
» result = get(hndl)
result =
                  Color: [0 0 1]
              EraseMode: 'normal'
              LineStyle: '-'
              LineWidth: 0.5000
                 Marker: 'none'
             MarkerSize: 6
        MarkerEdgeColor: 'auto'
        MarkerFaceColor: 'none'
                  XData: [1x21 double]
                  YData: [1x21 double]
```

```
            ZData: [1x0 double]
      BeingDeleted: 'off'
     ButtonDownFcn: []
          Children: [0x1 double]
          Clipping: 'on'
         CreateFcn: []
         DeleteFcn: []
        BusyAction: 'queue'
  HandleVisibility: 'on'
           HitTest: 'on'
     Interruptible: 'on'
          Selected: 'off'
SelectionHighlight: 'on'
               Tag: ''
              Type: 'line'
     UIContextMenu: []
          UserData: []
           Visible: 'on'
            Parent: 303.0004
       DisplayName: ''
        XDataMode: 'manual'
       XDataSource: ''
       YDataSource: ''
       ZDataSource: ''
```

Note that the current line width is 0.5 pixels and the current line style is a solid line. We can change the line width and the line style with the commands

```
» set(hndl,'LineWidth',4,'LineStyle','--')
```

The plot after this command is issued is shown in Figure 9.2*b*.

The get and set functions are especially useful for programmers, because they can be directly inserted into MATLAB programs to modify a figure based on a user's input. As is shown in Chapter 10, these functions are used extensively in GUI programming.

For the end user, however, it is often easier to change the properties of a MATLAB object interactively. The Property Editor is a GUI-based tool designed for this purpose. The Property Editor is started by first selecting the Edit Button (🔲) on the figure toolbar, and then clicking on the object that you want to modify with the mouse. Alternatively, the property editor can be started from the command line.

```
propedit(HandleList);
propedit;
```

For example, the following statements will create a plot containing the line $y = x^2$ over the range 0 to 2, and will open the Property Editor to allow the user to interactively change the properties of the line.

(a)

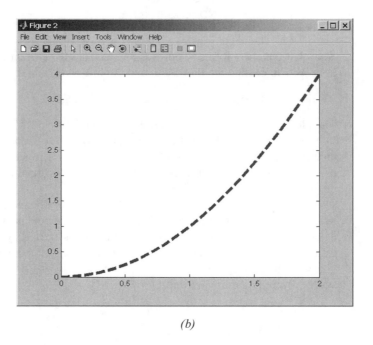

(b)

Figure 9.2 *(a)* Plot of the function $y = x^2$ using the default linewidth. *(b)* Plot of the function after modifying the LineWidth and LineStyle properties.

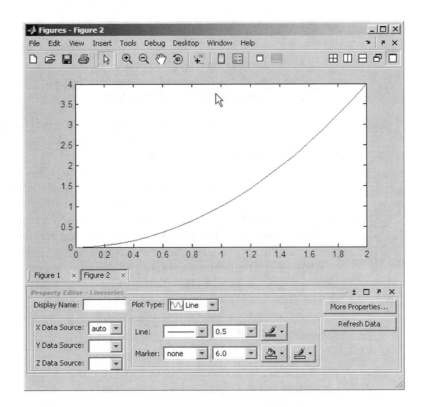

Figure 9.3 The Property Editor when editing a line object. Changes in style are immediately displayed on the figure as the object is edited.

```
figure(2);
x = 0:0.1:2;
y = x.^2;
hndl = plot(x,y);
propedit(hndl);
```

The Property Editor invoked by these statements is shown in Figure 9.3. The Property Editor contains a series of panes that vary depending on the type of object being modified.

Example 9.1—Using Low-Level Graphics Commands

The function sinc(x) is defined by the equation

$$\text{sinc } x = \begin{cases} \dfrac{\sin x}{x} & x \neq 0 \\ 1 & x = 0 \end{cases} \tag{9-1}$$

Plot this function from $x = -3\pi$ to $x = 3\pi$. Use handle graphics functions to customize the plot as follows:

1. Make the figure background pink.
2. Use *y*-axis grid lines only (no *x*-axis grid lines).
3. Plot the function as a 2-point-wide solid orange line.

SOLUTION To create this graph, we need to plot the function sinc x from $x = -3\pi$ to $x = 3\pi$ using the `plot` function. The plot function will return a handle for the line that we can save and use later.

After plotting the line, we need to modify the color of the *figure* object, the grid status of the *axes* object, and the color and width of the *line* object. These modifications require us to have access to the handles of the `figure`, `axes`, and `line` objects. The handle of the `figure` object is returned by the `gcf` function, the handle of the `axes` object is returned by the `gca` function, and the handle of the `line` object is returned by the `plot` function that created it.

The low-level graphics properties that need to be modified can be found by referring to the on-line MATLAB Help Browser documentation, under the topic "Handle Graphics." They are the `'Color'` property of the current figure, the `'YGrid'` property of the current axes, and the `'LineWidth'` and `'Color'` properties of the line.

1. **State the problem.**
   Plot the function sinc x from $x = -3\pi$ to $x = 3\pi$ using a figure with a pink background, *y*-axis grid lines only, and a 2-point-wide solid orange line.

2. **Define the inputs and outputs.**
   There are no inputs to this program, and the only output is the specified figure.

3. **Describe the algorithm.**
   This program can be broken down into three major steps:

   ```
   Calculate sinc(x)
   Plot sinc(x)
   Modify the required graphics object properties
   ```

   The first major step is to calculate sinc x from $x = -3\pi$ to $x = 3\pi$. This can be done with vectorized statements, but the vectorized statements will produce a NaN at $x = 0$, since the division of 0/0 is undefined. We must replace the NaN with a 1.0 before plotting the function. The detailed pseudocode for this step is

   ```
   % Calculate sinc(x)
   x = -3*pi:pi/10:3*pi
   y = sin(x) ./ x

   % Find the zero value and fix it up. The zero is
   % located in the middle of the x array.
   ```

```
index = fix(length(y)/2) + 1
y(index) = 1
```

Next, we must plot the function, saving the handle of the resulting line for further modifications. The detailed pseudocode for this step is

```
hndl = plot(x,y);
```

Now we must use handle graphics commands to modify the figure background, *y*-axis grid, and line width and color. Remember that the figure handle can be recovered with the function gcf, and the axis handle can be recovered with the function gca. The color pink can be created with the RGB vector [1 0.8 0.8], and the color orange can be created with the RGB vector [1 0.5 0]. The detailed pseudocode for this step is

```
set(gcf,'Color',[1 0.8 0.8])
set(gca,'YGrid','on')
set(hndl,'Color',[1 0.5 0],'LineWidth',2)
```

4. **Turn the algorithm into MATLAB statements.**
   The final MATLAB program is shown here.

```
% Script file: plotsinc.m
%
% Purpose:
%   This program illustrates the use of handle graphics
%   commands by creating a plot of sinc(x) from -3*pi to
%   3*pi, and modifying the characteristics of the figure,
%   axes, and line using the "set" function.
%
% Record of revisions:
%    Date          Programmer          Description of change
%    ====          ==========          =====================
%    04/02/07     S. J. Chapman        Original code
%
% Define variables:
%   hndl            -- Handle of line
%   x               -- Independent variable
%   y               -- sinc(x)

% Calculate sinc(x)
x = -3*pi:pi/10:3*pi;
y = sin(x) ./ x;

% Find the zero value and fix it up. The zero is
% located in the middle of the x array.
index = fix(length(y)/2) + 1;
y(index) = 1;
```

Figure 9.4 Plot of sinc x versus x. A colour plot of this appears in the insert.

```
% Plot the function.
hndl = plot(x,y);

% Now modify the figure to create a pink background,
% modify the axis to turn on y-axis grid lines, and
% modify the line to be a 2-point wide orange line.
set(gcf,'Color',[1 0.8 0.8]);
set(gca,'YGrid','on');
set(hndl,'Color',[1 0.5 0],'LineWidth',2);
```

5. **Test the program.**
   Testing this program is very simple—we just execute it and examine the resulting plot. The plot created is shown in Figure 9.4, and it does have the characteristics that we wanted.
   ◄

9.4 Using set to List Possible Property Values

The set function can be used to provide lists of possible property values. If a set function call contains a property name but not a corresponding value, set returns a list of all of the legal choices for that property. For example, the command set(hndl,'LineStyle') will return a list of all legal line styles with the default choice in brackets:

```
» set(hnd1,'LineStyle')
ans =
    '-'
    '--'
    ':'
    '-.'
    'none'
```

This function shows that the legal line styles are '-', '--', ':', '-.', and 'none', with the first choice as the default.

If the property does not have a fixed set of values, MATLAB returns an empty cell array:

```
» set(hnd1,'LineWidth')
ans =
    {}
```

The function set(hnd1) will return all of the possible choices for all of the properties of an object.

```
» xxx = set(hnd1)
xxx =
                 Color: {}
             EraseMode: {4x1 cell}
             LineStyle: {5x1 cell}
             LineWidth: {}
                Marker: {14x1 cell}
            MarkerSize: {}
       MarkerEdgeColor: {2x1 cell}
       MarkerFaceColor: {2x1 cell}
                 XData: {}
                 YData: {}
                 ZData: {}
         ButtonDownFcn: {}
              Children: {}
              Clipping: {2x1 cell}
             CreateFcn: {}
             DeleteFcn: {}
            BusyAction: {2x1 cell}
      HandleVisibility: {3x1 cell}
               HitTest: {2x1 cell}
         Interruptible: {2x1 cell}
              Selected: {2x1 cell}
    SelectionHighlight: {2x1 cell}
                   Tag: {}
         UIContextMenu: {}
              UserData: {}
```

```
            Visible: {2x1 cell}
             Parent: {}
        DisplayName: {}
          XDataMode: {2x1 cell}
        XDataSource: {}
        YDataSource: {}
        ZDataSource: {}
```

Any of the items in this list can be expanded to show the available list of options.

```
» xxx.EraseMode
ans =
    'normal'
    'background'
    'xor'
    'none'
```

9.5   User-Defined Data

In addition to the standard properties defined for a GUI object, a programmer can define special properties to hold program-specific data. These extra properties are a convenient way to store any kind of data that the programmer might wish to associate with the GUI object. Any amount of any type of data can be stored and used for any purpose.

User-defined data is stored in a manner similar to standard properties. Each data item has a name and a value. Data values are stored in an object with the setappdata function and retrieved from the object using the getappdata function.

The general form of setappdata is

```
setappdata(hndl,'DataName',DataValue);
```

where hndl is the handle of the object to store the data into, 'DataName' is the name given to the data, and DataValue is the value assigned to that name. Note that the data value can be either numeric or a character string.

For example, suppose that we wanted to define two special data values, one containing the number of errors that have occurred on a particular figure and the other containing a string describing the last error detected. Such data values could be given the names 'ErrorCount' and 'LastError'. If we assume that h1 is the handle of the figure, then the command to create these data items and initialize them would be

```
setappdata(h1,'ErrorCount',0);
setappdata(h1,'LastError','No error');
```

Application data can be retrieved at any time using the function `getappdata`. The two forms of `getappdata` are

```
value = getappdata(hndl,'DataName');
struct = getappdata(hndl);
```

where `hndl` is the handle of the object containing the data and `'DataName'` is the name of the data to be retrieved. If a `'DataName'` is specified, then the value associated with that data name will be returned. If it is not specified, then *all* user-defined data associated with that object will be returned in a structure. The names of the data items will be structure element names in the returned structure.

For the example given previously, `getappdata` will produce the following results:

```
» value = getappdata(h1,'ErrorCount')
value =
     0
» struct = getappdata(h1)
struct =
     ErrorCount: 0
     LastError: 'No error'
```

The functions associated with user-defined data are summarized in Table 9.1.

Table 9.1   Functions for Manipulating User-Defined Data

Function	Description
`setappdata(hndl,'DataName',Data Value)`	Stores `DataValue` in an item named `'DataName'` within the object specified by the handle `hndl`.
`value = getappdata(hndl,'DataName')` `struct = getappdata(hndl)`	Retrieves user-defined data from the object specified by the handle `hndl`. The first form retrieves the value associated with `'DataName'` only, and the second form retrieves all user-defined data.
`isappdata(hndl,'DataName')`	A logical function that returns a 1 if `'DataName'` is defined within the object specified by the handle `hndl`, and 0 otherwise.
`rmappdata(hndl,'DataName')`	Removes the user-defined data item named `'DataName'` from the object specified by the handle `hndl`.

9.6 Finding Objects

Each new graphics object that is created has its own handle, and that handle is returned by the creating function. If you intend to modify the properties of an object that you create, then it is a good idea to save the handle for later use with get and set.

☀ Good Programming Practice

If you intend to modify the properties of an object that you create, save the handle of that object for later use with get and set.

However, sometimes we might not have access to the handle. Suppose that we lost a handle for some reason. How can we examine and modify the graphics objects?

MATLAB provides four special functions to help find the handles of objects.

- gcf—Returns the handle of the current *figure*.
- gca—Returns the handle of the current *axes* in the current *figure*.
- gco—Returns the handle of the current *object*.
- findobj—Finds a graphics object with a specified property value.

The function gcf returns the handle of the current figure. If no figure exists, gcf *will create one* and return its handle. The function gca returns the handle of the current axes within the current figure. If no figure exists or if the current figure exists but contains no axes, gca *will create a set of axes* and return its handle. The function gco has the form

```
h_obj = gco;
h_obj = gco(h_fig);
```

where h_obj is the handle of the object and h_fig is the handle of a figure. The first form of this function returns the handle of the *current object in the current figure*, and the second form of the function returns the handle of the *current object in a specified figure*.

The current object is defined as the last object clicked on with the mouse. This object can be any graphics object except the root. There will not be a current object in a figure until a mouse click has occurred within that figure. Before the first mouse click, function gco will return an empty array []. Unlike gcf and gca, gco does not create an object if it does not exist.

Once the handle of an object is known, we can determine the type of the object by examining its 'Type' property. The 'Type' property will be a character string, such as 'figure', 'line', 'text', and so forth.

```
h_obj = gco;
type = get(h_obj,'Type')
```

The easiest way to find an arbitrary MATLAB object is with the `findobj` function. The basic form of this function is

```
hndls = findobj('PropertyName1',value1,...)
```

This command starts at the root object, and searches the entire tree for all objects that have the specified values for the specified properties. Note that multiple property/value pairs may be specified, and `findobj` returns only the handles of objects that match *all* of them.

For example, suppose that we have created Figures 1 and 3. Then the function `findobj('Type','figure')` will return the results

```
» h_fig = findobj('Type','figure')
h_fig =
        3
        1
```

This form of the `findobj` function is very useful, but it can be slow, since it must search through the entire object tree to locate any matches. If you must use an object multiple times, make only one call to `findobj` and save the handle for reuse.

Restricting the number of objects that must be searched can increase the execution speed of this function. This can be done with the following form of the function:

```
hndls = findobj(Srchhndls,'PropertyName1',value1,...)
```

Here, only the handles listed in array `Srchhndls` and their children will be searched to find the object. For example, suppose that you wanted to find all of the dashed lines in Figure 1. The command to do this would be

```
hndls = findobj(1,'Type','line','LineStyle','--');
```

✳ Good Programming Practice

If possible, restrict the scope of your searches with `findobj` to make them faster.

9.7 Selecting Objects with the Mouse

Function `gco` returns the handle of the current object, which is the last object clicked on by the mouse. Each object has a **selection region** associated with it, and any mouse click within that selection region is assumed to be a click on that object. This is very important for thin objects such as lines or points—the selection region allows the user to be slightly sloppy in mouse position and still select the line. The width and shape of the selection region varies for different types of objects. For instance, the selection region for a line is 5 pixels on either side of the line, whereas the selection region for a surface, patch, or text object is the smallest rectangle that can contain the object.

The selection region for an `axes` object is the area of the axes plus the area of the titles and labels. However, lines or other objects inside the axes have a higher priority, so to select the axes you must click on a point within the axes that is not near lines or text. Clicking on a figure outside of the axes region will select the figure itself.

What happens if a user clicks on a point that has two or more objects, such as the intersection of two lines? The answer depends on the **stacking order** of the objects. The stacking order is the order in which MATLAB selects objects. This order is specified by the order of the handles listed in the `'Children'` property of a figure. If a click is in the selection region of two or more objects, the one with the highest position in the `'Children'` list will be selected.

MATLAB includes a function called `waitforbuttonpress` that is sometimes used when selecting graphics objects. The form of this function is

```
k = waitforbuttonpress
```

When this function is executed, it halts the program until either a key is pressed or a mouse button is clicked. The function returns 0 if it detects a mouse button click or 1 if it detects a key press.

The function can be used to pause a program until a mouse click occurs. After the mouse click occurs, the program can recover the handle of the selected object using the `gco` function.

Example 9.2—Selecting Graphics Objects

The program that follows explores the properties of graphics objects and incidentally shows how to select objects using `waitforbuttonpress` and `gco`. The program allows objects to be selected repeatedly until a key press occurs.

```
% Script file: select_object.m
%
% Purpose:
%   This program illustrates the use of waitforbuttonpress
%   and gco to select graphics objects. It creates a plot
%   of sin(x) and cos(x), and then allows a user to select
%   any object and examine its properties. The program
%   terminates when a key press occurs.
%
% Record of revisions:
%     Date          Programmer          Description of change
%     ====          ==========          =====================
%   04/02/07      S. J. Chapman        Original code
%
% Define variables:
%   details        -- Object details
%   h1             -- handle of sine line
%   h2             -- handle of cosine line
%   handle         -- handle of current object
```

```matlab
%   k               -- Result of waitforbuttonpress
%   type            -- Object type
%   x               -- Independent variable
%   y1              -- sin(x)
%   y2              -- cos(x)
%   yn              -- Yes/No

% Calculate sin(x) and cos(x)
x = -3*pi:pi/10:3*pi;
y1 = sin(x);
y2 = cos(x);

% Plot the functions.
h1 = plot(x,y1);
set(h1,'LineWidth',2);
hold on;
h2 = plot(x,y2);
set(h2,'LineWidth',2,'LineStyle',':','Color','r');
title('\bfPlot of sin \itx \rm\bf and cos \itx');
xlabel('\bf\itx');
ylabel('\bfsin \itx \rm\bf and cos \itx');
legend('sine','cosine');
hold off;

% Now set up a loop and wait for a mouse click.
k = waitforbuttonpress;

while k == 0

   % Get the handle of the object
   handle = gco;

   % Get the type of this object.
   type = get(handle,'Type');

   % Display object type
   disp (['Object type = ' type '.']);

   % Do we display the details?
   yn = input('Do you want to display details? (y/n) ','s');

   if yn == 'y'
      details = get(handle);
      disp(details);
   end

   % Check for another mouse click
   k = waitforbuttonpress;
end
```

Figure 9.5 Plot of sin x and cos x.

When this program is executed, it produces the plot shown in Figure 9.5. Experiment by clicking on various objects on the plot and observing the resulting characteristics.

◀

9.8 Position and Units

Many MATLAB objects have a 'position' property, which specifies the size and position of the object on the computer screen. This property differs slightly for different kinds of objects, as described in the following subsections.

9.8.1 Positions of figure Objects

The 'position' property for a figure specifies the location of that figure on the computer screen using a four-element row vector. The values in this vector are [left bottom width height], where left is the leftmost edge of the figure, bottom is the bottom edge of the figure, width is the width of the figure, and height is the height of the figure. These position values are in the units specified in the 'Units' property for the object. For example, the position and units associated with a the current figure can be found as follows:

```
» get(gcf,'Position')
ans =
   176  204  672  504
» get(gcf,'Units')
ans =
pixels
```

This information specifies that the lower-left corner of the current figure window is 176 pixels to the right and 204 pixels above the lower-left corner of the screen, and the figure is 672 pixels wide by 504 pixels high. This is the drawable region of the figure, excluding borders, scrollbars, menus, and the figure title area.

The 'units' property of a figure defaults to pixels, but it can be inches, centimeters, points, characters, or normalized coordinates. Pixels are screen pixels, which are the smallest rectangular shape that can be drawn on a computer screen. Typical computer screens re at least 640 pixels wide × 480 pixels high, and screens can have more than 1000 pixels in each direction. Since the number of pixels varies from computer screen to computer screen, the size of an object specified in pixels will also vary.

Normalized coordinates are coordinates in the range 0 to 1, where the lower-left corner of the screen is at (0,0) and the upper-right corner of the screen is at (1,1). If an object position is specified in normalized coordinates, it will appear in the same relative position on the screen regardless of screen resolution. For example, the following statements create a figure and place it into the upper-left quadrant of the screen on any computer, regardless of screen size[2].

```
h1 = figure(1)
set(h1,'units','normalized','position',[0 .5 .5 .45])
```

✳ Good Programming Practice

If you would like to place a window in a specific location, it is easier to place the window at the desired location using normalized coordinates; the results will be the same regardless of the computer's screen resolution.

9.8.2 Positions of axes and uicontrol Objects

The position of axes and uicontrol objects is also specified by a four-element vector, but the object position is specified relative to the lower-left corner of the *figure* instead of the position of the screen. In general, the 'Position' property of a child object is relative to the position of its parent.

---

[2]The normalized height of this Figure is reduced to 0.45 to allow room for the Figure title and menu bar, both of which are above the drawing area.

By default, the positions of axes objects are specified in *normalized* units within a figure, with (0,0) representing the lower-left corner of the figure, and (1,1) representing the upper-right corner of the figure.

9.8.3 Positions of text Objects

Unlike other objects, `text` objects have a position property containing only two or three elements. These elements correspond to the *x*, *y*, and *z* values of the text object *within* an axes object. Note that these values are in the units being displayed on the axes themselves.

The position of the text object with respect to the specified point is controlled by the object's `HorizontalAlignment` and `VerticalAlignment` properties. The `HorizontalAlignment` can be `{Left}`, `Center`, or `Right`, and the `VerticalAlignment` can be `Top`, `Cap`, `{Middle}`, `Baseline`, or `Bottom`.

The size of `text` objects is determined by the font size and the number of characters being displayed, so there are no height and width values associated with them.

▶

Example 9.3—Positioning Objects within a Figure

As was mentioned previously, axis positions are defined relative to the lower-left corner of the frame they are contained in, whereas text object positions are defined within axes in the data units being displayed on the axes.

To illustrate the positioning of graphics objects within a figure, we will write a program that creates two overlapping sets of axes within a single figure. The first set of axes will display sin *x* versus *x* and will have a text comment attached to the display line. The second set of axes will display cos *x* versus *x* and will have a text comment in the lower-left corner.

A program to create the figure follows. Note that we are using the `figure` function to create an empty figure, and then two `axes` functions to create the two sets of axes within the figure. The position of the `axes` functions is specified in normalized units within the figure, so the first set of axes, which starts at (0.05, 0.05), is in the lower-left corner of the figure, and the second set of axes, which starts at (0.45,0.45), is in the upper-right corner of the figure. Each set of axes has the appropriate function plotted on it.

The first `text` object is attached to the first set of axes at position $(-\pi, 0)$, which is a point on the curve. The `'HorizontalAlignment'`, `'right'` property is selected, so the *attachment point* $(-\pi, 0)$ is on the *right-hand side* of the text string. As a result, the text appears to the *left* of the of the attachment point in the final figure. (This can be confusing for new programmers!)

The second `text` object is attached to the second set of axes at position $(-7.5, -0.9)$, which is near the lower-left corner of the axes. This string uses the default horizontal alignment, which is `'left'`, so the attachment point $(-7.5, -0.9)$ is on the *left-hand side* of the text string. As a result, the text appears to the right of the of the attachment point in the final figure. The final program is:

```
% Script file: position_object.m
%
% Purpose:
%   This program illustrates the positioning of graphics
%   graphics objects. It creates a figure, and then places
%   two overlapping sets of axes on the figure. The first
%   set of axes is placed in the lower left hand corner of
%   the figure, and contains a plot of sin(x). The second
%   set of axes is placed in the upper right hand corner of
%   the figure, and contains a plot of cos(x). Then two
%   text strings are added to the axes, illustrating the
%   positioning of text within axes.
%
% Record of revisions:
%     Date         Programmer         Description of change
%     ====         ==========         =====================
%   04/03/07     S. J. Chapman        Original code
%
% Define variables:
%   h1               -- Handle of sine line
%   h2               -- Handle of cosine line
%   ha1              -- Handle of first axes
%   ha2              -- Handle of second axes
%   x                -- Independent variable
%   y1               -- sin(x)
%   y2               -- cos(x)

% Calculate sin(x) and cos(x)
x = -2*pi:pi/10:2*pi;
y1 = sin(x);
y2 = cos(x);

% Create a new figure
figure;

% Create the first set of axes and plot sin(x).
% Note that the position of the axes is expressed
% in normalized units.
ha1 = axes('Position',[.05 .05 .5 .5]);
h1 = plot(x,y1);
set(h1,'LineWidth',2);
title('\bfPlot of sin \itx');
xlabel('\bf\itx');
ylabel('\bfsin \itx');
axis([-8 8 -1 1]);
```

```
% Create the second set of axes and plot cos(x).
% Note that the position of the axes is expressed
% in normalized units.
ha2 = axes('Position',[.45 .45 .5 .5]);
h2 = plot(x,y1);
set(h2,'LineWidth',2,'Color','r','LineStyle','--');
title('\bfPlot of cos \itx');
xlabel('\bf\itx');
ylabel('\bfsin \itx');
axis([-8 8 -1 1]);

% Create a text string attached to the line on the first
% set of axes.
axes(ha1);
text(-pi,0.0,'sin(x)\rightarrow','HorizontalAlignment','right');

% Create a text string in the lower left hand corner
% of the second set of axes.
axes(ha2);
text(-7.5,-0.9,'Test string 2');
```

When this program is executed, it produces the plot shown in Figure 9.6. You should execute this program again on your computer, changing the size and/or location of the objects being plotted and observing the results.

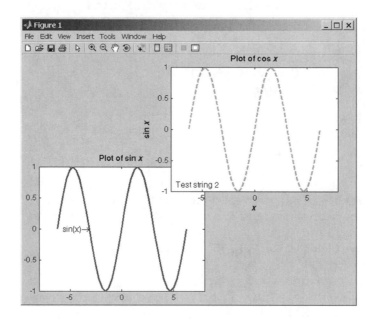

Figure 9.6   The output of program `position_object`.

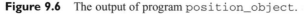

9.9   Printer Positions

The 'Position' and 'Units' properties specify the location of a figure on the *computer screen*. There are five other properties that specify the location of a figure on a sheet of paper *when it is printed*. These properties are summarized in Table 9.2.

For example, to set a plot to print out in landscape mode, on A4 paper, in normalized units, we could set the following properties:

```
set(hndl,'PaperType','A4')
set(hndl,'PaperOrientation','landscape')
set(hndl,'PaperUnits','normalized');
```

9.10   Default and Factory Properties

MATLAB assigns default properties to each object when it is created. If those properties are not what you want, then you must use set to select the desired values. If you wanted to change a property in every object that you have created, this process could become very tedious. For those cases, MATLAB allows you to modify the default property itself, so that all objects will inherit the correct value of the property when they are created.

When a graphics object is created, MATLAB looks for a default value for each property by examining the object's parent. If the parent sets a default value, that value

Table 9.2   Printing-Related Figure Properties

Option	Description
PaperUnits	Units for paper measurements: [{inches} | centimeters | normalized | points]
PaperOrientation	[{portrait} | landscape]
PaperPosition	A position vector of the form [left, bottom, width, height] where all units are as specified in PaperUnits
PaperSize	A two-element vector containing the power size, for example [8.5 11]
PaperType	Sets paper type. Note that setting this property automatically updates the PaperSize property. [{usletter} | uslegal | A0 | A1 | A2 | A3 | A4 | A5 | B0 | B1 | B2 | B3 | B4 | B5 | arch-A | arch-B | arch-C | arch-D | arch-E | A | B | C | D | E | tabloid | <custom>]

is used. If not, MATLAB examines the parent's parent to see whether that object sets a default value, and so on back to the root object. MATLAB uses the *first* default value that it encounters when working back up the tree.

Default properties may be set at any point in the graphics object hierarchy that is *higher* than level at which the object is created. For example, a default `figure` color would be set in the `root` object, and then all figures created after that time would have the new default color. On the other hand, a default `axes` color could be set in either the `root` object or the `figure` object. If the default `axes` color is set in the `root` object, it will apply to all new axes in all figures. If the default `axes` color is set in the `figure` object, it will apply to all new axes in the current figure only.

Default values are set using a string consisting of `'Default'` followed by the object type and the property name. Thus, the default figure color would be set with the property `'DefaultFigureColor'`, and the default axes color would be set with the property `'DefaultAxesColor'`. Some examples of setting default values are shown here:

`set(0,'DefaultFigureColor','y')`	Yellow figure background—all new figures
`set(0,'DefaultAxesColor','r')`	Red axes background—all new axes in all figures
`set(gcf,'DefaultAxesColor','r')`	Red axes background—all new axes in current figure only
`set(gca,'DefaultLineLineStyle',':')`	Set default line style to dashed, in current axes only.

If you are working with existing objects, it is always a good idea to restore them to their existing condition after they have been used. *If you change the default properties of an object in a function, save the original values and restore them before exiting the function.* For example, suppose that we wish to create a series of figures in normalized units. We could save and restore the original units as follows:

```
saveunits = get(0,'DefaultFigureUnits');
set(0,'DefaultFigureUnits','normalized');
...
<MATLAB statements>
...
set(0,'DefaultFigureUnits',saveunits);
```

If you want to customize MATLAB to use different default values at all times, then you should set the defaults in the `root` object every time that MATLAB starts up. The easiest way to do this is to place the default values into the `startup.m` file, which is automatically executed every time MATLAB starts. For example, suppose you always use A4 paper and you always want a grid displayed on your plots. Then you could set the following lines into `startup.m`:

```
set(0,'DefaultFigurePaperType','A4');
set(0,'DefaultFigurePaperUnits','centimeters');
set(0,'DefaultAxesXGrid','on');
set(0,'DefaultAxesYGrid','on');
set(0,'DefaultAxesZGrid','on');
```

There are three special value strings that are used with handle graphics: `'remove'`, `'factory'`, and `'default'`. If you have set a default value for a property, the `'remove'` value will remove the default that you set. For example, suppose that you set the default figure color to yellow:

```
set(0,'DefaultFigureColor','y');
```

The following function call will cancel this default setting and restore the previous default setting.

```
set(0,'DefaultFigureColor','remove');
```

The string `'factory'` allows a user to temporarily override a default value and use the original MATLAB default value instead. For example, the following figure is created with the factory default color despite a default color of yellow having been previously defined:

```
set(0,'DefaultFigureColor','y');
figure('Color','factory')
```

The string `'default'` forces MATLAB to search up the object hierarchy until it finds a default value for the desired property. It uses the first default value that it finds. If it fails to find a default value, then it uses the factory default value for that property. This use is illustrated here:

```
% Set default values
set(0,'DefaultLineColor','k');   % root default = black
set(gcf,'DefaultLineColor','g'); % figure default = green

% Create a line on the current axes. This line is green.
hndl = plot(randn(1,10));
set(hndl,'Color','default');
pause(2);

% Now clear the figure's default and set the line color to the new
% default. The line is now black.
set(gcf,'DefaultLineColor','remove');
set(hndl,'Color','default');
```

9.11 Graphics Object Properties

There are hundreds of different graphic object properties, far too many to discuss in detail here. The best place to find a complete list of graphics object properties is in the Help Browser distributed with MATLAB.

We have mentioned a few of the most important properties for each type of graphic object as we have needed them ('LineStyle', 'Color', and so forth). A complete set of properties is given in the MATLAB Help Browser documentation under the descriptions of each type of object.

9.12 Summary

Every element of a MATLAB plot is a graphics object. Each object is identified by a unique handle, and each object has many properties associated with it that affect the way the object is displayed.

MATLAB objects are arranged in a hierarchy with **parent objects** and **child objects**. When a child object is created, it inherits many of its properties from its parent.

The highest-level graphics object in MATLAB is the root, which can be thought of as the entire computer screen. Under the root there can be one or more Figure Windows. Each figure is a separate window on the computer screen that can display graphical data, and each figure has its own properties.

Each figure can contain seven types of objects: uimenus, uicontextmenus, uicontrols, uitoolbars, uipanels, uibuttongroups, and axes. Uimenus, uicontextmenus, uicontrols, uitoolbars, uipanels, and uibuttongroups are special graphics objects used to create graphical user interfaces—they are described in Chapter 10. Axes are regions within a figure where data is actually plotted. There can be more than one set of axes within a single figure.

Each set of axes can contain as many lines, text strings, patches, and the like as is necessary to create the plot of interest.

The handles of the current figure, current axes, and current object may be recovered with the gcf, gca, and gco functions, respectively. The properties of any object may be examined and modified using the get and set functions.

There are literally hundreds of properties associated with MATLAB graphics functions, and the best place to find the details of these of these functions is the MATLAB on-line documentation.

9.12.1 Summary of Good Programming Practice

The following guidelines should be adhered to when working with MATLAB handle graphics.

1. If you intend to modify the properties of an object that you create, save the handle of that object for later use with get and set.

2. If possible, restrict the scope of your searches with `findobj` to make them faster.
3. If you would like to place a window in a specific location, it is easier to place the window at the desired location using normalized coordinates, and the results will be the same regardless of the computer's screen resolution.

9.12.2   MATLAB Summary

The following summary lists all of the MATLAB commands and functions described in this chapter, along with a brief description of each one.

`axes`	Creates a new axes/makes axes current.
`figure`	Creates a new figure/makes figure current.
`findobj`	Finds an object based on one or more property values.
`gca`	Get handle of current axes.
`gcf`	Get handle of current figure.
`gco`	Get handle of current object.
`get`	Gets object properties.
`getappdata`	Gets user-defined data in an object.
`isappdata`	Tests to see if an object contains user-defined data with the specified name.
`rmappdata`	Removes user-defined data from an object.
`set`	Sets object properties.
`setappdata`	Stores user-defined data in an object.
`waitforbuttonpress`	Pauses program, waiting for a mouse click or keyboard input.

9.13   Exercises

9.1   What is meant by the term "handle graphics"? Sketch the hierarchy of MATLAB graphics objects.

9.2   Use the MATLAB Help Browser to learn about the `Name` and `NumberTitle` properties of a `figure` object. Create a figure containing a plot of the function $y(x) = e^x$ for $-2 \leq x \leq 2$. Change the properties mentioned here to suppress the figure number and to add the title "Plot Window" to the figure.

9.3 Write a program that modifies the default figure color to orange and the default line width to 3.0 points. Then create a figure plotting the ellipse defined by the equations

$$x(t) = 10 \cos t$$
$$y(t) = 6 \sin t$$

(9-1)

from $t = 0$ to $t = 2\pi$. What color and width was the resulting line?

9.4 Use the MATLAB Help Browser to learn about the CurrentPoint property of an axes object. Use this property to create a program that creates an axes object and plots a line connecting the locations of successive mouse clicks within the axes. Use the function waitforbuttonpress to wait for mouse clicks, and update the plot after each click. Terminate the plot when a keyboard press occurs.

9.5 Use the MATLAB Help Browser to learn about the CurrentCharacter property of a figure object. Modify the program created in Problem 9.4 by testing the CurrentCharacter property when a keyboard press occurs. If the character typed on the keyboard is a "c" or "C," change the color of the line being displayed. If the character typed on the keyboard is an "s" or "S," change the line style of the line being displayed. If the character typed on the keyboard is a "w" or "W," change the width of the line being displayed. If the character typed on the keyboard is an "x" or "X," terminate the plot. (Ignore all other input characters.)

9.6 Create a MATLAB program that plots the functions

$$x(t) = \cos \frac{t}{\pi}$$
$$x(t) = 2 \sin \frac{t}{2\pi}$$

(9-2)

for the range $-2 \leq t \leq 2$. The program should then wait for mouse clicks, and if the mouse has clicked on one of the two lines, the program should change the line's color randomly from a choice of red, green, blue, yellow, cyan, magenta, or black. Use the function waitforbuttonpress to wait for mouse clicks, and update the plot after each click. Use the function gco to determine the object clicked on, and use the Type property of the object to determine whether the click was on a line.

9.7 Create a MATLAB figure and store some user-defined data values in the figure. Then, save the figure in a figure file (*.fig) using the File/Save As menu option on the Figure Window. Next close down and restart MATLAB. Reload the figure using the File/Open menu option on a Figure Window, and try to recover the user-defined data using the getappdata function. Was the information preserved?

9.8 The `plot` function plots a line and returns a handle to that line. This handle can be used to get or set the line's properties after it has been created. Two of a line's properties are `XData` and `YData`, which contain the x- and y-values currently plotted. Write a program that plots the function

$$x(t) = \cos(2\pi t - \theta) \tag{9-3}$$

between the limits $-1.0 \leq t \leq 1.0$, and saves the handle of the resulting line. The angle θ is initially 0 radians. Then, re-plot the line over and over with $\theta = \pi/10$ rad, $\theta = 2\pi/10$ rad, $\theta = 3\pi/10$ rad, and so forth up to $\theta = 2\pi$ rad. To re-plot the line, use a `for` loop to calculate the new values of x and t, and update the line's `XData` and `YData` properties with `set` commands. Pause 0.5 seconds between each update, using MATLAB's `pause` command.

10

Graphical User Interfaces

A graphical user interface (GUI) is a pictorial interface to a program. A good GUI can make programs easier to use by providing them with a consistent appearance and with intuitive controls such as pushbuttons, edit boxes, list boxes, sliders, and menus. The GUI should behave in an understandable and predictable manner so that a user knows what to expect when he or she performs an action. For example, when a mouse click occurs on a pushbutton, the GUI should initiate the action described on the label of the button.

This chapter contains an introduction to the basic elements of the MATLAB GUIs. It does not contain a complete description of components or GUI features, but it does provide the basics required to create functional GUIs for your programs.

10.1 How a Graphical User Interface Works

A graphical user interface provides the user with a familiar environment in which to work. It contains pushbuttons, toggle buttons, lists, menus, text boxes, and so forth, all of which are already familiar to the user, so that he or she can concentrate on the purpose of the application instead of the mechanics involved in doing things. However, GUIs are harder for the programmer, because a GUI-based program must be prepared for mouse clicks (or possibly keyboard input) for any GUI element at any time. Such inputs are known as **events**, and a program that responds to events is said to be *event driven*.

The three principal elements required to create a MATLAB graphical user interface are

1. **Components.** Each item on a MATLAB GUI (pushbuttons, labels, edit boxes, etc.) is a graphical component. The types of components include graphical **controls** (pushbuttons, toggle buttons, edit boxes, lists, sliders, etc.) static elements (text boxes), **menus**, **toolbars**, and **axes**. Graphical controls and text boxes are created by the function `uicontrol`, and menus are created by the functions `uimenu` and `uicontextmenu`. Toolbars are created by function `uitoolbar`. Axes, which are used to display graphical data, are created by the function `axes`.

2. **Containers.** The components of a GUI must be arranged within a **container**, which is a window on the computer screen. The most common container is a **figure**. A figure is a window on the computer screen that has a title bar along the top and that can optionally have menus attached. In the past, figures have been created automatically whenever we plotted data. However, empty figures can be created with the function `figure`, and they can be used to hold any combination of components and other containers.

   The other types of containers are **panels** (created by the function `uipanel`) and **button groups** (created by the function `uibuttongroup`). Panels can contain components or other containers, but they do not have a title bar and cannot have menus attached. Button groups are special panels that can manage groups of radio buttons or toggle buttons to ensure that no more than one button in the group is on at any time.

3. **Callbacks.** Finally, there must be some way to perform an action if a user clicks a mouse on a button or types information on a keyboard. A mouse click or a key press is an **event**, and the MATLAB program must respond to each event if the program is to perform its function. For example, if a user clicks on a button, then that event must cause the MATLAB code that implements the function of the button to be executed. The code executed in response to an event is known as a **callback**. There must be a callback to implement the function of each graphical component on the GUI.

The basic GUI elements are summarized in Table 10.1, and some sample elements are shown in Figure 10.1. We will study examples of these elements and then build working GUIs from them.

10.2   Creating and Displaying a Graphical User Interface

MATLAB Graphical User Interfaces are created using a tool called `guide`, the GUI Development Environment. This tool allows a programmer to lay out the GUI, selecting and aligning the GUI components to be placed in it. Once the components are in place, the programmer can edit their properties: name, color, size, font, text to be displayed, and so forth. When `guide` saves the GUI, it creates a working program, including skeleton functions that the programmer can modify to implement the behavior of the GUI.

(*text continues on page 455*)

Table 10.1   Some Basic GUI Components

Component	Created By	Description
	Containers	
Figure	`figure`	Creates a figure, which is container that can hold components and other containers. Figures are separate windows that have title bars and can have menus.
Panel	`uipanel`	Creates a panel, which is container that can hold components and other containers. Unlike figures, panels do not have title bars or menus. Panels can be placed inside figures or other panels.
Button Group	`uibuttongroup`	Creates a button group, which is a special kind of panel. Button groups automatically manage groups of radio buttons or toggle buttons to ensure that only one item of the group is on at any given time.
	Graphical Controls	
Pushbutton	`uicontrol`	A graphical component that implements a pushbutton. It triggers a callback when clicked with a mouse.
Toggle Button	`uicontrol`	A graphical component that implements a toggle button. A toggle button is either "on" or "off," and it changes state each time it is clicked. Each mouse button click also triggers a callback.
Radio Button	`uicontrol`	A radio button is a type of toggle button that appears as a small circle with a dot in the middle when it is "on." Groups of radio buttons are used to implement mutually exclusive choices. Each mouse click on a radio button triggers a callback.
Checkbox	`uicontrol`	A checkbox is a type of toggle button that appears as a small square with a check mark in it when it is "on." Each mouse click on a checkbox triggers a callback.
Edit Box	`uicontrol`	An edit box displays a text string and allows the user to modify the information displayed. A callback is triggered when the user presses the Enter key, or when the user clicks on a different object with the mouse.
List Box	`uicontrol`	A list box is a graphical control that displays a series of text strings. A user may select one of the text strings by single- or double-clicking on them. A callback is triggered when the user selects a string.
Popup Menus	`uicontrol`	A popup menu is a graphical control that displays a series of text strings in response to a mouse click. When the popup menu is not clicked on, only the currently selected string is visible.
Slider	`uicontrol`	A slider is a graphical control to adjust a value in a smooth, continuous fashion by dragging the control with a mouse. Each slider change triggers a callback.

(continued)

Table 10.1   (continued)

Component	Created By	Description
		Static Elements
Frame	uicontrol	Creates a frame, which is a rectangular box within a figure. Frames are used to group sets of controls together. Frames never trigger callbacks. (This is a deprecated component, which should not be used in new GUIs.)
Text Field	uicontrol	Creates a label, which is a text string located at a point on the figure. Text fields never trigger callbacks.
		Menus, Toolbars, Axes
Menu Items	uimenu	Creates a menu item. Menu items trigger a callback when a mouse button is released over them.
Context Menus	uicontextmenu	Creates a context menu, which is a menu that appears over a graphical object when a user right-clicks the mouse on that object.
Toolbar	uitoolbar	Creates a toolbar, which is a bar across the top of the figure containing quick-access buttons.
Toolbar Pushbutton	uipushtool	Creates a pushbutton to go in a toolbar.
Toolbar Toggle Button	uitoggletool	Creates a toggle button to go in a toolbar.
Axes	axes	Creates a new set of axes to display data on. Axes never trigger callbacks.

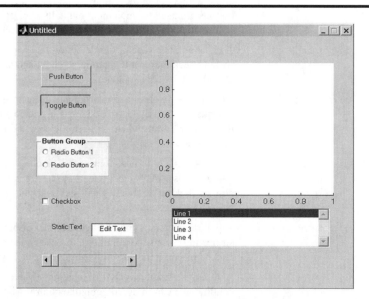

Figure 10.1 A Figure Window showing examples of MATLAB GUI elements. From top to bottom and left to right, the elements are: (1) a pushbutton; (2) a toggle button in the 'on' state; (3) two radio buttons within a button group; (4) a check box; (5) a label and an edit box; (6) a slider; (7) a set of axes; and (8) a list box.

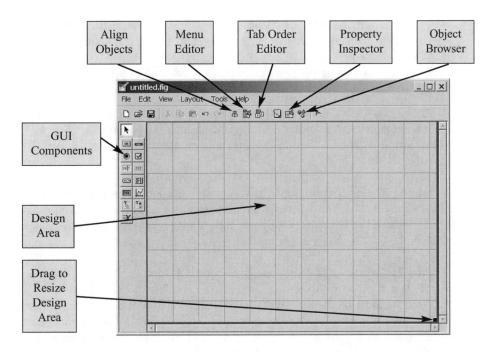

Figure 10.2   The guide tool window.

When guide is executed, it creates the Layout Editor, shown in Figure 10.2. The large gray area with grid lines is the *layout area*, where a programmer can lay out the GUI. The Layout Editor window has a palate of GUI components along the left-hand side of the layout area. A user can create any number of GUI components by first clicking on the desired component and then dragging its outline in the layout area. The top of the window has a toolbar with a series of useful tools that allow the user to distribute and align GUI components, modify the properties of GUI components, and add menus to GUIs, among other things.

The basic steps required to create a MATLAB GUI are as follows:

1. Decide what elements are required for the GUI and what the function of each element will be. Make a rough layout of the components by hand on a piece of paper.
2. Use the MATLAB tool called guide (GUI Development Environment) to lay out the components on a figure. The size of the figure and the alignment and spacing of components on the figure can be adjusted using the tools built into guide.
3. Use a MATLAB tool called the Property Inspector (built into guide) to give each component a name (a "tag") and to set the characteristics of each component, such as its color or the text it displays.
4. Save the figure to a file. When the figure is saved, two files will be created on disk with the same name but different extents. The fig file contains the

GUI layout and the components of the GUI; while the M-file contains the code to load the figure as well as skeleton callback functions for each GUI element.

5. Write code to implement the behavior associated with each callback function.

As an example of these steps, let's consider a simple GUI that contains a single pushbutton and a single text string. Each time the pushbutton is clicked, the text string will be updated to show the total number of clicks since the GUI started.

Step 1: The design of this GUI is very simple. It contains a single pushbutton and a single text field. The callback from the pushbutton will cause the number displayed in the text field to increase by one each time the button is pressed. A rough sketch of the GUI is shown in Figure 10.3.

Step 2: To lay out the components on the GUI, run the MATLAB function guide. When guide is executed, it creates the window shown in Figure 10.2.

First, we must set the size of the layout area, which will become the size of the final GUI. We do this by dragging the small square on the lower-right corner of the layout area until it has the desired size and shape. Then, click on the "pushbutton" button in the list of GUI components and create the shape of the pushbutton in the layout area. Finally, click on the "text" button in the list of GUI components and create the shape of the text field in the layout area. The resulting figure after performing these steps is shown in Figure 10.4. We could now adjust the alignment of these two elements using the Alignment Tool, if desired.

Step 3: To set the properties of the pushbutton, click on the button in the layout area and then select "Property Inspector" (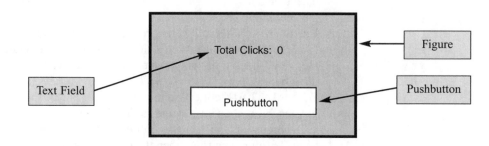) from the toolbar. Alternatively, right-click on the button and select "Property Inspector" from the popup menu. The Property Inspector window shown in Figure 10.5 will appear. Note that this window lists every property available for the pushbutton and allows us to set each value using a GUI interface. The Property Inspector performs the

(text continues on page 459)

Figure 10.3 Rough layout for a GUI containing a single pushbutton and a single label field.

Figure 10.4   The completed GUI layout within the `guide` window.

Figure 10.5   The Property Inspector showing the properties of the pushbutton. Note that the String is set to `'Click Here'`, and the Tag is set to `'MyFirstButton'`.

same function as the `get` and `set` functions introduced in Chapter 9 but in a much more convenient form.

For the pushbutton, we may set many properties such as color, size, font, text and alignment. However, we *must* set two properties: the `String` property, which contains the text to be displayed, and the `Tag` property, which is the name of the pushbutton. In this case, the `String` property will be set to `'Click Here'`, and the `Tag` property will be set to `MyFirstButton`.

For the text field, we *must* set two properties: the `String` property, which contains the text to be displayed, and the `Tag` property, which is the name of the text field. This name will be needed by the callback function to locate and update the text field. In this case, the `String` property will be set to `'Total Clicks: 0'`, and the `Tag` property defaults to `'MyFirstText'`. The layout area after performing these steps is shown in Figure 10.6.

It is possible to set the properties of the figure itself by clicking on a clear spot in the Layout Editor and using the Property Inspector to examine and set the figure's properties. Although not required, it is a good idea to set the figure's `Name` property. The string in the `Name` property will be displayed in the title bar of the resulting GUI when it is executed. In this program, we will set the `Name` to `'MyFirstGUI'`.

Step 4: We will now save the layout area under the name `MyFirstGUI`. Select the "File/Save As" menu item, type the name `MyFirstGUI` as the file name, and click "Save". This action will automatically create two files: `MyFirstGUI.fig` and `MyFirstGUI.m`. The figure file contains the actual GUI that we have created. The M-file contains code that loads the figure file and creates the GUI, plus a skeleton callback function for each active GUI component.

At this point, we have a complete GUI, but one that does not yet do the job it was designed to do. You can start this GUI by typing `MyFirstGUI` in the Command Window, as shown in Figure 10.7. If the button is clicked on this GUI, nothing happens.

Figure 10.6   The design area after the properties of the pushbutton and the text field have been modified.

Figure 10.7   Typing `MyFirstGUI` in the Command Window starts the GUI.

A portion of the M-file automatically created by `guide` is shown in Figure 10.8. This file contains the main function `MyFirstGUI`, plus subfunctions to specify the behavior of the active GUI components. The file contains a *dummy callback function for every active GUI component* that you defined. In this case, the only active GUI component was the pushbutton, so there is a callback function called `MyFirstButton_Callback`, which is executed when the user clicks on the button.

If function `MyFirstGUI` is called *without* arguments, then the function displays the GUI contained in file `MyFirstGUI.fig`. If function `MyFirstGUI` is called *with* arguments, then the function assumes that the first argument is the name of a subfunction, and it calls that subfunction using `feval`, passing the other arguments on to that subfunction.

Each callback function handles events from a single GUI component. If a mouse click (or keyboard input for edit fields) occurs on the GUI component, then the component's callback function will be automatically called by MATLAB. The name of the callback function will be the value in the `Tag` property of the GUI component plus the characters "`_Callback`". Thus, the callback function for `MyFirstButton` will be named `MyFirstButton_Callback`.

M-files created by `guide` contain callbacks for each active GUI component, but these callbacks don't do anything yet.

Step 5: Now, we need to write the callback subfunction code for the pushbutton. This function will include a `persistent` variable that can be used

```
function varargout = MyFirstGUI(varargin)
% MYFIRSTGUI M-file for MyFirstGUI.fig
%      MYFIRSTGUI, by itself, creates a new MYFIRSTGUI or raises the existing
%      singleton*.
%
%      H = MYFIRSTGUI returns the handle to a new MYFIRSTGUI or the handle to
%      the existing singleton*.
%
%      MYFIRSTGUI('CALLBACK',hObject,eventData,handles,...) calls the local
%      function named CALLBACK in MYFIRSTGUI.M with the given input arguments.
%
%      MYFIRSTGUI('Property','Value',...) creates a new MYFIRSTGUI or raises the
%      existing singleton*.  Starting from the left, property value pairs are
%      applied to the GUI before MyFirstGUI_OpeningFunction gets called.  An
%      unrecognized property name or invalid value makes property application
%      stop.  All inputs are passed to MyFirstGUI_OpeningFcn via varargin.
%
%      *See GUI Options on GUIDE's Tools menu.  Choose "GUI allows only one
%      instance to run (singleton)".
%
% See also: GUIDE, GUIDATA, GUIHANDLES

% Edit the above text to modify the response to help MyFirstGUI

% Last Modified by GUIDE v2.5 21-Feb-2004 16:17:45

% Begin initialization code - DO NOT EDIT
gui_Singleton = 1;
gui_State = struct('gui_Name',       mfilename, ...
                   'gui_Singleton',  gui_Singleton, ...
                   'gui_OpeningFcn', @MyFirstGUI_OpeningFcn, ...
                   'gui_OutputFcn',  @MyFirstGUI_OutputFcn, ...
                   'gui_LayoutFcn',  []  , ...
                   'gui_Callback',   []);
if nargin & isstr(varargin{1})
    gui_State.gui_Callback = str2func(varargin{1});
end

if nargout
    [varargout{1:nargout}] = gui_mainfcn(gui_State, varargin{:});
else
    gui_mainfcn(gui_State, varargin{:});
end
% End initialization code - DO NOT EDIT


% --- Executes just before MyFirstGUI is made visible.
function MyFirstGUI_OpeningFcn(hObject, eventdata, handles, varargin)
% This function has no output args, see OutputFcn.
% hObject    handle to figure
% eventdata  reserved - to be defined in a future version of MATLAB
% handles    structure with handles and user data (see GUIDATA)
% varargin   command line arguments to MyFirstGUI (see VARARGIN)

% Choose default command line output for MyFirstGUI
handles.output = hObject;

% Update handles structure
guidata(hObject, handles);
```

Main Function

Figure
Opening
Function

Figure 10.8 The M-file for MyFirstGUI, automatically created by guide.

```
% UIWAIT makes MyFirstGUI wait for user response (see UIRESUME)
% uiwait(handles.figure1);


% --- Outputs from this function are returned to the command line.
function varargout = MyFirstGUI_OutputFcn(hObject, eventdata, handles)
% varargout   cell array for returning output args (see VARARGOUT);
% hObject     handle to figure
% eventdata   reserved - to be defined in a future version of MATLAB
% handles     structure with handles and user data (see GUIDATA)

% Get default command line output from handles structure
varargout{1} = handles.output;


% --- Executes on button press in MyFirstButton.
function MyFirstButton_Callback(hObject, eventdata, handles)
% hObject     handle to MyFirstButton (see GCBO)
% eventdata   reserved - to be defined in a future version of MATLAB
% handles     structure with handles and user data (see GUIDATA)
```

Data Output Function

Button Callback Function

Figure 10.8   (*continued*)

to count the number of clicks that have occurred. When a click occurs on the pushbutton, MATLAB will call the function `MyFirstGUI` with `MyFirstButton_Callback` as the first argument. Then function `MyFirstGUI` will call subfunction `MyFirstButton_Callback`, as shown in Figure 10.9. This function should increase the count of clicks by one, create a new text string containing the count, and store the new string in the `String` property of the text field `MyFirstText`. A function to perform this step is shown here:

```
function MyFirstButton_Callback(hObject, eventdata, handles)
% hObject     handle to MyFirstButton (see GCBO)
% eventdata   reserved - to be defined in a future version of MATLAB
% handles     structure with handles and user data (see GUIDATA)

% Declare and initialize variable to store the count
persistent count
if isempty(count)
    count = 0;
end

% Update count
count = count + 1;

% Create new string
str = sprintf('Total Clicks: %d',count);

% Update the text field
set (handles.MyFirstText,'String',str);
```

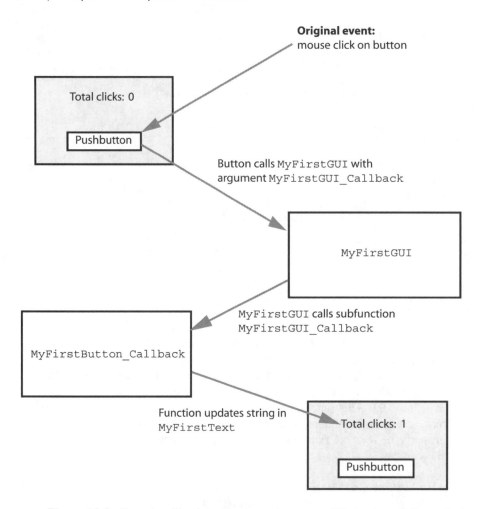

Figure 10.9 Event handling in program MyFirstGUI. When a user clicks on the button with the mouse, the function MyFirstGUI is called automatically with the argument MyFirstButton_Callback. Function MyFirstGUI in turn calls subfunction MyFirstButton_Callback. This function increments count, and then saves the new count in the text field on the GUI.

Note that this function declares a persistent variable count and initializes it to zero. Each time the function is called, it increments count by 1 and creates a new string containing the count. Then, it updates the string displayed in the text field MyFirstText.

The resulting program is executed by typing MyFirstGUI in the Command Window. When the user clicks on the button, MATLAB automatically calls function MyFirstGUI with MyFirstButton_Callback as the first argument,

Figure 10.10   The resulting program after three button pushes.

and function `MyFirstGUI` calls subfunction `MyFirstButton_Callback`. This function increments variable `count` by one and updates the value displayed in the text field. The GUI that result after three button pushes is shown in Figure 10.10.

※ **Good Programming Practice**

Use `guide` to lay out a new GUI, and use the Property Inspector to set the initial properties of each component such as the text displayed on the component, the color of the component, and the name of the callback function, if required.

※ **Good Programming Practice**

After creating a GUI with `guide`, manually edit the resulting M-file to add comments describing its purpose and components, and to implement the behavior of callbacks.

10.2.1   A Look Under the Hood

Figure 10.8 shows the M-file that was automatically generated by `guide` for `MyFirstGUI`. We will now examine this M-file more closely to understand how it works.

First, let's look at the main `function` declaration itself. Note that this function uses `varargin` to represent its input arguments and `varargout` to represent its output results. As we learned in Chapter 7, function `varargin` can represent an arbitrary number of input arguments, and function `varargout` can represent a varying number of output arguments. Therefore, *a user can call function* `MyFirstGUI` *with any number of arguments.*

The main function begins with a series of comments that serve as the help message displayed when the user types "`help MyFirstGUI`". You should edit these comments to reflect the actual function of your program.

Next, the main function creates a structure called `gui_State`. The code to create this structure is

```
gui_Singleton = 1;
gui_State = struct('gui_Name',        mfilename, ...
                   'gui_Singleton',   gui_Singleton, ...
                   'gui_OpeningFcn',  @MyFirstGUI_OpeningFcn, ...
                   'gui_OutputFcn',   @MyFirstGUI_OutputFcn, ...
                   'gui_LayoutFcn',   [] , ...
                   'gui_Callback',    []);
if nargin & isstr(varargin{1})
   gui_State.gui_Callback = str2func(varargin{1});
end
```

The structure contains some control information, plus function handles for some of the subfunctions in the file. Other MATLAB GUI functions use these function handles to call the subfunctions from outside of the M-file. Note that the first argument is converted into a callback function handle, if it exists.

The value `gui_Singleton` specifies whether there can be one or more simultaneous copies of the GUI. If `gui_Singleton` is 1, then there can be only one copy of the GUI. If `gui_Singleton` is 0, then there can be many simultaneous copies of the GUI.

The main function calls the MATLAB function `gui_mainfcn` and passes the `gui_State` structure and all of the input arguments to it. Function `gui_mainfcn` is a standard MATLAB function. It actually does the work of creating the GUI, or of calling the appropriate subfunction in response to a callback.

Calling the M-File without Arguments

If the user calls `MyFirstGUI` *without* arguments, function `gui_mainfcn` loads the GUI from the figure file `MyFirstGUI.fig` using the `openfig` function. The form of this function is

```
fig = openfig(mfilename,reuse);
```

where `mfilename` is the name of the figure file to load. The second argument in the function specifies whether there can be only one copy of the figure running at a given time or multiple copies can be run. If `gui_State.gui_Singleton` is 1, then the second argument is set to '`reuse`', and only one copy of the figure can be run. If `openfig` is called with the '`reuse`' option and the specified figure already exists, the preexisting figure will be brought to the top of the screen and reused. In contrast, if `gui_State.gui_Singleton` is 0, then the argument is set to '`new`', and a new copy of the figure will be created each time `MyFirstGUI` is called without arguments. By default, a GUI created by guide has the gui_State.gui_Singleton set to 1, so only one copy of the figure can exist at any time. If you wish to have multiple copies of the GUI, turn off the "GUI allows only one instance to run" flag in the GUI Options selection on guide's Tools menu.

After the figure has been loaded, the gui_mainfcn function executes the statement

```
set(fig,'Color',get(0,'defaultUicontrolBackgroundColor'));
```

This function sets the background color of the figure to match the default background color used by the computer that MATLAB is executing on. It makes the color of the GUI match the color of native windows on the computer. Therefore, a GUI can be written on a Windows-based PC and used on a Unix-based computer, and vice versa. It will look natural in either environment.

Then, function gui_mainfcn creates a structure containing the handles of all the objects in the current figure, and stores that structure as application data in the figure.

```
guidata(gui_hFigure, guihandles(gui_hFigure));
```

Function guihandles creates a structure containing handles to all of the objects within the specified figure. The element names in the structure correspond to the Tag properties of each GUI component, and the values are the handles of each component. For example, the handle structure returned in MyFirstGUI.m is

```
» handles = guihandles(fig)
handles =
            figure1: 99.0005
        MyFirstText: 3.0021
      MyFirstButton: 100.0007
```

There are three GUI components in this figure—the figure itself, plus a text field and a pushbutton. Function guidata saves the handles structure as application data in the figure, using the setappdata function that we studied in Chapter 9.

Finally, just before making the figure visible, function gui_mainfcn calls the function specified in gui_OpeningFcn. This function provides a way for the programmer to customize the GUI before showing it to the user. For example, a programmer could load initial data values here, change background colors, and so forth.

Calling the M-File with Arguments

When the user clicks on an active GUI element, MATLAB calls MyFirstGUI with the name of the GUI element's callback function in the first argument. If MyFirstGUI is called *with* arguments, the value returned by nargin will be greater than zero. In this case, function MyFirstGUI converts the callback function name into a function handle using the following code.

```
    if nargin & isstr(varargin{1})
        gui_State.gui_Callback = str2func(varargin{1});
    end
```

When function `gui_mainfcn` is called this time, it calls the callback function using this function handle. The callback executes and responds to the mouse click or keyboard input, as appropriate.

Figure 10.11 summarizes the operation of `MyFirstGUI` on first and subsequent calls.

10.2.2   The Structure of a Callback Subfunction

Every callback subfunction has the standard form

```
function ComponentTag_Callback(hObject, eventdata, handles)
```

where `ComponentTag` is the name of the component generating the callback (the string in its `Tag` property). The arguments of this subfunction are as follows:

- **hObject**—The handle of the parent figure.
- **eventdata**—A currently unused (in MATLAB 7) array.
- **Handles**—The handles structure contains the handles of all GUI components on the figure.

Note that each callback function has full access to the `handles` structure, and so each callback function can modify any GUI component in the figure. We took advantage of this structure in the callback function for the pushbutton in `MyFirstGUI`, where the callback function for the pushbutton modified the text displayed in the text field.

```
% Update the text field
set (handles.MyFirstText,'String',str);
```

10.2.3   Adding Appication Data to a Figure

It is possible to store any application-specific information needed by a GUI program in the `handles` structure instead of using global or persistent memory for that data. The resulting GUI design is more robust, since other MATLAB programs cannot accidentally modify the global GUI data and since multiple copies of the same GUI cannot interfere with each other.

To add local data to the `handles` structure, we must manually modify the M-file after it has been created by `guide`. A programmer first adds the required local data to the `handles` structure and then calls `guidata` to update the `handles` structure stored in the figure. For example, to add the number of mouse clicks `count` to the `handles` structure, we would modify the `MyFirstButton_Callback` function as follows:

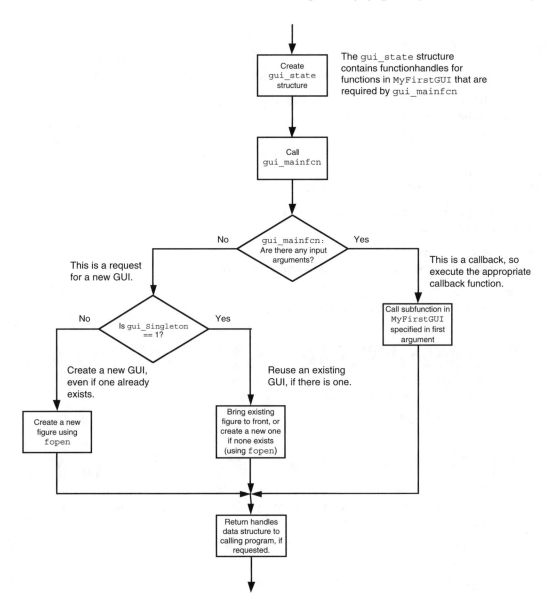

Figure 10.11 The operation of `MyFirstGUI`. If there are no calling arguments, it either creates a GUI or displays an existing GUI. If there are calling arguments, the first arguments is assumed to be a callback function name, and `MyFirstGUI` calls the appropriate callback function.

```
function MyFirstButton_Callback(hObject, eventdata, handles)
% hObject    handle to MyFirstButton (see GCBO)
% eventdata  reserved - to be defined in a future version of MATLAB
% handles    structure with handles and user data (see GUIDATA)

% Create the count field if it does not exist
if ~isfield(handles,'count')
   handles.count = 0;
end

% Update count
handles.count = handles.count + 1;

% Save the updated handles structure
guidata(hObject, handles);

% Create new string
str = sprintf('Total Clicks: %d',handles.count);

% Update the text field
set (handles.MyFirstText,'String',str);
```

☀ Good Programming Practice

Store GUI application data in the `handles` structure, so that it will automatically be available to any callback function.

☀ Good Programming Practice

If you modify any of the GUI application data in the `handles` structure, be sure to save the structure with a call to `guidata` before exiting the function where the modifications occurred.

10.2.4   A Few Useful Functions

Three special functions are occasionally used in the design of callback functions: `gcbo`, `gcbf`, and `findobj`. These functions are not actually needed with MATLAB 7 GUIs, because the same information is available in the `handles` data structure. However, they were commonly used in earlier versions of MATLAB, and a programmer is sure to encounter them.

Function `gcbo` (*get callback object*) returns the handle of the object that generated the callback, and function `gcbf` (*get callback figure*) returns the handle of the figure containing that object. These functions can be used by a callback function to determine the object and figure producing the callback, so that it can modify objects on that figure.

Function `findobj` searches through all of the child objects within a parent object, looking for those that have a specific value of a specified property. It returns a handle to any objects with the matching characteristics. The most common form of `findobj` is

```
Hndl = findobj(parent,'Property',Value);
```

where `parent` is the handle of a parent object such as a figure, `'Property'` is the property to examine, and `'Value'` is the `value` to look for.

For example, suppose that a programmer would like to change the color of all the lines in a plot on the callback figure. He or she could find the lines and change the line colors to red with the following statements:

```
Hndl = findobj(gcbf, 'Type', 'Line');
for ii = 1:length(Hndl)
    set( Hndl, 'Color','r' );
end
```

10.3   Object Properties

Every GUI object includes an extensive list of properties that can be used to customize the object. These properties are slightly different for each type of object (figures, axes, `uicontrols`, etc.). All of the properties for all types of objects are documented on the on-line Help Browser, but a few of the more important properties for `figure` and `uicontrol` objects are summarized in Tables 10.2 and 10.3.

Table 10.2   Important `figure` Properties

Property	Description
Color	Specifies the color of the figure. The value is either a predefined color such as `'r'`, `'g'`, or `'b'`, or else a three-element vector specifying the red, green, and blue components of the color on a 0–1 scale. For example, the color magenta would be specified by `[1  0  1]`.
CurrentCharacter	Contains the character corresponding to the last key pressed in this figure.
CurrentPoint	Location of the last button click in this figure, measured from the lower-left corner of the figure in units specified in the `Units` property.
Dockable	Specifies whether or not the figure can be docked to the desktop. Possible values are `'on'` or `'off'`.
MenuBar	Specifies whether or not the default set of menus appears on the figure. Possible values are `'figure'` to display the default menus, or `'none'` to delete them.
Name	A string containing the name that appears in the title bar of a figure.
NumberTitle	Specifies whether or not the figure number appears in the title bar. Possible values are `'on'` or `'off'`.

(continued)

Table 10.2 (continued)

Property	Description
Position	Specifies the position of a figure on the screen, in the units specified by the `'units'` property. This value accepts a four-element vector in which the first two elements are the *x* and *y* positions of the lower-left corner of the figure and the next two elements are the width and height of the figure.
SelectionType	Specifies the type of selection for the last mouse click on this figure. A single click returns type `'normal'`, while a double click returns type `'open'`. There are additional options; see the MATLAB on-line documentation.
Tag	The "name" of the figure, which can be used to locate it.
Units	The units used to describe the position of the figure. Possible choices are `'inches'`, `'centimeters'`, `'normalized'`, `'points'`, `'pixels'`, or `'characters'`. The default units are `'pixels'`.
Visible	Specifies whether or not this figure is visible. Possible values are `'on'` or `'off'`.
WindowStyle	Specifies whether this figure is normal or modal (see discussion of Dialog Boxes). Possible values are `'normal'` or `'modal'`.

Table 10.3 Important uicontrol Properties

Property	Description
BackgroundColor	Specifies the background color of the object. The value is either a predefined color such as `'r'`, `'g'`, or `'b'`, or else a three-element vector specifying the red, green, and blue components of the color on a 0–1 scale. For example, the color magenta would be specified by [1 0 1].
Callback	Specifies the name and parameters of the function to be called when the object is activated by a keyboard or text input.
Enable	Specifies whether or not this object is selectable. If it not enabled, it will not respond to mouse or keyboard input. Possible values are `'on'` or `'off'`.
FontAngle	A string containing the font angle for text displayed on the object. Possible values are `'normal'`, `'italic'`, and `'oblique'`.
FontName	A string containing the font name for text displayed on the object.
FontSize	A number specifying the font size for text displayed on the object.
FontUnits	The units in which the font size is defined. Possible choices are `'inches'`, `'centimeters'`, `'normalized'`, `'points'`, and `'pixels'`. The default font units are `'points'`.
FontWeight	A string containing the font weight for text displayed on the object. Possible values are `'light'`, `'normal'`, `'demi'`, and `'bold'`. The default font weight is `'normal'`.
ForegroundColor	Specifies the foreground color of the object.

(continued)

Table 10.3   (continued)

Property	Description
HorizontalAlignment	Specifies the horizontal alignment of a text string within the object. Possible values are 'left', 'center', and 'right'.
Max	The maximum size of the value property for this object.
Min	The minimum size of the value property for this object.
Parent	The handle of the figure containing this object.
Position	Specifies the position of the object on the screen, in the units specified by the 'units' property. This value accepts a four-element vector in which the first two elements are the *x* and *y* positions of the lower-left corner of the object *relative to the figure containing it*, and the next two elements are the width and height of the object.
Tag	The "name" of the object, which can be used to locate it.
TooltipString	Specifies the help text to be displayed when a user places the mouse pointer over an object.
Units	The units used to describe the position of the figure. Possible choices are 'inches', 'centimeters', 'normalized', 'points', 'pixels', or 'characters'. The default units are 'pixels'.
Value	The current value of the uicontrol. For toggle buttons, checkboxes, and radio buttons, the value is max when the button is on and min when the button is off. Other controls have different meanings for this term.
Visible	Specifies whether or not this object is visible. Possible values are 'on' or 'off'.

Object properties can be modified using either the Property Inspector or the get and set functions. Although the Property Inspector is a convenient way to adjust properties during GUI design, we must use get and set to adjust them dynamically from within a program, such as in a callback function.

10.4   Graphical User Interface Components

This section summarizes the basic characteristics of common graphical user interface components. It describes how to create and use each component, as well as the types of events each component can generate. The components discussed in this section are

- Static Text Fields
- Edit Boxes
- Pushbuttons

- Toggle Buttons
- Checkboxes
- Radio Buttons
- Popup Menus
- List Boxes
- Sliders

10.4.1 Static Text Fields

A **static text field** is a graphical object that displays one or more text strings, which are specified in the text field's `String` property. The `String` property accepts a string or a cell array of strings. If the input value is a string, it will be displayed on a single line. If the input value is a cell array of strings, the first element will be displayed on the first line of the text box, the second element will be displayed on the second line of the text box, and so forth. You can specify how the text is aligned in the display area by setting the horizontal alignment property. By default, text fields are horizontally centered. A text field is created by a `uicontrol` whose style property is `'text'`. A text field may be added to a GUI by using the text tool ( ) in the Layout Editor.

Text fields do not create callbacks, but the value displayed in the text field can be updated from another component's callback function by changing the text field's `String` property, as shown in program `MyFirstGUI` in Section 10.2.

10.4.2 Edit Boxes

An **edit box** is a graphical object that allows a user to enter one or more text strings. It is created by a `uicontrol` whose style property is `'edit'`. If the `min` property and `max` property are both set to 1, then *the edit box will accept a single line of text*, and it will generate a callback when the user presses the Enter key or the ESC key after typing the text.

Figure 10.12*a* shows a simple GUI containing an edit box named `'EditBox'` and a text field named `'TextBox'`. When a user presses Enter or ESC after typing a string into the edit box, the program automatically calls the function `EditBox_Callback`, which is shown in Figure 10.12*b*. This function locates the edit box using the `handles` structure, and recovers the string typed by the user. Then, it locates the text field and displays the string in the text field. Figure 10.13 shows this GUI just after it has started, and after the user has typed the word "Hello" in the edit box.

If the `max` property is set to a number greater than the `min` property, then *the edit box will accept as many lines of text as the user wishes to enter*. The textbox will include a vertical scrollbar to allow the user to move up and down through the data. Either the scrollbar or the up and down arrows can be used to move between the input lines. If the user presses the Enter key in a multi-line edit box, the current line is finished and the cursor moves down to the next line for

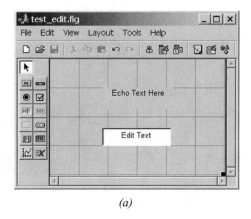

(a)

```
function EditBox_Callback(hObject, eventdata, handles)

% Find the value typed into the edit box
str = get (handles.EditBox,'String');

% Place the value into the text field
set (handles.TextBox,'String',str);
```
(b)

Figure 10.12 *(a)* Layout of a simple GUI with a single-line edit box and a text field. *(b)* The callback function for this GUI.

Figure 10.13 *(a)* The GUI produced by program `test_edit`. *(b)* The GUI after a user types `'Hello'` into the edit box and presses Enter.

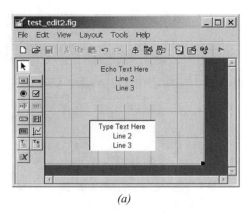

(a)

```
function EditBox2_Callback(hObject, eventdata, handles)

% Find the value typed into the edit box
str = get (handles.EditBox,'String');

% Place the value into the text field
set (handles.TextBox2,'String',str);
```

(b)

Figure 10.14 *(a)* Layout of a simple GUI with a multi-line edit box and a text field. *(b)* The callback function for this GUI.

additional input. If the user presses the ESC key or clicks a point on the figure background with the mouse, a callback will be generated and the data typed into the edit box will be available as a cell array of strings in the uicontrol's String property.

Figure 10.14*a* shows a simple GUI containing a multi-line edit box named 'EditBox2' and a text field named 'TextBox2'. When a user presses ESC after typing a set of lines into the edit box, the program automatically calls the function EditBox2_Callback, which is shown in Figure 10.14*b*. This function locates the edit box using the handles structure and recovers the strings typed by the user. Then, it locates the text field and displays the strings in the text field. Figure 10.15 shows this GUI just after it has started and after the user has typed four lines in the edit box.

10.4.3 Pushbuttons

A **pushbutton** is a component that a user can click on to trigger a specific action. The pushbutton generates a callback when the user clicks on it with the mouse. A pushbutton is created by creating a uicontrol whose style property

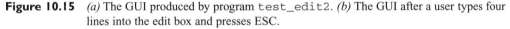

(a)   *(b)*

Figure 10.15   *(a)* The GUI produced by program `test_edit2`. *(b)* The GUI after a user types four lines into the edit box and presses ESC.

is `'pushbutton'`. It can be added to a GUI by using the pushbutton tool (⌗) in the Layout Editor.

Function `MyFirstGUI` in Figure 10.10 illustrates the use of pushbuttons.

10.4.4   Toggle Buttons

A **toggle button** is a type of button that has two states: on (depressed) and off (not depressed). A toggle button switches between these two states whenever the mouse clicks on it, and it generates a callback each time. The `'Value'` property of the toggle button is set to max (usually 1) when the button is on, and min (usually 0) when the button is off.

A toggle button is created by a `uicontrol` whose style property is `'togglebutton'`. It can be added to a GUI by using the toggle button tool (⌗) in the Layout Editor.

Figure 10.16*a* shows a simple GUI containing a toggle button named `'ToggleButton'` and a text field named `'TextBox'`. When a user clicks on the toggle button, it automatically calls the function `ToggleButton_Callback`, which is shown in Figure 10.16*b*. This function locates the toggle button using the `handles` structure and recovers its state from the `'Value'` property. Then, it locates the text field and displays the state in the text field. Figure 10.17 shows this GUI just after it has started, and after the user has clicked on the toggle button for the first time.

10.4.5   Checkboxes and Radio Buttons

Checkboxes and radio buttons are essentially identical to toggle buttons except that they have different shapes. Like toggle buttons, they have two states: on and off. They switch between these two states whenever the mouse clicks on them,

(a)

```
function ToggleButton_Callback(hObject, eventdata, handles)
% hObject       handle to ToggleButton (see GCBO)
% eventdata     reserved - to be defined in a future version of MATLAB
% handles       structure with handles and user data (see GUIDATA)

% Find the state of the toggle button
state = get(handles.ToggleButton,'Value');

% Place the value into the text field
if state == 0
    set (handles.TextBox,'String','Off');
else
    set (handles.TextBox,'String','On');
end
```
(b)

Figure 10.16   *(a)* Layout of a simple GUI with a toggle button and a text field. *(b)* The callback
function for this GUI.

(a)   *(b)*

Figure 10.17   *(a)* The GUI produced by program `test_togglebutton` when the toggle button is
off. *(b)* The GUI when the toggle button is on.

generating a callback each time. The `'Value'` property of the checkbox or radio button is set to max (usually 1) when it is on, and min (usually 0) when it is off. Both checkboxes and radio buttons are illustrated in Figure 10.1.

A checkbox is created by a `uicontrol` whose style property is `'checkbox'`, and a radio button is created by a `uicontrol` whose style property is `'radiobutton'`. A checkbox may be added to a GUI by using the checkbox tool (☑) in the Layout Editor, and a radio button may be added to a GUI by using the radio button tool (◉) in the Layout Editor.

Checkboxes are traditionally used to display on/off options, whereas groups of radio buttons are traditionally used to select among mutually exclusive options.

Figure 10.18*a* shows a simple GUI containing a checkbox named `'CheckBox'` and a text field named `'TextBox'`. When a user clicks on the checkbox, it automatically calls the function `CheckButton_Callback`, which is shown in Figure 10.18*b*. This function locates the checkbox using the `handles` structure and recovers its state from the `'Value'` property. Then, it

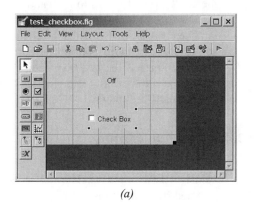

(a)

```
function CheckBox_Callback(hObject, eventdata, handles)

% Find the state of the checkbox
state = get(handles.CheckBox,'Value');

% Place the value into the text field
if state == 0
   set (handles.TextBox,'String','Off');
else
   set (handles.TextBox,'String','On');
end ;
```

(b)

Figure 10.18 *(a)* Layout of a simple GUI with a CheckBox and a text field. *(b)* The callback function for this GUI.

(a)   (b)

Figure 10.19 *(a)* The GUI produced by program `test_checkbox` when the toggle button is off.
(b) The GUI when the toggle button is on.

locates the text field and displays the state in the text field. Figure 10.19 shows this GUI just after it has started, and after the user has clicked on the toggle button for the first time.

Figure 10.20*a* shows an example of how to create a group of mutually exclusive options with radio buttons. The GUI in this figure creates three radio buttons, labeled "Option 1," "Option 2," and "Option 3," plus a text field to display the currently selected results.

The corresponding callback functions are shown in Figure 10.20*b*. When the user clicks on a radio button, the corresponding callback function is executed. That function sets the text box to display the current option, turns on that radio button, and turns off all other radio buttons.

Figure 10.21 shows this GUI after Option 2 has been selected.

10.4.6 Popup Menus

Popup menus are graphical objects that allow a user to select one of a mutually exclusive list of options. The list of options that the user can select among is specified by a cell array of strings, and the `'Value'` property indicates which of the strings is currently selected. A popup menu may be added to a GUI by using the popup menu tool (⌨) in the Layout Editor.

Figure 10.22*a* shows an example of a popup menu. This GUI in this figure creates a popup menu with five options, labeled "Option 1," "Option 2," and so forth.

The corresponding callback function is shown in Figure 10.22*b*. The callback function recovers the selected option by checking the `'Value'` parameter of the popup menu and creates and displays a string containing that value in the text field. Figure 10.23 on page 468 shows this GUI after Option 4 has been selected.

10.4.7 List Boxes

List boxes are graphical objects that display many lines of text and allow a user to select one or more of those lines. If there are more lines of text than can fit in

(a)

```
function Option1_Callback(hObject, eventdata, handles)

% Display the radio button clicked in the text field
set (handles.TextBox,'String','Option 1');

% Update all text fields
set (handles.Option1,'Value',1);
set (handles.Option2,'Value',0);
set (handles.Option3,'Value',0);

function Option2_Callback(hObject, eventdata, handles)

% Display the radio button clicked in the text field
set (handles.TextBox,'String','Option 2');

% Update all text fields
set (handles.Option1,'Value',0);
set (handles.Option2,'Value',1);
set (handles.Option3,'Value',0);

function Option3_Callback(hObject, eventdata, handles)

% Display the radio button clicked in the text field
set (handles.TextBox,'String','Option 3');

% Update all text fields
set (handles.Option1,'Value',0);
set (handles.Option2,'Value',0);
set (handles.Option3,'Value',1);
```

(b)

Figure 10.20   *(a)* Layout of a simple GUI with three radio buttons and a text field. *(b)* The callback functions for this GUI. When a user clicks on a radio button, it is set to 'on' and all other radio buttons are set to 'off'.

Figure 10.21 The GUI produced by program `test_radio_button` when Option 2 has been selected.

(a)

```
function Popup1_Callback(hObject, eventdata, handles)

% Find the value of the popup menu
value = get(handles.Popup1,'Value');

% Place the value into the text field
str = ['Option ' num2str(value)];
set (handles.Label1,'String',str);
```
(b)

Figure 10.22 *(a)* Layout of a simple GUI with a popup menu and a text field to display the current selection. *(b)* The callback functions for this GUI.

Figure 10.23 The GUI produced by program `test_popup_menu`.

the list box, a scroll bar will be created to allow the user to scroll up and down within the list box. The lines of text that the user can select among are specified by a cell array of strings, and the `'Value'` property indicates which of the strings are currently selected.

A list box is created by a `uicontrol` whose style property is `'list-box'`. A list box may be added to a GUI by using the listbox tool (▤) in the Layout Editor.

List boxes can be used to select a single item from a selection of possible choices. In normal GUI usage, a single mouse click on a list item selects that item but does not cause an action to occur. Instead, the action waits on some external trigger, such as a pushbutton. However, a mouse double-click causes an action to happen immediately. Single-click and double-click events can be distinguished using the `SelectionType` property of the figure in which the clicks occurred. A single mouse click will place the string `'normal'` in the `SelectionType` property, while a double mouse click will place the string `'open'` in the `SelectionType` property.

It is also possible for a list box to allow multiple selections from the list. If the difference between the `max` and `min` properties of the list box is greater than one, then multiple selection is allowed. Otherwise, only one item may be selected from the list.

Figure 10.24*a* shows an example of a single-selection list box. The GUI in this figure creates a list box with eight options, labeled "Option 1," "Option 2," and so forth. In addition, it creates a pushbutton to perform selection and a text field to display the selected choice. Both the list box and the pushbutton generate callbacks.

The corresponding callback functions are shown in Figure 10.24*b*. If a selection is made in the list box, then function `Listbox1_Callback` will be executed. This function will check the *figure producing the callback* (using function `gcbf`) to determine whether the selecting action was a single-click or a double-click. If it were a single-click, the function does nothing. If it was a double-click, then the function gets the selected value from the listbox, and writes an appropriate string into the text field.

If the pushbutton is selected, then function `Button1_Callback` will be executed. This function gets the selected value from the listbox and writes an appropriate string into the text field.

In an end-of-chapter exercise, you will be asked to modify this example to allow multiple selections in the list box.

10.4.8   Sliders

Sliders are graphical objects that allow a user to select values from a continuous range between a specified minimum value and a specified maximum value by moving a bar with a mouse. The `'Value'` property of the slider is set to a value between `min` and `max` depending on the position of the slider.

A slider is created by a `uicontrol` whose style property is `'slider'`. A slider may be added to a GUI by using the slider tool (▬) in the Layout Editor.

(a)

```
function Button1_Callback(hObject, eventdata, handles)

% Find the value of the popup menu
value = get(handles.Listbox1,'Value');

% Update text label
str = ['Option ' num2str(value)];
set (handles.Label1,'String',str);



function Listbox1_Callback(hObject, eventdata, handles)

% If this was a double click, update the label.
selectiontype = get(gcbf,'SelectionType');
if selectiontype(1) == 'o'

   % Find the value of the popup menu
   value = get(handles.Listbox1,'Value');

   % Update text label
   str = ['Option ' num2str(value)];
   set (handles.Label1,'String',str);
end
```

(b)

Figure 10.24 *(a)* Layout of a simple GUI with a list box, a pushbutton, and a text field.
(b) The callback functions for this GUI.

Figure 10.26*a* shows the layout for a simple GUI containing a slider and a text field. The 'Min' property for this slider is set to zero, and the 'Max' property is set to one. When a user drags the slider, it automatically calls the function Slider1_Callback, which is shown in Figure 10.26*b*. This function gets the value of the slider from the 'Value' property and displays the value in the text field. Figure 10.27 shows this GUI with the slider at some intermediate position in its range.

Figure 10.25   The GUI produced by program `test_listbox`.

(a)

```
function Slider1_Callback(hObject, eventdata, handles)

% Find the value of the slider
value = get(handles.Slider1,'Value');

% Place the value in the text field
str = sprintf('%.2f',value);
set (handles.Label1,'String',str);
```

(b)

Figure 10.26   *(a)* Layout of a simple GUI with a slider and a text field. *(b)* The callback function for this GUI.

Figure 10.27   The GUI produced by program `test_slider`.

▶

Example 10.1—Temperature Conversion

Write a program that converts temperature from degrees Fahrenheit to degrees Celsius and vice versa over the range 0–100°C, using a GUI to accept data and display results. The program should include an edit box for the temperature in degrees Fahrenheit, an edit box for the temperature in degrees Celsius, and a slider to allow for the continuous adjustment of temperature. The user should be able to enter temperatures in either edit box or by moving the slider, and all GUI elements should adjust to the corresponding values.

SOLUTION   To create this program, we will need a text field and an edit box for the temperature in degrees Fahrenheit, another text field and an edit box for the temperature in degrees Celsius, and a slider. We will also need a function to convert degrees Fahrenheit to degrees Celsius, and a function to convert degrees Celsius to degrees Fahrenheit. Finally, we will need to write a callback function to support user inputs.

The range of values to convert will be 32–212°F or 0–100°C, so it will be convenient to set up the slider to cover the range 0–100, and to treat the value of the slider as a temperature in degrees C.

The first step in this process is to use guide to design the GUI. We can use guide to create the five required GUI elements and locate them in approximately the correct positions. Then we can use the Property Inspector to perform the following steps:

1. Select appropriate names for each GUI element and store them in the appropriate Tag properties. The names will be 'Label1', 'Label2', 'Edit1', 'Edit2', and 'Slider1'.
2. Store 'Degrees F' and 'Degrees C' in the String properties of the two labels.
3. Set the slider's minimum and maximum limits to 0 and 100, respectively.
4. Store initial values in the String property of the two edit fields and in the Value property of the slider. We will initialize the temperature to 32°F or 0°C, which corresponds to a slider value of 0.
5. Set the Name property of the figure containing the GUI to 'Temperature Conversion'.

Once these changes have been made, the GUI should be saved to file temp_conversion.fig. This will produce both a figure file and a matching M-file. The M-file will contain stubs for the three callback functions needed by the edit fields and the slider. The resulting GUI is shown during the layout process in Figure 10.28.

The next step in the process is to create the functions to convert degrees Fahrenheit to degrees Celsius. Function to_c will convert temperature from degrees Fahrenheit to degrees Celsius. It must implement the equation

$$\deg C = \frac{5}{9}(\deg F - 32) \tag{10-1}$$

Figure 10.28 Layout of the temperature conversion GUI.

The code for this function is

```
function deg_c = to_c(deg_f)

% Convert degrees Fahrenheit to degrees C.
deg_c = (5/9) * (deg_f - 32);
end % function deg_c
```

Function `to_f` will convert temperature from degrees Celsius to degrees Fahrenheit. It must implement the equation

$$\deg F = \frac{9}{5} \deg C + 32 \qquad (10\text{-}2)$$

The code for this function is

```
function deg_f = to_f(deg_c)

% Convert degrees Celsius to degrees Fahrenheit.
deg_f = (9/5) * deg_c + 32;
```

Finally, we must write the callback functions to tie it all together. The functions must respond either to the edit box or to the slider, and must update all three components. (Note that we will update even the edit box that the user types into, so that the data can be displayed with a consistent format at all times and to correct errors if the user types an out-of-range input value.)

There is an extra complication here, since the values entered into edit boxes are *strings*, and we wish to treat them as *numbers*. If a user types the value 100 into an edit box, he or she has really created the string `'100'`, not the number 100. The callback function must convert the strings into numbers so that the conversion can be calculated. This conversion is done with the `str2num` function, which converts a string into a numerical value.

In addition, the callback function will have to limit user entries to the valid temperature ranges, which are 0–100°C and 32–212°F.

The resulting callback functions are shown in Figure 10.29.

```
function Edit1_Callback(hObject, eventdata, handles)

% Update all temperature values
deg_f = str2num( get(hObject,'String') );
deg_f = max( [ 32 deg_f] );
deg_f = min( [212 deg_f] );
deg_c = to_c(deg_f);

% Now update the fields
set (handles.Edit1,'String',sprintf('%.1f',deg_f));
set (handles.Edit2,'String',sprintf('%.1f',deg_c));
set (handles.Slider1,'Value',deg_c);


function Edit2_Callback(hObject, eventdata, handles)

% Update all temperature values
deg_c = str2num( get(hObject,'String') );
deg_c = max( [  0 deg_c] );
deg_c = min( [100 deg_c] );
deg_f = to_f(deg_c);

% Now update the fields
set (handles.Edit1,'String',sprintf('%.1f',deg_f));
set (handles.Edit2,'String',sprintf('%.1f',deg_c));
set (handles.Slider1,'Value',deg_c);


function Slider1_Callback(hObject, eventdata, handles)

% Update all temperature values
deg_c = get(hObject,'Value');
deg_f = to_f(deg_c);

% Now update the fields
set (handles.Edit1,'String',sprintf('%.1f',deg_f));
set (handles.Edit2,'String',sprintf('%.1f',deg_c));
set (handles.Slider1,'Value',deg_c);
```

Figure 10.29 Callback functions for the temperature conversion GUI.

The program is now complete. Execute it and enter several different values using both the edit boxes and the sliders. Be sure to use some out-of-range values. Does it appear to be functioning properly?

◀

10.5   Additional Containers: Panels and Button Groups

MATLAB GUIs include two other types of containers: **panels** (created by the function `uipanel`) and **button groups** (created by the function `uibuttongroup`).

10.5.1   Panels

Panels are containers that can contain components or other containers, but they do *not* have a title bar and cannot have menus attached. A panel can contain GUI elements such as uicontrols, axes, other panels, or button groups. Any elements placed in a panel will be positioned relative to the panel. If the panel is moved on the GUI, then all of the elements within it are moved as well. Panels are a great way to group related controls on a GUI.

A panel is created by a `uipanel` function. It can be added to a GUI by using the panel tool (🔳) in the Layout Editor.

Each panel has a title and is usually surrounded by an etched or beveled line marking the edges of the panel. The title of a panel can be located at the left, center, or right side of either the top or bottom of the panel. Samples of panels with several combinations of title positions and edge styles are shown in Figure 10.30.

Figure 10.30   Examples of various panel styles.

Let's look at a simple example using panels. Suppose that we wanted to create a GUI to plot the function $y = ax^2 + bx + c$ between two specified values x_{min} and x_{max}. The GUI should allow the user to specify the values a, b, c, x_{min}, and x_{max}. In addition, it should allow the user to specify the style, color, and thickness of the line being plotted. These two sets of values (the ones specifying the line and the ones specifying what the line looks like) are logically distinct, so we can group them together in two panels on the GUI. One possible layout is shown in Figure 10.31. (You will be asked to finish this GUI and create an operational program in Exercise 10.7 at the end of the chapter.)

Table 10.4 contains a list of some important `uipanel` properties. These properties can be modified by the Property Inspector during the design phase, or they can be modified during execution with `get` and `set` functions.

10.5.2 Button Groups

Button groups are a special type of panel that can manage groups of radio buttons or toggle buttons to ensure that *no more than one button in the group is on at any time*. A button group is just like any other panel, except that the button group ensures that at most one radio button or toggle button is on at any given time. If one of them is turned on, then the button group turns off any buttons that were already on.

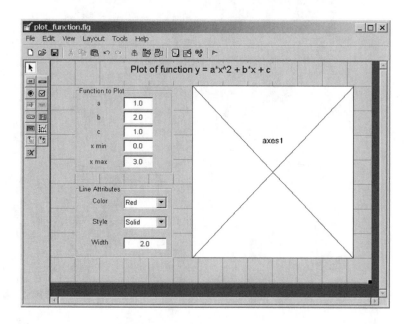

Figure 10.31 Layout of the Plot Function GUI, using panels to group related characteristics together.

Table 10.4   Important `uipanel` and `uibuttongroup` Properties

Property	Description
BackgroundColor	Specifies the color of the `uipanel` background. The value is either a predefined color such as `'r'`, `'g'`, or `'b'`, or else a three-element vector specifying the red, green, and blue components of the color on a 0–1 scale. For example, the color magenta would be specified by `[1  0  1]`.
BorderType	Type of border around the `uipanel`. Options are `'none'`, `'etchedin'`, `'etchedout'`, `'beveledin'`, `'beveledout'`, or `'line'`. The default border type is `'etchedin'`.
BorderWidth	Width of border around the `uipanel`.
FontAngle	A string containing the font angle for the title text. Possible values are `'normal'`, `'italic'`, and `'oblique'`.
FontName	A string containing the font name for the title text.
FontSize	A number specifying the font size for the title text.
FontUnits	The units in which the font size is defined. Possible choices are `'inches'`, `'centimeters'`, `'normalized'`, `'points'`, and `'pixels'`. The default font units are `'points'`.
FontWeight	A string containing the font weight for the title text. Possible values are `'light'`, `'normal'`, `'demi'`, and `'bold'`. The default font weight is `'normal'`.
ForegroundColor	Specifies the color of the title font and the border.
HighlightColor	Specifies the three-dimensional border highlight color.
Position	Specifies the position of a panel relative to its parent `figure`, `uipanel`, or `uibuttongroup`, in the units specified by the `'units'` property. This value accepts a four-element vector in which the first two elements are the *x* and *y* positions of the lower-left corner of the panel, and the next two elements are the width and height of the panel.
ShadowColor	Specifies the color of the three-dimensional border shadow.
Tag	The 'name' of the `uipanel`, which can be used to access it.
Title	The title string.
TitlePosition	Location of the title string on the `uipanel`. Possible values are `'lefttop'`, `'centertop'`, `'righttop'`, `'leftbottom'`, `'centerbottom'`, and `'rightbottom'`. The default value is `'lefttop'`.
Units	The units used to describe the position of the `uipanel`. Possible choices are `'inches'`, `'centimeters'`, `'normalized'`, `'points'`, `'pixels'`, or `'characters'`. The default units are `'normalized'`.
Visible	Specifies whether or not this `uipanel` is visible. Possible values are `'on'` or `'off'`.

Figure 10.32 A button group controlling three radio buttons.

A button group is created by a `uibuttongroup` function. It can be added to a GUI by using the button group tool ( ) in the Layout Editor.

If a radio button or a toggle button is controlled by a button group, then the user must attach the name of the function to execute when that button is selected in a special property called `SelectionChangeFcn`. This callback is executed by the GUI whenever a radio button or toggle button changes state. Do *not* place the function in the usual `Callback` property, since the button group overwrites the callback property for every radio button or toggle button that it controls.

Figure 10.32 shows a simple GUI containing a button group and three radio buttons, labeled `'Option 1'`, `'Option 2'`, and `'Option 3'`. When a user clicks on one radio button in the group, the button is turned on, and all other buttons in the group are turned off.

10.6 Dialog Boxes

A **dialog box** is a special type of figure that is used to display information or to get input from a user. Dialog boxes are used to display errors, provide warnings, ask questions, or get user input. They are also used to select files or printer properties.

Dialog boxes may be **modal** or **non-modal**. A modal dialog box does not allow any other window in the application to be accessed until it is dismissed, while a normal dialog box does not block access to other windows. Modal dialog boxes are typically used for warning and error messages that need urgent attention and cannot be ignored. By default, most dialog boxes are non-modal.

MATLAB includes many types of dialog boxes, the more important of which are summarized in Table 10.5. We will examine only a few types of dialog box here, but you can consult the MATLAB on-line documentation for the details of the others.

10.6.1 Error and Warning Dialog Boxes

Error and warning dialog boxes have similar calling parameters and behavior. In fact, the only difference between them is the icon displayed in the dialog box. The most common calling sequence for these dialog boxes is

Table 10.5   Selected Dialog Boxes

Property	Description
dialog	Create a generic dialog box.
errordlg	Displays an error message in a dialog box. The user must click the OK button to continue.
helpdlg	Displays a help message in a dialog box. The user must click the OK button to continue.
inputdlg	Displays a request for input data and accepts the user's input values.
listdlg	Allows a user to make one or more selections from a list.
printdlg	Displays a printer selection dialog box.
questdlg	Asks a question. This dialog box can contain either two or three buttons, which by default are labeled Yes, No, and Cancel.
uigetfile	Displays a file open dialog box. This box allows a user to select a file to open, *but does not actually open the file.*
uiputfile	Displays a file save dialog box. This box allows a user to select a file to save, *but does not actually save the file.*
uisetcolor	Displays a color selection dialog box.
uisetfont	Displays a font selection dialog box.
warndlg	Displays a warning message in a dialog box. The user must click the OK button to continue.

```
errordlg(error_string,box_title,create_mode);
warndlg(warning_string,box_title,create_mode);
```

The error_string or warning_string is the message to display to the user, and the box_title is the title of the dialog box. Finally, create_mode is a string that can be 'modal' or 'non-modal', depending on the type of dialog box you wish to create.

For example, the following statement creates a modal error message that cannot be ignored by the user. The dialog box produced by this statement is shown in Figure 10.33.

```
errordlg('Invalid input values!','Error Dialog Box','modal');
```

Figure 10.33   An error dialog box.

10.6.2 Input Dialog Boxes

Input dialog boxes prompt a user to enter one or more values that may be used by a program. They may be created with one of the following calling sequences:

```
answer = inputdlg(prompt)
answer = inputdlg(prompt,title)
answer = inputdlg(prompt,title,line_no)
answer = inputdlg(prompt,title,line_no,default_answer)
```

Here, `prompt` is a cell array of strings, with each element of the array corresponding to one value that the user will be asked to enter. The parameter `title` specifies the title of the dialog box, while `line_no` specifies the number of lines to be allowed for each answer. Finally, `default_answer` is a cell array containing the default answers that will be used if the user fails to enter data for a particular item. Note that there must be as many default answers as there are prompts.

When the user clicks the `OK` button on the dialog box, his or her answers will be returned as a cell array of strings in variable `answer`.

As an example of an input dialog box, suppose that we wanted to allow a user to specify the position of a figure using an input dialog. The code to perform this function would be

```
prompt{1} = 'Starting x position:';
prompt{2} = 'Starting y position:';
prompt{3} = 'Width:';
prompt{4} = 'Height:';
title = 'Set Figure Position';
default_ans = {'50','50','180','100'};
answer = inputdlg(prompt,title,1,default_ans);
```

The resulting dialog box is shown in Figure 10.34.

Figure 10.34 An input dialog box.

10.6.3 The `uigetfile`, `uisetfile`, and `uigetdir` Dialog Boxes

The `uigetfile` and `uisetfile` dialog boxes allow a user to interactively pick files to open or save. These functions use the standard file open or file save dialog boxes for the particular operating system that MATLAB is running on. They return strings containing the name and the path of the file but do not actually read or save it. The programmer is responsible for writing additional code for that purpose.

The form of these two dialog boxes is

```
[filename, pathname] = uigetfile(filter_spec,title);
[filename, pathname] = uisetfile(filter_spec,title);
```

Parameter `filter_spec` is a string specifying the type of files to display in the dialog box, such as `'*.m'`, `'*.mat'`, and so forth. Parameter `title` is a string specifying the title of the dialog box. After the dialog box executes, `filename` contains the name of the selected file and `pathname` contains the path of the file. If the user cancels the dialog box, `filename` and `pathname` are set to zero.

The following script file illustrates the use of these dialog boxes. It prompts the user to enter the name of a Mat-file, and then reads the contents of that file. The dialog box created by this code on a Windows 2000 Professional system is shown in Figure 10.35. (This is the standard open file dialog for Windows 2000 Professional. It will appear slightly different on Windows NT 4.0, Windows XP, or Unix/Linux systems).

```
[filename, pathname] = uigetfile('*.mat','Load MAT File');
if filename ~= 0
   load([pathname filename]);
end
```

Figure 10.35 A file open dialog box created by `uigetfile`.

The `uigetdir` dialog box allows a user to interactively select a directory. This function uses the standard directory selection dialog box for the particular operating system that MATLAB is running on. It returns the name of the directory but does not actually do anything with it. The programmer is responsible for writing additional code to use the directory name.

The form of this dialog box is

```
directoryname = uigetdir(start_path, title);
```

Parameter `start_path` is the path of the initially selected directory. If it is not valid, the dialog box opens with the base directory selected. Parameter `title` is a string specifying the title of the dialog box. After the dialog box executes, `directoryname` contains the name of the selected directory. If the user cancels the dialog box, `directoryname` is set to zero.

The following script file illustrates the use of this dialog box. It prompts the user to select a directory starting with the current MATLAB working directory. This dialog box created by this code on a Windows 2000 Professional system is shown in Figure 10.36. (The style of the box varies for different operating systems. It will appear slightly different on a Windows NT 4.0, Windows XP, or Unix/Linux systems).

```
dir1 = uigetdir('d:\matlab7\work','Select a directory');
if dir1 ~= 0
   cd(dir1);
end
```

10.6.4   The `uisetcolor` and `uisetfont` Dialog Boxes

The `uisetcolor` and `uisetfont` dialog boxes allow a user to interactively select colors or fonts using the standard dialog boxes for the computer on which

Figure 10.36   A file open dialog box created by `uigetdir`.

MATLAB is executing. The appearances of these boxes will vary for different operating systems. They provide a standard way to select colors or fonts within a MATLAB GUI.

Consult the MATLAB on-line documentation to learn more about these special-purpose dialog boxes. We will use them in some of the end-of-chapter exercises.

✳ Good Programming Practice

Use dialog boxes to provide information or request input in GUI-based programs. If the information is urgent and should not be ignored, make the dialog boxes modal.

10.7 Menus

Menus can also be added to MATLAB GUIs. A menu allows a user to select actions without additional components appearing on the GUI display. They are useful for selecting less commonly used options without cluttering up the GUI with a lot of extra buttons.

There are two types of menu in MATLAB: **standard menus**, which are pulled down from the menu bar at the top of a figure, and **context menus**, which pop up over the figure when a user right-clicks the mouse over a graphical object. We will learn how to create and use both types of menu in this section.

Standard menus are created with uimenu objects. Each item in a menu is a separate uimenu object, including items in submenus. These uimenu objects are similar to uicontrol objects, and they have many of the same properties such as Parent, Callback, Enable, and so forth. A list of the more important uimenu properties is given in Table 10.6.

Each menu item is attached to a parent object, which is a figure for the top-level menus, or another menu item for submenus. All of the uimenus connected to the same parent appear on the same menu, and the cascade of items forms a tree of submenus. Figure 10.37a shows a typical MATLAB menu in operation, and Figure 10.37b shows the relationship among the objects making up the menu.

MATLAB menus are created using the Menu Editor, which can be selected by clicking the (📝) icon on the toolbar in the guide Layout Editor. Figure 10.37c shows the Menu Editor with the menu items that generate this menu structure. The additional properties in Table 10.6 that are not shown in the Menu Editor can be set with the Property Editor (propedit).

Top-level context menus are created by uicontextmenu objects, and the lower-level items within context menus are created by uimenu objects. Context

Table 10.6 Important uimenu Properties

Property	Description
Accelerator	A single character specifying the keyboard equivalent for the menu item. The keyboard combination CTRL+key allows a user to activate the menu item from the keyboard.
Callback	Specifies the name and parameters of the function to be called when the menu item is activated. It the menu item has a submenu, the callback executes *before the submenu is displayed*. If the menu item does not have submenus, then the callback executes when the mouse button is *released*.
Checked	When this property is 'on', a checkmark is placed to the left of the menu item. This property can be used to indicate the status of menu items that toggle between two states. Possible values are 'on' or 'off'.
Enable	Specifies whether or not this menu item is selectable. If it not enabled, the menu item will not respond to mouse clicks or accelerator keys. Possible values are 'on' or 'off'.
ForegroundColor	Set color of text in the menu item.
Label	Specifies the text to be displayed on the menu. The ampersand character (&) can be used to specify a keyboard mnemonic for this menu item; it will not appear on the label. For example, the string '&File' will create a menu item displaying the text 'File' and responding to the F key.
Parent	The handle of the parent object for this menu item. The parent object could be a figure or another menu item.
Position	Specifies the position of a menu item on the menu bar or within a menu. Position 1 is the left-most menu position for a top-level menu, and the highest position within a submenu.
Separator	When this property is 'on', a separating line is drawn above this menu item. Possible values are 'on' or 'off'.
Tag	The "name" of the menu item, which can be used to access it.
Visible	Specifies whether or not this menu item is visible. Possible values are 'on' or 'off'.

menus are basically the same as standard menus, except that they can be associated with any GUI object (e.g., axes, lines, text, figures).

10.7.1 Suppressing the Default Menu

Every MATLAB figure comes with a default set of standard menus. If you wish to delete these menus from a figure and create your own menus, you must first turn the default menus off. The display of default menus is controlled by the figure's MenuBar property. The possible values of this property are 'figure'

Table 10.7 Important `uicontextmenu` Properties

Property	Description
Callback	Specifies the name and parameters of the function to be called when the context menu is activated. The callback executes before the context menu is displayed.
Parent	The handle of the parent object for this context menu.
Tag	The "name" of the context menu, which can be used to access it.
Visible	Specifies whether or not this context menu is visible. This property is set automatically and should normally not be modified.

and `'none'`. If the property is set to `'figure'`, then the default menus are displayed. If the property is set to `'none'`, then the default menus are suppressed. You can use the Property Inspector to set the `MenuBar` property for your GUIs when you create them.

10.7.2 Creating Your Own Menus

Creating your own standard menus for a GUI is basically a three-step process.

1. First, create a new menu structure with the Menu Editor. Use the Menu Editor to define the structure, giving each menu item a `Label` to display and a unique `Tag` value. You can also specify whether or not there is a separator bar between menu items, and whether or not each menu item has a check mark by it. A dummy callback function will be generated automatically for each menu item.

(a)

Figure 10.37 *(a)* A typical menu structure.

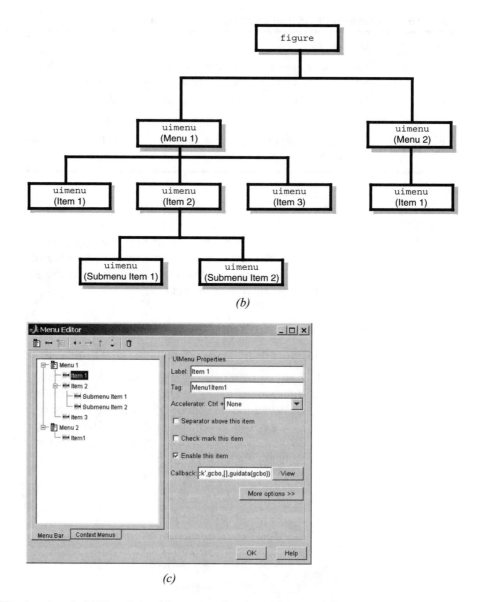

(b)

(c)

Figure 10.37 *(continued)* *(b)* The relationships among the uimenu items creating the menu. *(c)* The Menu Editor structure that generated these menus.

2. If necessary, edit the properties of each menu item using the Property Inspector. The Property Inspector can be started by clicking the "More Options" button on the Menu Editor. The most important menu item properties (Label, Tag, Callback, Checked, and Separator) can be set on the Menu Editor, so the Property Inspector is usually not needed. However, if you must set any of the other properties listed in Table 10.6, you will need to use the Property Inspector.

3. Third, implement a callback function to perform the actions required by your menu items. The prototype function is created automatically, but you must add the code to make each menu item behave properly.

The process of building menus is illustrated in an example at the end of this section.

⚫ Programming Pitfalls

Only the `Label`, `Tag`, `Callback`, `Checked`, and `Separator` properties of a menu item can be set from the Menu Editor. If you need to set any of the other properties, you will have to use the Property Inspector on the figure and select the appropriate menu item to edit.

10.7.3   Accelerator Keys and Keyboard Mnemonics

MATLAB menus support accelerator keys and keyboard mnemonics. **Accelerator keys** are "CTRL+key" combinations that cause a menu item to be executed *without opening the menu first*. For example, the accelerator key "o" might be assigned to the File/Open menu item. In that case, the keyboard combination CTRL+o will cause the File/Open callback function to be executed.

A few CRTL+key combinations are reserved for the use of the host operating system. These combinations differ between PC and Unix systems; consult the MATLAB on-line documentation to determine which combinations are legal for your type of computer.

Accelerator keys are defined by setting the `Accelerator` property in a `uimenu` object.

Keyboard mnemonics are single letters that can be pressed to cause a menu item to execute once the menu is open. The keyboard mnemonic letter for a given menu item is underlined[1]. For top-level menus, the keyboard mnemonic is

---

[1]On some versions of Windows, the underlines are hidden until the ALT key is held down. This behavior can be modified. For example, the underlines can be made visible all the time in Windows 2000 by going to the Display item in the Control Panel, selecting the Effects tab, and clearing the item "Hide keyboard navigation indicators until I use the Alt key."

Figure 10.38   An example showing keyboard mnemonics. The menu shown was opened by typing the keys ALT+f, and the Exit option could be executed by simply typing 'x'.

executed by pressing ALT plus the mnemonic key at the same time. Once the top-level menu is open, simply pressing the mnemonic key will cause a menu item to execute.

Figure 10.38 illustrates the use of keyboard mnemonics. The File menu is opened with the keys ALT+f, and once it is opened, the Exit menu item can be executed by simply typing "x."

Keyboard mnemonics are defined by placing the ampersand character (&) before the desired mnemonic letter in the `Label` property. The ampersand will not be displayed, but the following letter will be underlined and it will act as a mnemonic key. For example, the `Label` property of the Exit menu item in Figure 10.29 is `'E&xit'`.

10.7.4   Creating Context Menus

Context menus are created in the same fashion as ordinary menus, except that the top-level menu item is a `uicontextmenu`. The parent of a `uicontextmenu` must be a figure, but the context menu can be associated with and respond to right mouse clicks on any graphical object. Context menus are created using the "Context Menu" selection on the Menu Editor. Once the context menu is created, any number of menu items can be created under it.

To associate a context menu with a specific object, you must set the object's `UIContextMenu` property to the handle of the `uicontextmenu`. This is normally done using the Property Inspector, but it can be done with the `set` command, which follows. If `Hcm` is the handle to a context menu, the following statements will associate the context menu with a line created by a `plot` command:

```
H1 = plot(x,y);
set (H1,'UIContextMenu',Hcm);
```

We will create a context menu and associate it with a graphical object in the following example.

►

Example 10.2—Plotting Data Points

Write a program that opens a user-specified data file and plots the line specified by the points in the file. The program should include a File menu, with Open and Exit menu items. The program should also include a context menu attached to the line, with options to change the line style. Assume that the data in the file is in the form of (x, y) pairs, with one pair of data values per line.

SOLUTION This program should include a standard menu with Open and Exit menu items, plus a set of axes on which to plot the data. It should also include a context menu specifying various line styles, which can be attached to the line after it is plotted. The options should include solid, dashed, dotted, and dash-dot line styles.

The first step in creating this program is to use guide to create the required GUI, which is only a set of axes in this case (see Figure 10.39a). Then, we must use the Menu Editor to create the File menu. This menu will contain Open and Exit menu items, as shown in Figure 10.39b. Note that we must use the Menu Editor to set the Label and Tag and strings for each of these menu items. We will also define keyboard mnemonics " "F" for File, "O" for Open, and "x" for Exit, and place a separator between the Open and Exit menu items. Figure 10.39b shows the Exit menu item. Note that "x" is the keyboard mnemonic and that the separator switch is turned on.

Next, we must use the Menu Editor to create the context menu. This menu starts with a uicontextmenu object, with four menu items attached to it (see Figure 10.39c). Again, we must set the Label and Tag strings for each of these menu items.

At this point, the GUI should be saved as plot_line.fig, and plot_line.m will be automatically created. Dummy callback functions will be automatically created for the menu items.

After the GUI is created, we must implement six callback functions for the Open, Exit, and linestyle menu items. The most difficult callback function is the response to the File/Open menu item. This callback must prompt the user for the name of the file (using a uigetfile dialog box) and to open the file, read the data, save it into x and y arrays, and close the file. Then, it must plot the line and save the line's handle as application data so that we use it to modify the line style later. Finally, it must associate the context menu with the line. The FileOpen_Callback function is shown in Figure 10.40. Note that the function uses a dialog box to inform the user of file open errors.

Figure 10.39 *(a)* The layout for plot_line. *(b)* The File menu in the Menu Editor.

(c)

Figure 10.39 *(continued) (c)* The context menu in the Menu Editor.

```
function varargout = FileOpen_Callback(h, eventdata, ...
                    handles, varargin)

% Get the file to open
[filename, pathname] = uigetfile('*.dat','Load Data');
if filename ~= 0

   % Open the input file
   filename = [pathname filename];
   [fid,msg] = fopen(filename,'rt');

   % Check to see of the open failed.
   if fid < 0

      % There was an error--tell user.
      str = ['File ' filename ' could not be opened.'];
      title = 'File Open Failed';
      errordlg(str,title,'modal');

   else
```

Get file
name to open

Open file

Error message
if open fails

Figure 10.40 The File/Open callback function.

```
% File opened successfully. Read the (x,y) pairs from
% the input file. Get first (x,y) pair before the
% loop starts.
[in,count] = fscanf(fid,'%g',2);                          Read data
ii = 0;

while ~feof(fid)
  ii = ii + 1;
   x(ii) = in(1);
   y(ii) = in(2);

  % Get next (x,y) pair
  [in,count] = fscanf(fid,'%g',2);
end

% Data read in. Close file.
fclose(fid);                                              Plot line

% Now plot the data.
 hline = plot(x,y,'LineWidth',3);
 xlabel('x');
 ylabel('y');
grid on;

% Associate the context menu with line                    Set context
set(hline,'Uicontextmenu',handles.ContextMenu1);          menu

% Save the line's handle as application data
handles.hline = hline;                                    Save handle
guidata(gcbf, handles);                                   as app data

end
```

Figure 10.40   *(continued)*

The remaining callback functions are very simple. The `FileExit_ Callback` function simply closes the figure, and the line style functions simply set the line style. When the user right-clicks a mouse button over the line, the context menu will appear. If the user selects an item from the menu, the resulting callback will use the line's saved handle to change its properties. These five functions are shown in Figure 10.41.

The output of the final program is shown in Figure 10.42. Experiment with it on your own computer to verify that it behaves properly.

```
function varargout = FileExit_Callback(h, eventdata, ...
                         handles, varargin)
close(gcbf);


function varargout = LineSolid_Callback(h, eventdata, ...
                         handles, varargin)
set(handles.hline,'LineStyle','-');function varargout =
LineDashed_Callback(h, eventdata, ...
                         handles, varargin)
set(handles.hline,'LineStyle','--');


function varargout = LineDotted_Callback(h, eventdata, ...
                         handles, varargin)
set(handles.hline,'LineStyle',':');


function varargout = LineDashDot_Callback(h, eventdata, ...
                         handles, varargin)
set(handles.hline,'LineStyle','-.');
```

Figure 10.41 The remaining callback functions in plot_line.

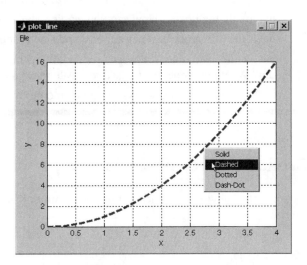

Figure 10.42 The GUI produced by program plot_line.

Quiz 10.1

This quiz provides a quick check to see if you have understood the concepts introduced in Sections 10.1 through 10.7. If you have trouble with the quiz, reread the section, ask your instructor, or discuss the material with a fellow student. The answers to this quiz are found in the back of the book.

1. List the types of graphical components discussed in this chapter. What is the purpose of each one?

2. List the types of containers discussed in this chapter. What are the differences among them?

3. What is a callback function? How are callback functions used in MATLAB GUIs?

4. Describe the steps required to create a GUI-based program.

5. Describe the purpose of the handles data structure.

6. How is application data saved in a MATLAB GUI? Why would you want to save application data in a GUI?

7. How can you make a graphical object invisible? How can you turn a graphical object off so that it will not respond to mouse clicks or keyboard input?

8. Which of the GUI components described in this chapter respond to mouse clicks? Which ones respond to keyboard inputs?

9. What are dialog boxes? How can you create a dialog box?

10. What is the difference between a modal and a non-modal dialog box?

11. What is the difference between a standard menu and a context menu? What components are used to create these menus?

12. What are accelerator keys? What are mnemonics?

10.8   Tips for Creating Efficient GUIs

This section lists a few miscellaneous tips for creating efficient graphical user interfaces.

10.8.1   Tool Tips

MATLAB GUIs support **tool tips**, which are small help windows that pop up beside a `uicontrol` GUI object whenever the mouse is held over the object for a while. Tool tips are used to provide a user with quick help about the purpose of each object on a GUI.

A tool tip is defined by setting an object's `TooltipString` property to the string that you wish to display. You will be asked to create tool tips in the end-of-chapter exercises.

✳ **Good Programming Practice**

Define tool tips to provide users with helpful hints about the functions of your GUI components.

10.8.2 Pcode

The first time that MATLAB executes a function during a program's execution, it compiles (or parses) the function into an intermediate code called **pcode** (which is short for pseudocode), and then executes the pcode in its run-time interpreter. Once a function has been compiled, it remains in MATLAB's memory, and it can be executed over and over again without having to be recompiled. However, the next time MATLAB executes, the function will have to be compiled again.

The penalty associated with this initial compilation is relatively small, but it becomes more and more significant as function sizes get larger. Since the functions that define a GUI are typically quite large, the compilation penalty for GUI-based programs is relatively larger than for other types of programs. In other words, the GUIs run more slowly because of this initial compilation time.

Fortunately, there is a way to avoid this extra penalty. It is possible to pre-compile your MATLAB functions and script files into pcode and save the pcode in files for immediate execution in the future. Executing the pcode files saves the initial compilation time and so makes your programs faster.

MATLAB creates pcode files with the command `pcode`. This command takes one of the following forms:

```
pcode fun1.m fun2.m fun3.m ...
pcode *.m
```

The first form compiles the named files, whereas the second form compiles every M-file in the current directory. The compiled output is saved with the file extent "p." For example, if you compile the file `foo.m`, the compiled output will be stored in file `foo.p`.

If a function exists as both an M-file and a p-file, MATLAB will automatically execute the p-file version since it will be faster. However, if you modify the M-file, you *must* remember to manually recompile it, or the program will continue to execute your old code!

Compiling files to pcode can have an additional advantage as well. You can protect your investment in your source code by distributing the program to others in the form of pcode files. They may be freely executed, but it is not easy for someone to reengineer the files and take your ideas.

✳ Good Programming Practice

Once a program is working properly, you may use the `pcode` command to pre-compile its M-files and increase the program's speed.

✸ Programming Pitfalls

If you change an M-file that has been compiled into pcode, always remember to recompile the file. Otherwise, you will continue to execute your old, unmodified code!

10.8.3   Toolbars

MATLAB GUIs can also support *toolbars*. A toolbar is a row of special pushbuttons or toggle buttons along the top of a figure, just below the menu bar. Each button has a small figure or icon on it, representing its function. We have seen examples of toolbars in most of the MATLAB figures produced in this book. For example, Figure 10.43 shows a simple plot displaying the default toolbar.

Every figure has a `ToolBar` property, which determines whether or not the default figure toolbar is displayed. The possible values of this property are `'none'`, `'auto'`, and `'figure'`. If the property is `'none'`, the default

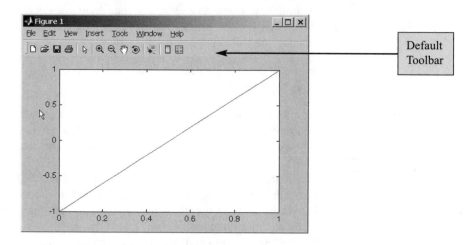

Figure 10.43   A MATLAB figure showing the default toolbar.

toolbar is not displayed. If the property is `'figure'`, the default toolbar is displayed. If the property is `'auto'`, the default toolbar is displayed unless the user defines a custom toolbar. If the property is `'auto'` and the user has defined a custom toolbar, then it will be displayed instead of the default toolbar.

A programmer can create his or her own toolbar using the `uitoolbar` function, and can add the toolbar equivalent of pushbuttons and toggle buttons to the toolbar using the `uipushtool` and `uitoggletool` functions. The user-defined toolbar can be displayed in addition to or instead of the default figure toolbar.

Unfortunately, toolbars are not yet integrated into `guide`. If you want to use toolbars in your GUIs, you must create them manually. More details about these functions can be found in the MATLAB on-line help documentation.

10.8.4 Additional Enhancements

GUI-based programs can be much more sophisticated than we have described in this introductory chapter. In addition to the `Callback` property that we have been using in the chapter, `uicontrols` support four other types of callbacks: `CreateFcn`, `DeleteFcn`, `ButtonDownFcn`, and `KeyPressFcn`. MATLAB figures also support three important types of callbacks: `WindowButtonDownFcn`, `WindowButtonMotionFcn`, and `WindowButtonUpFcn`.

The **CreateFcn** property defines a callback that is automatically called whenever an object is created. It allows a programmer to customize his or her objects as they are created during program execution. Since this callback is executed before the object is completely defined, a programmer must specify the function to execute as a default property of the root before the object is created. For example, the following statement will cause the function `function_name` to be executed each time a `uicontrol` is created. The function will be called after MATLAB creates the object's properties, so they will be available to the function when it executes.

```
set(0,'DefaultUicontrolCreateFcn','function_name')
```

The **DeleteFcn** property defines a callback that is automatically called whenever an object is destroyed. It is executed before the object's properties are destroyed, so they will be available to the function when it executes. This callback provides the programmer with an opportunity to do custom clean-up work.

The **ButtonDownFcn** property defines a callback that is automatically called whenever a mouse button is pressed within a five-pixel border around a `uicontrol`. If the mouse button is pressed on the `uicontrol`, the `Callback` is executed. Otherwise, if it is near the border, the `ButtonDownFcn` is executed. If the `uicontrol` is not enabled, the `ButtonDownFcn` is executed even for clicks on the control.

The **KeyPressFcn** property defines a callback that is automatically called whenever a key is pressed while the specified object is selected or highlighted.

This function can find out which key was pressed by checking the `CurrentCharacter` property of the enclosing figure, or else by checking the contents of the event data structure passed to the callback. It can use this information to change behavior, depending on which key was pressed.

The figure-level callback functions **WindowButtonDownFcn, WindowButtonMotionFcn,** and **WindowButtonUpFcn** allow a programmer to implement features such as animations and drag-and-drop, since the callbacks can detect the initial, intermediate, and final locations at which the mouse button is pressed. They are beyond the scope of this book, but are well worth learning about. Refer to the *Creating Graphical User Interfaces* section in the MATLAB user documentation for a description of these callbacks.

▶

Example 10.3—Creating a Histogram GUI

Write a program that opens a user-specified data file and calculates a histogram of the data in the file. The program should calculate the mean, median, and standard deviation of the data in the file. It should include a File menu, with Open and Exit menu items. It should also include a means to allow the user to change the number of bins in the histogram.

Select a color other than the default color for the figure and the text label backgrounds, use keyboard mnemonics for menu items, and add tool tips where appropriate.

SOLUTION   This program should include a standard menu with Open and Exit menu items, a set of axes on which to plot the histogram, and a set of six text fields for the mean, median, and standard deviation of the data. Three of these text fields will hold labels, and three will hold the read-only mean, median and standard deviation values. It must also include a label and an edit field to allow the user to selected the number of bins to display in the histogram.

We will select a light blue color [0.6 1.0 1.0] for the background of this GUI. To make the GUI have a light-blue background, this color vector must be loaded into the `'Color'` property of the figure and into the `'BackgroundColor'` property of each text label with the Property Inspector during the GUI layout. Since MATLAB GUIs automatically reset their figure color to match the system default, we will also need to set this figure color in the `histGUI_OpeningFcn` callback, which is called just before the figure becomes visible.

The first step in creating this program is to use `guide` to lay out the required GUI (see Figure 10.44*a*). Then, use the Property Inspector to set the properties of the seven text fields and the edit field. The fields must be given unique tags so that we can locate them from the callback functions. Next, use the Menu Editor to create the File menu (see Figure 10.44*b*). Finally, the resulting GUI should be saved as `histGUI`, creating `histGUI.fig` and `histGUI.m`.

(a)

(b)

Figure 10.44   *(a)* The layout for `histGUI`. *(b)* The File menu in the Menu Editor.

After `histGUI.m` has been saved, the function `histGUI_OpeningFcn` must be edited to initialize the background color of the figure and to save the initial number of histogram bins in the `handles` structure. The modified code for the opening function is

```
% --- Executes just before histGUI is made visible.
function histGUI_OpeningFcn(hObject, eventdata, handles, varargin)
% This function has no output args, see OutputFcn.
% hObject     handle to figure
% eventdata   reserved - to be defined in a future version of MATLAB
% handles     structure with handles and user data (see GUIDATA)
% varargin    command line arguments to histGUI (see VARARGIN)

% Choose default command line output for histGUI
handles.output = hObject;
```

```
% Set the figure background color
set(hObject,'Color',[0.6 1 1]);
```

```
% Set the initial number of bins
handles.nbins = 11;
```

```
% Update handles structure
guidata(hObject, handles);
```

Next, we must create callback functions for the File/Open menu item, the File/Exit menu item, and the "number of bins" edit box.

The File/Open callback must prompt the user for a file name and then read the data from the file. It must calculate and display the histogram and update the statistics text fields. Note that the data in the file must also be saved in the `handles` structure, so that it will be available for recalculation if the user changes the number of bins in the histogram. The callback function to perform these steps is shown here:

```
function Open_Callback(hObject, eventdata, handles)
% hObject     handle to Open (see GCBO)
% eventdata   reserved - to be defined in a future version of MATLAB
% handles     structure with handles and user data (see GUIDATA)
```

```
% Get file name
[filename,path] = uigetfile('*.dat','Load Data File');
if filename ~= 0
```

```
   % Read data
   x = textread([path filename],'%f');
```

```
   % Save in handles structure
   handles.x = x;
   guidata(gcbf, handles);
```

```
   % Create histogram
   hist(handles.x,handles.nbins);
```

```
   % Set axis labels
   xlabel('\bfValue');
   ylabel('\bfCount');
```

```
   % Calculate statistics
   ave = mean(x);
```

```
med = median(x);
sd  = std(x);
n   = length(x);

% Update fields
set (handles.MeanData,'String',sprintf('%7.2f',ave));
set (handles.MedianData,'String',sprintf('%7.2f',med));
set (handles.StdDevData,'String',sprintf('%7.2f',sd));
set (handles.TitleString,'String',['Histogram (N = 'int2str(n)')']);
end
```

The File/Exit callback is trivial. All it has to do is close the figure.

```
function Exit_Callback(hObject, eventdata, handles)
% hObject      handle to Exit (see GCBO)
% eventdata    reserved - to be defined in a future version of MATLAB
% handles      structure with handles and user data (see GUIDATA)

close(gcbf);
```

The NBins callback must read a numeric input value, round it off to the nearest integer, display that integer in the Edit Box, and recalculate and display the histogram. Note that the number of bins must also be saved in the handles structure, so that it will be available for recalculation if the user loads a new data file. The callback function to perform these steps is shown here:

```
function NBins_Callback(hObject, eventdata, handles)
% hObject      handle to NBins (see GCBO)
% eventdata    reserved - to be defined in a future version of MATLAB
% handles      structure with handles and user data (see GUIDATA)

% Get number of bins, round to integer, and update field
nbins = str2num(get(hObject,'String'));
nbins = round(nbins);
if nbins < 1
   nbins = 1;
end
set (handles.NBins,'String',int2str(nbins));

% Save in handles structure
handles.nbins = nbins;
guidata(gcbf, handles);

% Re-display data, if available
if handles.nbins > 0 & ~isempty(handles.x)

   % Create histogram
   hist(handles.x,handles.nbins);

end
```

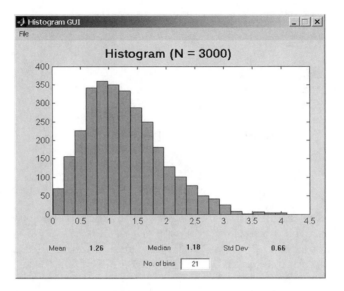

Figure 10.45   The GUI produced by program histGUI.

The final program is shown in Figure 10.45. Experiment with it on your own computer to verify that it behaves properly.   ◀

10.9   Summary

In Chapter 10, we learned how to create MATLAB graphical user interfaces. The three fundamental parts of a GUI are components (uicontrols, uimenus, uicontextmenus, toolbars, and axes), containers to contain them, and callbacks to implement actions in response to mouse clicks or keyboard inputs.

The standard GUI components created by uicontrol include text fields, edit boxes, pushbuttons, toggle buttons, checkboxes, radio buttons, popup menus, list boxes, and sliders. The standard GUI components created by uimenu and uicontextmenu are standard menus and context menus.

MATLAB containers consist of figures, panels, and button groups. Figures are created by the figure function. They are separate windows, complete with title bars, menus, and toolbars. Panels are created by the uipanel function. They are containers that reside within figures or other containers and do not have title bars, menus, or toolbars. Panels can contain uicontrol components and other panels or button groups, and those items will be laid out with respect to the panel itself. If the panel is moved, all of its contents move with it. Button groups are created by the `uibuttongroup` function. They are special types of panels that control any radio buttons or toggle buttons contained within them to ensure that at most one of them can be on at any time.

Any of these components and containers can be placed on a figure using `guide` (the GUI Development Environment tool). Once the GUI layout has been completed, the user must edit the object properties with the Property Inspector, and then write a callback function to implement the actions associated with each GUI object.

Dialog boxes are special figures used to display information or to get input from a user. Dialog boxes are used to display errors, provide warnings, ask questions, or get user input. They are also used to select files or printer properties.

Dialog boxes may be modal or non-modal. A modal dialog box does not allow any other window in the application to be accessed until it is dismissed, while a normal dialog box does not block access to other windows. Modal dialog boxes are typically used for warning and error messages that need urgent attention and cannot be ignored.

Menus can also be added to MATLAB GUIs. A menu allows a user to select actions without additional components appearing on the GUI display. They are useful for selecting less commonly used options without cluttering up the GUI with a lot of extra buttons. Menus are created with the Menu Editor, and then the programmer must write a callback function to implement the actions associated with each menu item. For each menu item, the user must use the Menu Editor to set at least the `Label` and `Tag` properties.

Accelerator keys and keyboard mnemonics can be used to speed the operation of windows.

Compiling MATLAB functions to pcode can speed the execution of a program. It also protects your investment in your source code by allowing you to distribute the program to others in the form of pcode files. They may be freely executed, but it is not easy for someone to reengineer the files and take your ideas.

MATLAB uicontrol components have several additional properties for specifying less common types of callbacks, including `CreateFcn`, `DeleteFcn`, `ButtonDownFcn`, and `KeyPressFcn`. MATLAB figures also have several properties for specifying types of callbacks, including `WindowButtonDownFcn`, `WindowButtonMotionFcn`, and `WindowButtonUpFcn`. These various callbacks allow a user to customize the appearance and response of these MATLAB GUIs to various user inputs.

10.9.1 Summary of Good Programming Practice

The following guidelines should be adhered to when working with MATLAB GUIs.

1. Use `guide` to lay out a new GUI, and use the Property Inspector to set the initial properties of each component, such as the text displayed on the

    component, the color of the component, and the name of the callback function, if required.

2. After creating a GUI with `guide`, manually edit the resulting function to add comments describing its purpose and components, and to implement the function of callbacks.

3. Store GUI application data in the `handles` structure so that it will automatically be available to any callback function.

4. If you modify any of the GUI application data in the `handles` structure, be sure to save the structure with a call to `guidata` before exiting the function where the modifications occurred.

5. Use dialog boxes to provide information or request input in GUI-based programs. If the information is urgent and should not be ignored, make the dialog boxes modal.

6. Define tool tips to provide users with helpful hints about the functions of your GUI components.

7. One a program is working properly, you may use the `pcode` command to precompile its M-files and increase the program's speed.

10.9.2   MATLAB Summary

The following summary lists all of the MATLAB commands and functions described in this chapter, along with a brief description of each one. Also, refer to the summaries of graphical object properties in Tables 10.2, 10.3, 10.4, 10.6, and 10.7.

`axes`	Function to create a set of axes.
`dialog`	Create a generic dialog box.
`errordlg`	Display an error message.
`helpdlg`	Display a help message.
`findobj`	Find a GUI object by matching one or more of its properties.
`gcbf`	Get callback figure.
`gcbo`	Get callback object.
`guidata`	Save GUI application data in a figure.
`guihandles`	Get the `handles` structure from the application data stored in a figure.
`guide`	GUI Development Environment tool.
`inputdlg`	Dialog to get input data from the user.
`printdlg`	Print dialog box.
`questdlg`	Dialog box to ask a question.
`uibuttongroup`	Create a button group container.

uicontrol	Function to create a GUI object.
uicontextmenu	Function to create a context menu.
uigetdir	Dialog box to select a directory.
uigetfile	Dialog box to select an input file.
uimenu	Function to create a standard menu, or a menu item on either a standard menu or a context menu.
uipanel	Create a panel.
uipushtool	Create a pushbutton on a user-defined toolbar.
uiputfile	Dialog box to select an output file.
uisetcolor	Displays a color selection dialog box.
uisetfont	Displays a font selection dialog box.
uitoggletool	Create a toggle button on a user-defined toolbar.
uitoolbar	Create a user-defined toolbar.
warndlg	Displays a warning message.

10.10 Exercises

10.1 Explain the steps required to create a GUI in MATLAB.

10.2 What types of components can be used in MATLAB GUIs? What functions create them, and how do you select a particular component type?

10.3 What types of containers can be used in MATLAB GUIs? What function creates each of them?

10.4 How does a callback function work? How can a callback function locate the figures and objects that it needs to manipulate?

10.5 Create a GUI that uses a standard menu to select the background color displayed by the GUI. Include accelerator keys and keyboard mnemonics in the menu design. Design the GUI so that it defaults to a green background. (Note: You will need to use the figure's OpeningFcn to set the background color, since MATLAB defaults the figure color to the system default color.)

10.6 Create a GUI that uses a context menu to select the background color displayed by the GUI. Design the GUI so that it defaults to a yellow background.

10.7 Write a GUI program that plots the equation $y(x) = ax^2 + bx + c$. The program should include a set of axes for the plot and should include a panel containing GUI elements to input the values of a, b, c, and the minimum and maximum x to plot. A separate panel should contain controls to set the style, color, and thickness of the line being plotted. Include tool tips for each of your GUI elements.

10.8   Modify the GUI of Exercise 10.7 to include a menu. The menu should include two submenus to select the color and line style of the plotted line, with a check mark beside the currently selected menu choices. The menu should also include an "Exit" option. If the user selects this option, the program should create a modal question dialog box asking "Are You Sure?", with the appropriate responses. Include accelerator keys and keyboard mnemonics in the menu design. (Note that the menu items duplicate some GUI elements; so if a menu item is selected, the corresponding GUI elements must be updated too, and vice versa.)

10.9   Modify the List Box example in Section 10.4.7 to allow for multiple selections in the list box. The text field should be expanded to multiple lines, so that it can display a list of all selections whenever the "Select" button is clicked.

10.10   **Random Number Distributions**   Create a GUI to display the distributions of different types of random numbers. The program should create the distributions by generating an array of 1,000,000 random values from a distribution and using function `hist` to create a histogram. Be sure to label the title and axes of the histogram properly.

The program should support uniform, Gaussian, and Rayleigh distributions, with the distribution selection made by a popup menu. In addition, it should have an edit box to allow the user to select the number of bins in the histogram. Make sure that the values entered in the edit box are legal (the number of bins must be a positive integer).

10.11   Modify the temperature conversion GUI of Example 10.1 to add a "thermometer". The thermometer should be set of rectangular axes with a red "fluid" level corresponding to the current temperature in degrees Celsius. The range of the thermometer should be 0°–100°C.

10.12   Modify the temperature conversion GUI of Exercise 10.11 to allow you to adjust the displayed temperature by clicking the mouse. (**Warning:** This exercise requires material not discussed in this chapter. Refer to the `CurrentPoint` property of `axes` objects in the on-line MATLAB documentation.)

10.13   Create a GUI that contains a title, and four pushbuttons grouped within a panel. The pushbuttons should be labeled "Title Color," "Figure Color," "Panel Color," and "Title Font." If the "Title Color" button is selected, open a `uisetcolor` dialog box, and change the title text to be in the selected color. If the "Figure Color" button is selected, open a `uisetcolor` dialog box, and change the figure color and the title text background color to be the selected color. If the "Panel Color" button is selected, open a `uisetcolor` dialog box and change the panel background to be in the selected color. If the "Title Font" button is selected, open a `uisetfont` dialog box and change the title text to be in the selected font.

10.14   Create a GUI that contains a title and a button group. The button group will be titled "Style," and it should contain four radio buttons labeled "Plain," "Italic," "Bold," and "Bold Italic". Design the GUI so that the style in the currently selected radio button is applied to the title text.

10.15   **Least-Squares Fit** Create a GUI that can read an input data set from a file and perform a least-squares fit to the data. The data will be stored in a disk file in (x, y) format, with one x and one y value per line. Perform the least-squares fit with the MATLAB function `polyfit`, and plot both the original data and the least-squares fitted line. Include two menus: File and Edit. The File menu should include File/Open and File/Exit menu items, and the user should receive an "Are You Sure?" prompt before exiting. The Edit menu item should allow the user to customize the display, including line style, line color, and grid status.

10.16   Modify the GUI of the previous exercise to include an Edit/Preferences menu item that allows the user to suppress the "Are You Sure?" exit prompt.

10.17   Modify the GUI of the previous exercise to read and write an initialization file. The file should contain the line style, line color, grid choice (on/off), and exit prompt choice made by the user on previous runs. These choices should be automatically written out and saved when the program exits via the File/Exit menu item, and they should be read in and used whenever the program is started again.

CHAPTER 11

The MATLAB Compiler

The MATLAB Compiler is a tool that can convert your MATLAB programs into stand-alone applications that can be distributed to users who do not have MATLAB installed on their computers. It can also be used to convert collections of MATLAB functions into shared libraries that can be called from C or C++ programs.

The programs produced by the MATLAB compiler are similar to Java programs. Java programs need a Java Virtual Machine to be installed on the host computer before they can be executed. Similarly, compiled MATLAB programs need a MATLAB Component Runtime (MCR) to be installed on the host computer before they can be executed. The MCR is similar in function to a Java Virtual Machine; it is a free special-purpose version of MATLAB that is used to execute compiled MATLAB programs.

When a MATLAB program is compiled, the MATLAB compiler creates two output files. One of the files is an executable file, and the other one is a Component Technology File (CTF), which contains compressed and encrypted versions of the M-files and related data files that are used by the application. When the MATLAB program is started for the first time on a new computer, it unpacks the contents of the CTF into a subdirectory and then calls the MATLAB runtime to execute the M-files in exactly the same manner as the original program would have run.

The MATLAB compiler provides a number of important advantages for programmers and users:

1. Compiled programs can be executed by any number of users on their own computers without requiring MATLAB licenses.
2. The algorithms inside the programs are hidden, so a programmer can protect his or her intellectual property while allowing the program to be used by various users.

3. Programmers can take advantage of the many built-in MATLAB functions to make their programming tasks easier.

The MATLAB compiler also has a number of important limitations that a programmer must be aware of:

1. The starting M-file of any compiled program must be a *function*, not a script file. If an existing program starts with a script file, then that file must be converted into a function before attempting to compile it.
2. Not all built-in MATLAB functions and tool boxes are available in compiled MATLAB programs. The missing functions are mostly related to the MATLAB Command Window, the help subsystem, or third-party licensed packages.
3. The MATLAB Component Runtime (MCR) must be installed before a compiled program can be executed.

11.1 Setting Up the MATLAB Compiler

When the MATLAB compiler is installed on a computer used to develop applications, it must be set up before it can be used. This is accomplished by typing the command "mbuild -setup" in the MATLAB Command Window. The mbuild function will search for available C and C++ compilers on your computer and ask you which one to use when compiling MATLAB programs.

The example that follows is taken from the computer used to write this textbook. This computer has four compilers on it: Microsoft Visual C++ 6.0, Microsoft Visual C++ .NET 2003, Microsoft Visual C++ 2005, and the LCC C compiler. Note that the LCC C compiler is distributed with the MATLAB compiler, so any computer that has the MATLAB compiler installed will at least have that choice available. In this example, the MATLAB compiler has been configured to use the LCC compiler.

```
» mbuild -setup
Please choose your compiler for building standalone MATLAB
applications:

Would you like mbuild to locate installed compilers [y]/n? y
Select a compiler:
[1] Lcc-win32 C 2.4.1 in C:\PROGRA~1\MATLAB\R2007a\sys\lcc
[2] Microsoft Visual C++ 2005 in C:\PROGRA~1\Microsoft Visual Studio 8
[3] Microsoft Visual C++ .NET 2003 in C:\PROGRA~1\Microsoft Visual
Studio .NET 2003
[4] Microsoft Visual C++ 6.0 in C:\ PROGRA~1\Microsoft Visual Studio

[0] None

Compiler:
Please verify your choices: 1
```

```
Compiler: Lcc-win32 C 2.4.1
Location: C:\PROGRA~1\MATLAB\R2007a\sys\lcc

Are these correct?([y]/n): y
Trying to update options file: C:\Documents and Settings\
schapman\Application Data\MathWorks\MATLAB\R2007a\compopts.bat
From template:
C:\PROGRA~1\MATLAB\R2007a\bin\win32\mbuildopts\lcccompp.bat

Done...
```

✳ Good Programming Practice

Always set up the MATLAB compiler using the `mbuild -setup` command before trying to compile programs.

11.2   Setting Up Computers that Run Compiled Applications

Before a computer is used to execute compiled MATLAB applications, the MATLAB Component Runtime (MCR) must be installed. On Windows-based computers, the MCR is installed using file `MCRInstaller.exe`. The application creator needs to distribute this file to all users along with the application itself.

✳ Good Programming Practice

The MATLAB Component Runtime (MCR) must be installed before a computer can be used to run a compiled MATLAB application. Always distribute the runtime together with your compiled applications.

11.3   Using the MATLAB Compiler

We will now learn how to use the MATLAB compiler to create a stand-alone application program. The steps involved in creating and running compiled MATLAB programs are as follows:

1. Copy all of the M-files and supporting data files for the program into a single directory.
2. If the starting point for the program is a script file, convert it into a function by adding a function statement to the top of the file.

3. Compile the program using the `mcc` command. This will produce an executable file to start the program and a CTF file containing compressed and encrypted copies of the MATLAB M-files and supporting data.
4. Copy the executable file and CTF file to the computer where the program is to be executed.
5. Install the MATLAB Component Runtime if it is not already present on that computer.
6. Run the program.

11.3.1 A Simple Example

To illustrate these steps, we will reexamine the selection sort function `ssort` and the driver program `test_ssort` developed in Chapter 5. These M-files are reproduced in Figures 11.1 and 11.2.

The first step in building a compiled version of this sort program is to create a directory called `test_ssort`, and then to copy the `test_ssort.m` and `ssort.m` files from Chapter 5 into that directory. After copying, the contents of the directory are as shown here:

```
D:\work\test_ssort>dir
 Volume in drive D is DATA
 Volume Serial Number is E46A-233F

 Directory of D:\work\test_ssort

26/05/2007  03:09 PM    <DIR>          .
26/05/2007  03:09 PM    <DIR>          ..
19/05/2007  04:57 PM             1,293 ssort.m
19/05/2007  04:58 PM             1,132 test_ssort.m
               2 File(s)          2,425 bytes
               2 Dir(s)  18,683,101,184 bytes free
```

The next step is to convert the script file `test_ssort.m` into a function by adding the line "`function test_ssort()`" at the top of the file.

Next, the program should be compiled using the `mcc` command. This command can be executed either in the MATLAB Command Window or at an ordinary command prompt outside of MATLAB. The syntax required to compile this program is

```
mcc -m test_ssort.m
```

The −m switch means that we are creating a stand-alone C application which will be compiled with the LCC compiler, and `test_ssort.m` is the name of the function that will be executed first in the program. Note that we do *not* have to include the file `ssort.m` in the command, because the MATLAB compiler automatically checks for all referenced M-files and includes them with the main function in the compilation.

```
function out = ssort(a)
%SSORT Selection sort data in ascending order
% Function SSORT sorts a numeric data set into
% ascending order.  Note that the selection sort
% is relatively inefficient.  DO NOT USE THIS
% FUNCTION FOR LARGE DATA SETS.  Use MATLAB's
% "sort" function instead.

% Define variables:
%   a        -- Input array to sort
%   ii       -- Index variable
%   iptr     -- Pointer to min value
%   jj       -- Index variable
%   nvals    -- Number of values in "a"
%   out      -- Sorted output array
%   temp     -- Temp variable for swapping

% Record of revisions:
%     Date          Programmer           Description of change
%     ====          ==========           =====================
%   02/02/07      S. J. Chapman          Original code

% Get the length of the array to sort
nvals = size(a,2);

% Sort the input array
for ii = 1:nvals-1

    % Find the minimum value in a(ii) through a(n)
    iptr = ii;
    for jj = ii+1:nvals
       if a(jj) < a(iptr)
          iptr = jj;
       end
    end

    % iptr now points to the minimum value, so swap a(iptr)
    % with a(ii) if ii ~= iptr.
    if ii ~= iptr
       temp     = a(ii);
       a(ii)    = a(iptr);
       a(iptr) = temp;
    end
  end

% Pass data back to caller
out = a;
```

Figure 11.1   The selection sort function ssort.

```
function out = ssort(a)
%SSORT Selection sort data in ascending order
% Function SSORT sorts a numeric data set into
% ascending order.  Note that the selection sort
% is relatively inefficient.  DO NOT USE THIS
% FUNCTION FOR LARGE DATA SETS.  Use MATLAB's
% "sort" function instead.

% Define variables:
%   a          -- Input array to sort
%   ii         -- Index variable
%   iptr       -- Pointer to min value
%   jj         -- Index variable
%   nvals      -- Number of values in "a"
%   out        -- Sorted output array
%   temp       -- Temp variable for swapping

%   Record of revisions:
%     Date         Programmer          Description of change
%     ====         ==========          =====================
%     02/02/07     S. J. Chapman       Original code

% Get the length of the array to sort
nvals = size(a,2);

% Sort the input array
for ii = 1:nvals-1

    % Find the minimum value in a(ii) through a(n)
    iptr = ii;
    for jj = ii+1:nvals
       if a(jj) < a(iptr)
            iptr = jj;
       end
    end

    % iptr now points to the minimum value, so swap a(iptr)
    % with a(ii) if ii ~= iptr.
    if ii ~= iptr
       temp     = a(ii);
       a(ii)    = a(iptr);
       a(iptr)  = temp;
    end
 end

% Pass data back to caller
out = a;
```

Figure 11.2   The test driver program `test_ssort`.

After the program is compiled, the directory will contain the following files:

```
D:\work\test_ssort>dir
 Volume in drive D is DATA
 Volume Serial Number is E46A-233F

 Directory of D:\work\test_ssort

26/05/2007    04:05 PM    <DIR>             .
26/05/2007    04:05 PM    <DIR>             ..
26/05/2007    04:05 PM             1,081 mccExcludedFiles.log
26/05/2007    04:05 PM             6,999 readme.txt
19/05/2007    04:57 PM             1,293 ssort.m
26/05/2007    04:05 PM            81,750 test_ssort.ctf
26/05/2007    04:05 PM            11,032 test_ssort.exe
26/05/2007    04:00 PM             1,155 test_ssort.m
26/05/2007    04:05 PM            10,605 test_ssort.prj
26/05/2007    04:05 PM             3,079 test_ssort_main.c
26/05/2007    04:05 PM             6,574 test_ssort_mcc_
                                          component_data.c
              9 File(s)       123,568 bytes
              2 Dir(s)   18,682,978,304 bytes free
```

The MATLAB compiler created the C function test_ssort_main.c, which was compiled by the LCC C compiler to create the program test_ssort.exe. In addition, the MATLAB compiler created the CTF file test_ssort.ctf, which contains compressed and encrypted versions of test_ssort.m and ssort.m, as well as any standard MATLAB functions that M-files called. The readme.txt file that was generated describes how to distribute this program to other users, and the file mccExcludedFiles.log lists any functions referred to in the program that could not be included in the stand-alone program (usually files relating to the help subsystem, licensed components, Simulink, and so forth). If one of these excluded functions is called by the compiled program, the program will not work properly.

The files test_ssort.exe, test_ssort.ctf, and the MATLAB runtime installer MCRInstaller.exe can now be distributed to anyone who may need to use the program.

To use the compiled application, place files test_ssort.exe and test_ssort.ctf in a directory and then execute test_ssort.exe. The first time it is started, this program will unpack the files from the CTF into a sub-directory called test_ssort_mcr, and the runtime will then execute the encrypted M-files in that subdirectory. An example execution of this application is shown here.

```
D:\work\test_ssort>test_ssort
Extracting CTF archive. This may take a few seconds,
 depending on the size of your application. Please
 wait ...
```

```
...CTF archive extraction complete.
Enter number of values to sort: 6
Enter value 1: 2
Enter value 2: -6
Enter value 3: 5
Enter value 4: 9
Enter value 5: 4
Enter value 6: -99

Sorted data:
-99.0000
 -6.0000
  2.0000
  4.0000
  5.0000
  9.0000
```

11.3.2 Additional Details

Some programs need to include data files for the program to work properly. If extra data files are required for a particular application, they can be included in the CTF file by adding them to the mcc command line with the −a option. For example, if program MyProg.m needs the data file MyData.dat in order to function correctly, the program would be compiled with the following command:

```
mcc -m MyProg.m -a MyData.dat
```

Another complication occurs in the case of programs with graphical user interfaces. The MATLAB compiler is not able to detect that a program needs to include a function if the only reference to it is in an eval function, an feval function, or a handle graphics callback. To make such a program work, we must force the compiler to include all referenced M-files in the final program. This can be done by listing all of the callback M-files with the −a switch in the mcc command, or else it can be done using a %#function pragma in the main function. A *pragma* is an instruction to the compiler to do something. In this case, it is an instruction to include the specified file in the final compiled program. A series of these pragmas can be placed in the main function, and the compiler will detect that the named files have to be included in the program. For example, if we needed to include functions fun1, fun2, and fun3 in program MyProg, we would include the following lines in the program:

```
program MyProg(...)
...
%#function fun1 fun2 fun3
...
```

Since this line starts with a # symbol, it will be ignored by MATLAB when it executes. However, the MATLAB compiler will read it and know that the additional functions need to be included in the compiled program.

▶
━━

Example 11.1—Creating a Histogram GUI

Convert the histogram GUI created in Example 10.3 into a stand-alone application program, and test that program.

```
D:\work\histGUI>dir
 Volume in drive D is DATA
 Volume Serial Number is E46A-233F

Directory of D:\work\histGUI

26/05/2007  05:42 PM    <DIR>          .
26/05/2007  05:42 PM    <DIR>          ..
29/03/2004  07:57 AM            21,688 histGUI.fig
28/03/2004  07:40 AM             5,811 histGUI.m
02/05/1999  10:59 PM            54,000 test_hist.dat
               3 File(s)        81,499 bytes
               2 Dir(s)  18,682,626,048 bytes free

D:\work\histGUI>mcc -m histGUI.m -a histGUI.fig

D:\work\histGUI>dir
 Volume in drive D is DATA
 Volume Serial Number is E46A-233F

 Directory of D:\work\histGUI

26/05/2007  05:46 PM    <DIR>          .
26/05/2007  05:46 PM    <DIR>          ..
26/05/2007  05:46 PM            84,090 histGUI.ctf
26/05/2007  05:46 PM            10,992 histGUI.exe
29/03/2004  07:57 AM            21,688 histGUI.fig
28/03/2004  07:40 AM             5,811 histGUI.m
26/05/2007  05:46 PM            10,641 histGUI.prj
26/05/2007  05:46 PM             3,041 histGUI_main.c
26/05/2007  05:46 PM             6,530 histGUI_mcc_
                                       component_data.c
26/05/2007  05:46 PM             1,081 mccExcludedFiles.log
26/05/2007  05:46 PM             6,942 readme.txt
02/05/1999  10:59 PM            54,000 test_hist.dat
              10 File(s)       204,816 bytes
               2 Dir(s)  18,682,490,880 bytes free
```
(a)

Figure 11.3   *(a)* The steps required to compile program histGUI.m.

(b)

Figure 11.3 *(continued)* *(b)* The GUI produced by the stand-alone nprogram `histGUI`.

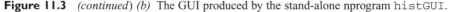

SOLUTION   This program is implemented by file `histGUI.m` and the figure file `histGUI.fig`. To make a stand-alone program, we must copy these files into a separate directory and then run the compiler with a command-line option that includes the figure file in the compiled applications. The command to compile this application would be

```
mcc -m histGUI.m -a histGUI.fig
```

After compiling, the program will run on any computer with the MATLAB runtime installed. The results are a shown in Figure 11.3*b*.

◀

11.3.3   The Deployment Tool

The MATLAB compiler also includes a GUI-based tool to make the creation of stand-alone MATLAB programs easier. It is called the Deployment Tool. This tool is started by typing `deploytool` in the MATLAB Command Window. When the tool starts, it opens the window shown in Figure 11.4.

To create a compiled program with this tool, first click the New Project Icon (⬜) to start a new project. When this icon is selected, a window will appear that allows the user to select both the type of project to create (see Figure 11.5*a*) and the directory in which the project will be built. After these steps have been taken, the Deployment Tool shows a blank project with no files in it yet (see Figure 11.5*b*).

To re-create the histogram GUI using the Deployment Tool, right-click the "Main function" folder and add the main M-file (the one where the program starts). Then, right-click the "Other files" folder and add all the other files needed by this

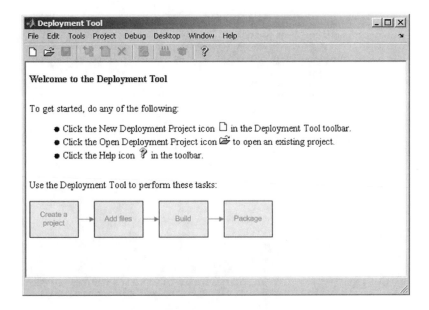

Figure 11.4   The Deployment Tool Window when the tool is first opened.

Figure 11.5   *(a)* Selecting a new project type.

(b)

Figure 11.5 *(continued)* *(b)* The Deployment Tool Window after a new project has been created. Note that there are no files in this project yet.

Figure 11.6 The Deployment Tool Window containing project `histGUI`.

Figure 11.7   *(a)* Contents of the `src` directory after compiling. *(b)* Contents of the `distrib` directory after packaging.

project. Figure 11.6 shows the Deployment Tool after the `histGUI` program has been inserted into it. The project can then be compiled by clicking the Build Icon ( ) and packaged for delivery to the customer by clicking the Package Icon ( ).

When the Build and Package Icons have been clicked, two directories are created under the project directory. The `src` directory contains all of the compilation files produced by running `mcc`, and the `distrib` directory contains the executable program, the CTF file, and an install batch file for distribution to customers (see Figure 11.7).

11.4   Summary

This chapter has introduced the MATLAB Compiler, which is a great tool for converting MATLAB programs into stand-alone applications. The compiler can be executed either from the MATLAB Command Windows or from a command prompt outside of MATLAB. It can also be executed using the Deployment Tool.

The entry point in a compiled program must be a MATLAB function, not a script file. If the entry point is a script file, convert it into a function before attempting to compile the program.

When the compiler is executed, it examines the M-files named on the command line and attempts to determine all of the other M-files called from the named files. It includes those additional files in the final program. However, the compiler cannot identify files that are only referenced in the `eval` function, the `feval` function, or callbacks. In that case, the files must be either explicitly included after the `-a` option on the command line or else included in a `%#function` pragma in the main function.

11.4.1 Summary of Good Programming Practice

The following guidelines should be adhered to when working with MATLAB GUIs.

1. Always set up the MATLAB compiler using the `mbuild  -setup` command before trying to compile programs.
2. The MATLAB Component Runtime (MCR) must be installed before a computer can be used to run a compiled MATLAB application. Always distribute the runtime together with your compiled applications.

11.4.2 MATLAB Summary

The following summary lists all of the MATLAB commands and functions described in this chapter, along with a brief description of each one.

`deploytool`	GUI-based tool to compile and deploy MATLAB programs
`#%function`	Pragma to tell compiler to include the named M-files in a compiled program
`mbuild`	Function used to set up MATLAB compiler
`mcc`	MATLAB compiler execution command

11.5 Exercises

11.1 What are the principal advantages and disadvantages of compiling a MATLAB program?

11.2 Compile the Least-Squares Fit GUI created in Exercises 10.15 through 10.17, and test the resulting program by installing and running it on another computer that does not have MATLAB installed.

APPENDIX A

ASCII
Character Set

MATLAB strings use the ASCII character set, which consists of the 127 characters shown in the table that follows. The results of MATLAB string comparison operations depend on the *relative lexicographic positions* of the characters being compared. For example, the character 'a' in the ASCII character set is at position 97 in the table, while the character 'A' is at position 65. Therefore, the relational operator 'a' > 'A' will return a 1 (true), since 97 > 65.

Each MATLAB character is stored in a 16-bit field, which means that in the future MATLAB can support the entire Unicode character set.

The table shows the ASCII character set, with the first two digits of the character number defined by the row, and the third digit defined by the column. Thus, the letter 'R' is on row 8 and column 2, so it is character 82 in the ASCII character set.

	0	1	2	3	4	5	6	7	8	9
0	nul	soh	stx	etx	eot	enq	ack	bel	bs	ht
1	nl	vt	ff	cr	so	si	dle	dc1	dc2	dc3
2	dc4	nak	syn	etb	can	em	sub	esc	fs	gs
3	rs	us	sp	!	"	#	$	%	&	'
4	(	)	*	+	,	-	.	/	0	1
5	2	3	4	5	6	7	8	9	:	;
6	<	=	>	?	@	A	B	C	D	E
7	F	G	H	I	J	K	L	M	N	O
8	P	Q	R	S	T	U	V	W	X	Y
9	Z	[	\	]	^	_	`	a	b	c
10	d	e	f	g	h	I	j	k	l	m
11	n	o	p	q	r	s	t	u	v	w
12	x	y	z	{			}	~	del		

APPENDIX

Answers
to
Quizzes

This appendix contains the answers to all of the quizzes in the book.

Quiz 1.1, page 18

1. The MATLAB Command Window is the window where a user enters commands. A user can enter interactive commands at the command prompt (») in the Command Window, and they will be executed on the spot. The Command Window is also used to start M-files executing. The Edit/Debug Window is an editor used to create, modify, and debug M-files. The Figure Window is used to display MATLAB graphical output.

2. You can get help in MATLAB by:

   - Typing `help <command_name>` in the Command Window. This command will display information about a command or function in the Command Window.

   - Typing `lookfor <keyword>` in the Command Window. This command will display in the Command Window a list of all commands or functions containing the keyword in their first comment line.

   - Starting the Help Browser by typing `helpwin` or `helpdesk` in the Command Window, by selecting "Help" from the Start menu, or by clicking on the question mark icon (⸮) on the desktop. The Help Browser contains an extensive hypertext-based description of all of the features in MATLAB, plus a complete copy of all manuals on-line in HTML and Adobe PDF formats. It is the most comprehensive source of help in MATLAB.

3. A workspace is the collection of all the variables and arrays that can be used by MATLAB when a particular command, M-file, or function is executing. All commands executed in the Command Window (and all script files executed from the Command Window) share a common workspace, so they can all share variables. The contents of the workspace can be examined with the `whos` command, or graphically with the Workspace Browser.

4. To clear the contents of a workspace, type `clear` or `clear variables` in the Command Window.

5. The commands to perform this calculation are

```
» t = 5;
» x0 = 10;
» v0 = 15;
» a = -9.81;
» x = x0 + v0 * t + 1/2 * a * t^2
x =
   -37.6250
```

6. The commands to perform this calculation are

```
» x = 3;
» y = 4;
» res = x^2 * y^3 / (x - y)^2
res =
   576
```

Questions 7 and 8 are intended to get you to explore the features of MATLAB. There is no single "right" answer for them.

Quiz 2.1, page 30

1. An array is a collection of data values organized into rows and columns and known by a single name. Individual data values within an array are accessed by including the name of the array followed by subscripts in parentheses that identify the row and column of the particular value. The term "vector" is usually used to describe an array with only one dimension, whereas the term "matrix" is usually used to describe an array with two or more dimensions.

2. (*a*) This is a 3×4 array; (*b*) $c(2, 3) = -0.6$; (*c*) The array elements whose value is 0.6 are $c(1, 4), c(2, 1)$, and $c(3, 2)$.

3. (*a*) 1×3; (*b*) 3×1; (*c*) 3×3; (*d*) 3×2; (*e*) 3×3; (*f*) 4×3; (*g*) 4×1.

4. w(2, 1) = 2
5. x(2, 1) = −20*i*
6. y(2, 1) = 0
7. v(3) = 3

Quiz 2.2, page 39

1. (*a*) c(2, :) = [0.6   1.1   −0.6   3.1]

(*b*) c(:, end) = $\begin{bmatrix} 0.6 \\ 3.1 \\ 0.0 \end{bmatrix}$

(*c*) c(1:2, 2:end) = $\begin{bmatrix} -3.2 & 3.4 & 0.6 \\ 1.1 & -0.6 & 3.1 \end{bmatrix}$

(*d*) c(6) = 0.6
(*e*) c(4, end) = [−3.2 1.1 0.6 3.4 −0.6 5.5 0.6 3.1 0.0]

(*f*) c(1:2, 2:end) = $\begin{bmatrix} -3.2 & 3.4 & 0.6 \\ 1.1 & -0.6 & 3.1 \end{bmatrix}$

(*g*) c([1 3], 2) = $\begin{bmatrix} -3.2 \\ 0.6 \end{bmatrix}$

(*h*) c([2 2], [3 3]) = $\begin{bmatrix} -0.6 & -0.6 \\ -0.6 & -0.6 \end{bmatrix}$

2. (*a*) a = $\begin{bmatrix} 7 & 8 & 9 \\ 4 & 5 & 6 \\ 1 & 2 & 3 \end{bmatrix}$ (*b*) a = $\begin{bmatrix} 4 & 5 & 6 \\ 4 & 5 & 6 \\ 4 & 5 & 6 \end{bmatrix}$ (*c*) a = $\begin{bmatrix} 4 & 5 & 6 \\ 4 & 5 & 6 \end{bmatrix}$

3. (*a*) a = $\begin{bmatrix} 1 & 0 & 0 \\ 1 & 2 & 3 \\ 0 & 0 & 1 \end{bmatrix}$ (*b*) a = $\begin{bmatrix} 1 & 0 & 4 \\ 0 & 1 & 5 \\ 0 & 0 & 6 \end{bmatrix}$ (*c*) a = $\begin{bmatrix} 1 & 0 & 0 \\ 0 & 1 & 0 \\ 9 & 7 & 8 \end{bmatrix}$

Quiz 2.3, page 46

1. The required command is "format long e".
2. (*a*) These statements get the radius of a circle from the user, and calculate and display the area of the circle. (*b*) These statements display the value of π as an integer, so they display the string: "The value is 3!".

3. The first statement outputs the value 12345.67 in exponential format; the second statement outputs the value in floating-point format; the third statement outputs the value in general format; and the fourth statement outputs the value in floating-point format in a field 12 characters wide, with four places after the decimal point. The results of these statements are

```
value = 1.234567e+004
value = 12345.670000
value = 12345.7
value = 12345.6700
```

Quiz 2.4, page 53

1. (*a*) This operation is illegal. Array multiplication must be between arrays of the same shape, or between an array and a scalar. (*b*) Legal matrix multiplication: result = $\begin{bmatrix} 4 & 4 \\ 3 & 3 \end{bmatrix}$ (*c*) Legal array multiplication: result = $\begin{bmatrix} 2 & 1 \\ -2 & 4 \end{bmatrix}$ (*d*) This operation is illegal. The matrix multiplication b * c yields a 1 × 2 array, and a is a 2 × 2 array, so the addition is illegal. (*e*) This operation is illegal. The array multiplication b .* c is between two arrays of different sizes, so the multiplication is illegal.

2. This result can be found from the operation x = A\B: $x = \begin{bmatrix} -0.5 \\ 1.0 \\ -0.5 \end{bmatrix}$

Quiz 3.1, page 103

Expression	Result	Comment
1. a > b	1	
	(logical true)	
2. b > d	0	
	(logical false)	
3. a > b && c > d	0	
	(logical false)	

4. a == b   0

(logical false)

5. a & b > c   0

(logical false)

6. ~~b   1

(logical true)

7. ~(a > b)
$$\begin{bmatrix} 0 & 0 \\ 0 & 1 \end{bmatrix}$$

(logical array)

8. a > c && b > c   Illegal   The && and || operators work only between *scalar* operands.

9. c <= d   Illegal   The <= operator must be between arrays of the same size, or between an array and a scalar.

10. logical(d)
$$\begin{bmatrix} 1 & 1 & 1 \\ 0 & 1 & 0 \end{bmatrix}$$

(logical array)

11. a * b > c
$$\begin{bmatrix} 1 & 0 \\ 0 & 1 \end{bmatrix}$$

(logical array)   The expression a * b is evaluated first, producing the double array $\begin{bmatrix} 2 & -4 \\ 0 & 20 \end{bmatrix}$, and the logical operation is evaluated second, producing the final answer.

12. a * (b > c)
$$\begin{bmatrix} 2 & 0 \\ 0 & 2 \end{bmatrix}$$

(double array)   The expression b > c produced the logical array $\begin{bmatrix} 1 & 0 \\ 0 & 1 \end{bmatrix}$, and multiplying that logical array by 2 converted the results back into a double array.

13. a*b^2 > a*c   0

(logical false)

14. d || b > a   1

(logical true)

15. `(d | b) > a`   0
    (`logical` false)

16. `isinf(a/b)`   0
    (`logical` false)

17. `isinf(a/c)`   1
    (`logical` true)

18. `a > b &&`
    `ischar(d)`   1
    (`logical` true)

19. `isempty(c)`   0
    (`logical` false)

20. `(~a) & b`   0
    (`logical` false)

21. `(~a) + b`   −2
    (`double` value)   ~a is a logical 0. When added to b, the result is converted back to a `double` value.

Quiz 3.2, page 118

1.
```
if x >= 0
   sqrt_x = sqrt(x);
else
   disp('ERROR: x < 0');
   sqrt_x = 0;
end
```

2.
```
if abs(denominator) < 1.0E-300
   disp('Divide by 0 error.');
else
   fun = numerator / denominator;
   disp(fun)
end
```

3.
```
if distance <= 100
   cost = 0.50 * distance;
elseif distance <= 300
   cost = 50 + 0.30 * (distance - 100);
else
   cost = 110 + 0.20 * (distance - 300);
end
```

4. These statement are incorrect. For this structure to work, the second `if` statement would need to be an `elseif` statement.

5. These statement are legal. They will display the message "`Prepare to stop.`"

6. These statement will execute, but they will not do what the programmer intended. If the `temperature` is 150, these statements will print out "`Human body temperature exceeded.`" instead of "`Boiling point of water exceeded.`", because the `if` structure executes the first `true` condition and skips the rest. To get proper behavior, the order of these tests should be reversed.

Quiz 3.3, page 135

1.
```
x = 0:pi/10:2*pi;
x1 = cos(2*x);
y1 = sin(x);
plot(x1,y1,'-ro','LineWidth',2.0,'MarkerSize',6,...
    'MarkerEdgeColor','b','MarkerFaceColor','b')
```
3. `'\itf\rm(\itx\rm) = sin \theta cos 2\phi'`
4. `'\bfPlot of \Sigma \itx\rm\bf^{2} versus \itx'`
5. This string creates the characters: τ_m
6. This string creates the characters: $x_1^2 + x_2^2$ (units: $\mathbf{m^2}$)
7. The backslash character is displayed using a double backslash (`'\\'`).

Quiz 4.1, page 175

1. 4 times
2. 0 times
3. 1 time
4. 2 times
5. 2 times
6. `ires = 10`
7. `ires = 55`
8. `ires = 25;`
9. `ires = 49;`
10. With loops and branches:
```
for ii = -6*pi:pi/10:6*pi
   if sin(ii) > 0
      res(ii) = sin(ii);
   else
      res(ii) = 0;
   end
end
```

With vectorized code:
```
arr1 = sin(-6*pi:pi/10:6*pi);
res = zeros(size(arr1));
res(arr1>0) = arr1(arr1>0);
```

Quiz 5.1, page 226

1. Script files are collections of MATLAB statements that are stored in a file. Script files share the Command Window's workspace, so any variables that were defined before the script file starts are visible to the script file, and any variables created by the script file remain in the workspace after the script file finishes executing. A script file has no input arguments and returns no results, but script files can communicate with other script files through the data left behind in the workspace. In contrast, each MATLAB function runs in its own independent workspace. It receives input data through an input argument list and returns results to the caller through an output argument list.

2. The `help` command displays all of the comment lines in a function until either the first blank line or the first executable statement is reached.

3. The H1 comment line is the first comment line in the file. This line is searched by and displayed by the `lookfor` command. It should always contain a one-line summary of the purpose of a function.

4. In the pass-by-value scheme, a *copy* of each input argument is passed from a caller to a function instead of the original argument itself. This practice contributes to good program design, because the input arguments may be freely modified in the function without causing unintended side effects in the caller.

5. A MATLAB function can have any number of arguments, and not all arguments need to be present each time the function is called. Function `nargin` is used to determine the number of input arguments actually present when a function is called, and function `nargout` is used to determine the number of output arguments actually present when a function is called.

6. This function call is incorrect. Function `test1` must be called with two input arguments. In this case, variable `y` will be undefined in function `test1`, and the function will abort.

7. This function call is correct. The function can be called with either one or two arguments.

Quiz 6.1, page 292

1. (*a*) `result = 1` (true), because the comparison is made between the real parts of the numbers. (*b*) `result = 0` (false), because the absolute values of the two numbers are identical. (*c*) `result = 25`.

2. The function `plot(array)` plots the imaginary part of the array versus the real part of the array, with the real part on the *x* axis and the imaginary part on the *y* axis.

3. The vector can be converted using the `double` function.

4. These statements concatenate the two lines together, and variable res contains the string `'This is a test!This line, too.'`.

5. These statements are illegal—there is no function `strcati`.

6. These statements are illegal—the two strings must have the same number of columns, and these strings are of different lengths.

7. These statements are legal, producing the result
$$res = \begin{bmatrix} \text{This is another test!} \\ \text{This line, too.} \end{bmatrix}$$. Note that each line is now 21 characters long, with line 2 padded out to that length.

8. These statements are legal, and the result `res = 1`, since the two strings are identical in their first five characters.

9. These statements are legal, and the result is `res = [4  7  13]`, since the letter "s" is at those locations in the string.

10. These statements are legal. Each space in the original string is replaced by an `'x'`, and the final string is `'Thisxisxaxtest!xx'`.

11. These statements are legal. The function isstrprop returns a 1 (`true`) for alphanumeric characters, and a 0 (`false`) for other characters. The result is

    ```
    res =
         1  1  1  1  0  1  1  1  1  0  0  0
    ```

12. These statements are legal, with the result `res = 'ThiS IS a test!'`.

13. These statements are legal. The results are $11 = 9$, $12 = 9$, $13 = 18$, $14 = 6$, and $15 = 12$.

14. These statements are illegal—you must specify the number of characters to compare in the two strings when using function `strncmp`.

1. A sparse array is a special type of array in which memory is allocated only for the nonzero elements in the array. Memory values are allocated for both the subscripts and the value of each element in a sparse array. By contrast, a memory location is allocated for every value in a full array, whether the value is 0 or not. Sparse arrays can be converted to full arrays using the `full` function, and full arrays can be converted to sparse arrays using the `sparse` function.

2. A cell array is an array of "pointers." each element of which can point to any type of MATLAB data. It differs from an ordinary array in that each element of a cell array can point to a different type of data, such as a numeric array, a string, another cell array, or a structure. Also, cell arrays use braces `{ }` instead of parentheses `( )` for selecting and displaying the contents of cells.

3. *Content indexing* involves placing braces `{ }` around the cell subscripts, together with cell contents in ordinary notation. This type of indexing defines the contents of the data structure contained in a cell. *Cell indexing* involves placing braces `{ }` around the data to be stored in a cell, together with cell subscripts in ordinary subscript notation. This type of indexing creates a data structure containing the specified data and then assigns that data structure to a cell.

4. A structure is a data type in which each individual element is given a name. The individual elements of a structure are known as fields, and each field in a structure may have a different type. The individual fields are addressed by combining the name of the structure with the name of the field, separated by a period. Structures differ from ordinary arrays and cell arrays in that ordinary arrays and cell array elements are addressed by subscript, but structure elements are addressed by name.

5. Function `varargin` appears as the last item in an input argument list, and it returns a cell array containing all of the actual arguments specified when the function is called, each in an individual element of a cell array. This function allows a MATLAB function to support any number of input arguments.

6. A function handle is a MATLAB data type that holds information to be used in referencing a function. A function handle is created using the `@` operator or the `str2func` function. A function is called using a function handle by naming the function handle, followed by the arguments in parentheses. If the function has no arguments, the handle must be followed by empty parentheses `( )`.

7. (*a*) a(1, 1) = [3x3 double]. The contents of cell array element
a(1, 1) is a 3 × 3 double array, and this data structure is displayed.

(*b*) a{1, 1} = $\begin{bmatrix} 1 & 2 & 3 \\ 4 & 5 & 6 \\ 7 & 8 & 9 \end{bmatrix}$. This statement displays the *value* of

the data structure stored in element a(1, 1).

(*c*) These statements are illegal, since you can not multiply a data
structure by a value.

(*d*) These statements are legal, since you *can* multiply the contents of

the data structure by a value. The result is $\begin{bmatrix} 2 & 4 & 6 \\ 8 & 10 & 12 \\ 14 & 16 & 18 \end{bmatrix}$.

(*e*) a{2, 2} = $\begin{bmatrix} -4 & -3 & -2 \\ -1 & 0 & 1 \\ 2 & 3 & 4 \end{bmatrix}$.

(*f*) This statement is legal.It initializes cell array element a(2, 3)

to be a 2 × 1 double array containing the values $\begin{bmatrix} -17 \\ 17 \end{bmatrix}$.

(*g*) a{2, 2}(2, 2) = 0.

8. (*a*) b(1).a - b(2).a = $\begin{bmatrix} -3 & 1 & -1 \\ -2 & 0 & -2 \\ -3 & 3 & 5 \end{bmatrix}$.

(*b*) strncmp (b(1).b, b(2).b, 6) = 1, because the two structure elements contain character strings that are identical in their first
six characters.

(*c*) mean (b(1).c) = 2

(*d*) This statement is illegal, because you cannot treat individual elements of a structure array as though it were an array itself.

(*e*) b = 1x2 struct array with fields:

            a

            b

            c

(*f*) b(1).('b') = 'Element 1'

(*g*) b(1) =

        a: [3x3 double]

        b: 'Element 1'

        c: [1 2 3]

9. (*a*) The result of this function is the name of function whose handle is passed as a calling argument:

```
» myfun(@cosh)
ans =
Element 1
```

Quiz 8.1, page 384

1. The `textread` function is designed to read ASCII files that are formatted into columns of data, where each column can be of a different type. This command is *very* useful for importing tables of data printed out by other applications, since it can handle data of mixed types within a single file.

2. MAT-files are relatively efficient users of disk space, and they store the full precision of every variable; no precision is lost due to conversion to and from ASCII format. (If compression is used, MAT-files take up even less space.) In addition, MAT-files preserve all of the information about each variable in the workspace, including its class, name, and whether or not it is global. A disadvantage of MAT-files is that they are unique to MATLAB and cannot be used to share data with other programs.

3. Function `fopen` is used to open files, and function `fclose` is used to close files. On PCs (but not on Linux or Unix computers), there is a difference between the format of a text file and a binary file. In order to open files in text mode on a PC, a `'t'` must be appended to the permission string in the `fopen` function.

4. `fid = fopen('myinput.dat','at')`

5.
```
fid = fopen('input.dat','r');
if fid < 0;
   disp('File input.dat does not exist.');
end
```

6. These statements are incorrect. They open a file as a text file, but then read the data in binary format. (Function `fscanf` should be used to read text data, as is shown later in this chapter.)

7. These statements are correct. They create a 10-element array x, open a binary output file `file1`, write the array to the file, and close the file. Next, they open the file again for reading, and read the data into array `array` in a `[2 Inf]` format. The resulting contents of the array are $\begin{bmatrix} 1 & 3 & 5 & 7 & 9 \\ 2 & 4 & 6 & 8 & 10 \end{bmatrix}$.

Quiz 8.2, page 400

1. Formatted I/O operations produce formatted files. A formatted file contains recognizable characters, numbers, and so forth stored as ASCII text. Formatted files have the advantages that we can readily see what sort of data they contain and it is easy to exchange data between different types of programs using them. However, formatted I/O operations take longer to read and write, and formatted files take up more space than unformatted files. Unformatted I/O operations copy the information from a computer's memory directly to the disk file with no conversions at all. These operations are much faster than formatted I/O operations because there is no conversion. In addition, the data occupies a much smaller amount of disk space. However, unformatted data cannot be examined and interpreted directly by humans.

2. Formatted I/O should be used whenever we need to exchange data between MATLAB and other programs, or when a person needs to be able to examine and/or modify the data in the file. Otherwise, unformatted I/O should be used.

3.
```
fprintf('Table of Cosines and Sines\n\n');
fprintf(' theta    cos(theta)    sin(theta)\n');
fprintf(' =====    ==========    ==========\n');
for ii = 0:0.1:1
    theta = pi * ii;
    fprintf('%7.4f %11.5f %11.5f\n',
            theta,cos(theta),sin(theta));
end
```

4. These statements are incorrect. The `%s` descriptor must correspond to a character string in the output list.

5. These statements are technically correct, but the results are undesirable. It is possible to mix binary and formatted data in a single file the way that these statements do, but the file is then very hard to use for any purpose. Normally, binary data and formatted data should be written to separate files.

Quiz 10.1, page 506

1. The types of graphical components discussed in this chapter are listed here, together with their purposes.

Table B.1   GUI Components Discussed in Chapter 10

Component	Created by	Description
		Graphical Controls
Pushbutton	uicontrol	A graphical component that implements a pushbutton. It triggers a callback when clicked with a mouse.
Toggle Button	uicontrol	A graphical component that implements a toggle button. A toggle button is either "on" or "off," and it changes state each time it is clicked. Each mouse button click also triggers a callback.
Radio Button	uicontrol	A radio button is a type of toggle button that appears as a small circle with a dot in the middle when it is "on." Groups of radio buttons are used to implement mutually exclusive choices. Each mouse click on a radio button triggers a callback.
Checkbox	uicontrol	A checkbox is a type of toggle button that appears as a small square with a check mark in it when it is "on." Each mouse click on a check box triggers a callback.
Edit Box	uicontrol	An edit box displays a text string and allows the user to modify the information displayed. A callback is triggered when the user presses the Enter key.
List Box	uicontrol	A list box is a graphical control that displays a series of text strings. A user may select one of the text strings by single- or double-clicking on them. A callback is triggered when the user selects a string.
Popup Menus	uicontrol	A popup menu is a graphical control that displays a series of text strings in response to a mouse click. When the popup menu is not clicked on, only the currently-selected string is visible.
Slider	uicontrol	A slider is a graphical control to adjust a value in a smooth, continuous fashion by dragging the control with a mouse. Each slider change triggers a callback.
		Static Elements
Frame	uicontrol	Creates a frame, which is a rectangular box within a figure. Frames are used to group sets of controls together. Frames never trigger callbacks. (This is a deprecated component, which should not be used in new GUIs.)
Text Field	uicontrol	Creates a label, which is a text string located at a point on the figure. Text fields never trigger callbacks.
		Menus, Toolbars, Axes
Menu Items	uimenu	Creates a menu item. Menu items trigger a callback when a mouse button is released over them.

(continued)

Table B.1 *(continued)*

Component	Created by	Description
Context Menus	`uicontextmenu`	Creates a context menu, which is a menu that appears over a graphical object when a user right-clicks the mouse on that object.
Toolbar	`uitoolbar`	Creates a toolbar, which is a bar across the top of the figure containing quick-access buttons.
Toolbar Pushbutton	`uipushtool`	Creates a pushbutton to go in a toolbar.
Toolbar Toggle Button	`uitoggletool`	Creates a toggle button to go in a toolbar.
Axes	`axes`	Creates a new set of axes to display data on. Axes never trigger callbacks.

2. The types of containers discussed in this chapter are listed here, together with their differences.

Table B.2 GUI Components Discussed in Chapter 10

Component	Created by	Description
		Containers
Figure	`uicontrol`	Creates a figure, which is a container that can hold components and other containers. Figures are separate windows that have title bars and can have menus.
Panel	`uipanel`	Creates a panel, which is a container that can hold components and other containers. Unlike figures, panels do not have title bars or menus. Panels can be placed inside figures or other panels.
Button Group	`uibuttongroup`	Creates a button group, which is a special kind of panel. Button groups automatically manage groups of radio buttons or toggle buttons to ensure that only one item of the group is on at any given time.

3. A callback function is a function that is executed whenever an action (mouse click, keyboard input, etc.) occurs on a specific GUI component. They are used to perform an action when a user clicks on or types in a GUI component. Callback functions are specified by the `'Callback'` property in a `uicontrol`, `uimenu`, `uicontextmenu`, `uipushtool`, or `uitoggletool`. When a new GUI is created, the callbacks are set automatically by `guide` to be `xxx_Callback`, where `xxx` is the value of the `Tag` property of the corresponding GUI component.

4. The basic steps required to create a MATLAB GUI are
   1. Decide what elements are required for the GUI and what the function of each element will be. Make a rough layout of the components by hand on a piece of paper.
   2. Use a MATLAB tool called `guide` (GUI Development Environment) to lay out the components on a figure. The size of the figure, and the alignment and spacing of components on the figure can be adjusted using the tools built into guide.
   3. Use a MATLAB tool called the Property Inspector (built into `guide`) to give each component a name (a "tag"), and to set the characteristics of each component, such as its color, the text it displays, and so forth.
   4. Save the figure to a file. When the figure is saved, two files will be created on disk with the same name but different extents. The `fig` file contains the actual GUI that you have created, and the M-file contains the code to load the figure, along with skeleton callbacks for each GUI element.
   5. Write code to implement the behavior associated with each callback function.

5. The `handles` data structure is a structure containing the handles of all components within a figure. Each structure element has the name of a component and the value of the component's handle. This structure is passed to every callback function, allowing each function to have access to every component in the figure.

6. Application data can be saved in a GUI by adding it to the handles structure and saving that structure after it has been modified using function `guidata`. Since the `handles` structure is automatically passed to every callback function, any additional data added to the structure will be available to any callback function in the GUI. (Each function that modifies the handles structure must be sure to save the modified version with a call to `guidata` before the function exits.)

7. A graphical object can be made invisible by setting its `'Visible'` property to `'off'`. A graphical object can be disabled so that it will not respond to mouse clicks or keyboard input by setting its `'Enable'` property to `'off'`.

8. Pushbuttons, toggle buttons, radio buttons, checkboxes, list boxes, popup menus, and sliders all respond to mouse clicks. Edit boxes respond to keyboard inputs.

9. A dialog box is a special type of figure that is used to display information or to get input from a user. Dialog boxes are used to display errors, provide warnings, ask questions, or get user input. Dialog

boxes can be created by any of the functions listed in Table 10.5, including `errordlg, warndlg, inputdlg, uigetfile,` and so on.

10. A modal dialog box does not allow any other window in the application to be accessed until it is dismissed, whereas a normal dialog box does not block access to other windows.

11. A standard menu is tied to a menu bar running across the top of a figure, but a context menu can be attached to any GUI component. Standard menus are activated by normal mouse clicks on the menu bar, whereas context menus are activated by mouse right-clicks over the associated GUI component. Menus are built out of `uimenu` components. Context menus are built out of both `uicontextmenu` and `uimenu` components.

12. Accelerator keys are keys that may be typed on the keyboard to cause a menu item to be selected. Keyboard mnemonic keys are CTRL+key combinations that cause a menu item to be executed. The principal difference between accelerator keys and keyboard mnemonics is that accelerator keys work to select a menu item only if a menu has already been opened, whereas keyboard mnemonics can trigger an action even if a menu has not been opened.

Index